UNREASONABLE
MATHEMATICS

JAMES R WARREN

BLOXWICH
2022

First Published in the United Kingdom in 2022 by Midland Tutorial Productions

First Edition: 1 August 2022

File Prefix Code: UNMATH

ISBN 978 1 7396296 5 6

Printed and Bound by IngramSpark

Midland Tutorial Productions Publishers
31 Victoria Avenue
Bloxwich
Walsall
WS3 3HS
United Kingdom

MI**DLAND**

TUTORIAL

UNREASONABLE MATHEMATICS

An Album of Research Reports

First Edition

James R Warren

MIDLAND TUTORIAL PRODUCTIONS
BLOXWICH

To The Glory of The Loving God

Who Made Our Minds Free

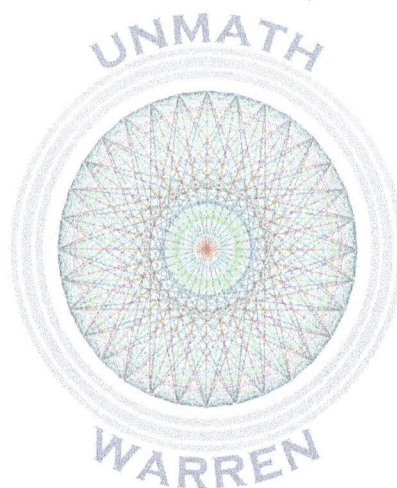

TABLE OF CONTENTS

Page

CHAPTER ONE

Pedometer Calibration

How I Calibrated My Pedometer

By
James R Warren BSc MSc PhD PGCE

Reports that say that something hasn't happened are always interesting to me, because as we know, there are known knowns; there are things we know we know. We also know there are known unknowns; that is to say we know there are some things we do not know. But there are also unknown unknowns – the ones we don't know we don't know.

- Donald Rumsfeld
12 February 2002

Mr Rumsfeld was much mocked at the time, and since. Perhaps it is little wonder, given the enormity of the context, and the inimitability of his delivery. But perhaps also it is because of, or despite, the severest epistemology of his startling oration.

The mind of a machine is a known unknown[1], at any event if you did not design and program it yourself. Yet to calibrate an instrument we need somehow to assess its behaviour against a known known, a fiducial which may itself be a shadow of reality, or even like the Ptolemaic Universe an utter misconception.

For temporal knowledge is impossible, and surmise temporary, or even transient. So our conception of known and unknown things reflects the fashions, prejudices and exploratory techniques of our time. This is not because we die, or even because we are stupid. It is a necessity of Action set in Time. Therefore, any scheme of knowledge is meaningless, and indeed itself impossible in its own paradox.

Only God knows, and is constant, and only the knowing constant can be eternal.

Without undue digression, we may perhaps clarify Mr Rumsfeld's dense prose with examples set helpfully in a pair of Johari windows[2]: One for the date 1 January 1901, the other for 1 January 2001. Such are respectively shown in Figures One and Two.

You are of course entitled to think the examples inapt or misclassified. Indeed you may think that some of the concepts cited are not so much unrecognised as actively deprecated. But the tabulations given are only illustrative, and the author hopes that they assist rather than hinder understanding of our tiny problems, notwithstanding the grandeur of the issues categorised.

We think we know, for example, that although uniformitarianism was a known known on both dates its flavor has changed, since we now surmise on evidence that the breeze which caressed the stromatolites differed from the airs that bathe today, and that the planet's chemistry and geology have changed. But such only goes to confirm the essential provisionality of what we know, and the mutability of Action in Time.

(Okay: Living stromatolites have now been found on the remotest coasts of Australia. And this goes to show that we cannot take anything for granted, especially not our own preconceptions).

	Things Described	Things Not Described
Things Cognised	**KNOWN KNOWNS** Newtonian Physics Darwinian Ecology Huttonian Uniformitarianism	**KNOWN UNKNOWNS** Life After Death The Extent of the Universe The Integrity of Probabilistic Assessments
Things Not Cognised	**UNKNOWN KNOWNS** Paranormal Effects Radioactivity Psychology of the Subconscious	**UNKNOWN UNKNOWNS** States Prior to the "Big Bang" Relativity Quantum Physics

Figure One
A Representation of the State of Knowledge on
1 January 1901

	Things Described	Things Not Described
Things Cognised	**KNOWN KNOWNS** Relativity Radioactivity Quantum Physics Darwinian Ecology Huttonian Uniformitarianism	**KNOWN UNKNOWNS** Life After Death The Extent of the Universe The Integrity of Probabilistic Assessments
Things Not Cognised	**UNKNOWN KNOWNS** Paranormal Effects Psychology of the Subconscious	**UNKNOWN UNKNOWNS** **?**

Figure Two
A Representation of the State of Knowledge on
1 January 2001

Around the year 2012, when I was sixty, a cardiologist opined that in view of my incipient angina I was in need of coronary artery surgery. He said that I risked having a heart attack and if I had one it would almost invariably be fatal in my case. I asked him to estimate the annual probability of my heart attack. He said "between seven and ten percent year-on-year". I did a swift piece of mental arithmetic and estimated that I had a half-life of ten years, or if you prefer a fifty percent chance of surviving ten years. I told him I would take the risk for the time being.

In December 2015 I dug two trenches in my back garden, and on returning to my Wife's kitchen I suffered a mild heart attack. Within ten minutes I rose from my passant position and carried on normally. On 7 January 2016 I was walking in Coventry when I tripped over a folded temporary road sign that was virtually the same color as the wet flagstones and partially hidden behind a large bollard. As I fell forward to the pavement I extended my right arm to save my head. The right shoulder dislocated as its tendons parted, and the upper right humerus shattered. I hailed a taxi to the station and returned to Walsall by train. Two days later corrective surgery was attempted (for the broken shoulder) and in the course of the operation the axillary artery tore. I lost four and a half liters of blood which was of course replaced immediately, but possibly I was seriously weakened by the total effect, because the next day when I was in intensive care I suffered a second heart attack. I was again referred to cardiologists, who made it clear that unless I had a coronary bypass I would almost certainly be dead by the end of the year.

On 13 April 2016 I therefore had a cardiac triple bypass using vascular tissue from my legs. This cured the angina completely, and I could walk steep slopes with the same speed and comfort as I did as a thirty-year-old, or so it seemed to me. I realised, however, that during the angina years I had fallen into a kind of decadence (forgive the pleonasm) in which I had largely ceased my habitual walks by the canals and up in the hills, and that I needed to walk again, though to do so alone in remote places would be unwise.

So I decided to walk every day around my large garden. There is hard pavement around the Western half of my back garden, and I decided to measure the arrowed circuit and perform multiple laps twice a day. The four sides of the walkway are shown by the colored arrows in Figure Three.

Figure Three
The Quadrilateral Lap of the Western Part of
My Back Garden at Bloxwich

The measurements were taken piecewise using a five-meter Robone-Chesterman retractable steel tape of excellent English manufacture. The total Lap Length was 49.34 meters as itemised in Table One:-

Path Segment	Length (meters)	Length (miles)	Color Code	
North	7.25	0.00450494	Yellow	
East	17.53	0.01089264	Blue	
West	17.10	0.01062545	Red	
South	7.46	0.00463543	Green	
	49.34	0.03065845		

Table One
Path Segments and their Lengths

This measurement of 49.34 meters I adopted as *fiducial*. It furnished the supposedly objective and infallible criterion against which the precision of the pedometer would be judged.

But to use the pedometer in later circumstances in which my walked distance would be unknown I also needed reliable knowledge of my personal Pace Length. Clearly this would be a function of my height, or more precisely my leg length, but would also vary arbitrarily as well as according to the terrain.

And as speed trialers paced their measured mile in one direction and then the other in order to mitigate the effects of wind and tide, so also I needed to pace my laps dextrally and sinistrally to cancel any topographical effects on my pacing behaviour, especially as stairs and a ramp were present on the Northern segment.

Figure Four is a photograph of the little Sportline electronic pedometer I bought shown against a ruler and a coin for scale. The coin is an English *"Decus et Tutamen"* Pound Coin 3mm thick by 22mm diameter. You may just be able to discern that the dull liquid crystal display registers the time-of-day as 15:46. The chrome-plated clip behind the instrument is designed to facilitate attachment to a trousers belt: That is how I wore it in use.

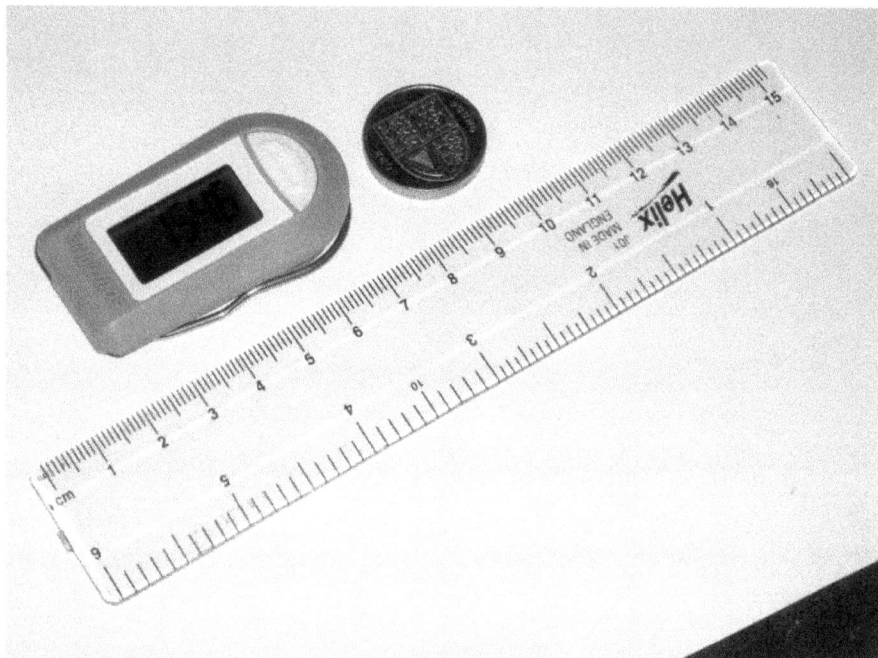

Figure Four
My Sportline Electronic Pedometer with a
Coin and Ruler for Scale

Pace Calibration

The programming of mean pace length is essential to the operation of an inertial pedometer such as my little Sportline instrument.

That pace length is of course both personal and environmental, subject to the interaction of psychology, physiology and the topography.

After several shorter experiments that inferred that I had a Mean Stride Length around 0.69 to 0.75 meters I adopted the outcome of a 32-lap exercise divided into 16 sinistral and 16 dextral laps.

I counted the paces mentally to the edge of a broken tile on the Northern straight.

The results are given in Table Two:-

Serial	Sinistral	Dextral
1	64	65
2	64	64.1
3	63	65.5
4	64	66.5
5	64	66.5
6	63	65.5
7	63	66.5
8	65	65.5
9	65.2	66.5
10	65	66.6
11	64	64.5
12	63	65.5
13	63	65.2
14	63.2	64.3
15	63.1	65.6
16	65	65.3

63.84375	65.5375	64.69063	**Mean**
0.836635	0.815578	1.183587	**s**
0.810068	0.78968	1.164946	**σ**

49.34	**Circuit (meters)**
0.762707	**Mean Stride Length (meters)**
0.697418	**Best Data 22 Aug STEP/KMS Implied Pace Length**
0.66	**Pedometer Assumed Stride Length**
0.865339	**Stride Multiplier**
0.914398	**STEP/KMS Regression Stride Multiplier**

Table Two
The Laps Pace Counts and their Means

As you can see there is a good 2.6% difference between the dextral and sinistral means of paces per lap, but their mean of means is 64.69063 paces per lap, which gives a Mean Stride Length of 0.762707 meters.

At once several anomalies manifest.

On a previous occasion I thought I had set the instrument's firmware to assume a Personal Pace Length of 75 cms (it only registers integral centimeters). When, however, I checked through the personal settings it always quoted the default Pace Length of 66 cms. Continued attempts to amend this we not successful.

Table Three lists the pedometer readouts as they appeared on 11 September 2016.

PERSONAL SETTINGS

Parameter	Unit	Value	Description
KG	Kg	89	Personal Weight
CM	cm	165	Personal Height
STRIDE	cm	66	Persisting Default Stride Length
GOAL	pace	9000	Pace Goal

INSTANTANIOUS READING

Parameter	Unit	Value	Description
TIME	hh:mm	09:29	Local Time (BST)
STEP	pace	116	Paces Unpaced since Last Reset
KMS	kilometers	0.08	Equivalent Paced Kilometers
CAL	kilocalories	6	Equivalent Calories Expended
TMP	hh/mm/ss	00:01:14	Activity Time
FAT	grams	0.7	Grams of Fat Expended
GOAL	pace	8884	Remaining Goal Paces

Table Three
Pedometer Settings as Retained on
11 September 2016

Personal Weight, Personal Height and Pace Goal were all accurate as I had set them in August, but the Personal Stride persisted to be 66 cms, the default that came with the machine.

It was clear that even if the chorological and machine pace counts were linearly related there would have to be some multiplier calculated to bring the pedometer readings nearer to physical reality.

Data Garnering Experiments

Table Four presents the pedometer data gathered in thirty-five lap pacing trials of 5-7 August 2016, together with three zero records included for regressive purposes.

Table Four
Consolidated Best Data for the Pedometer Calibration

Serial	Laps	Meters	Miles	TMA	Decimal Time	Comp Paces	Pedo Paces STEPS	Paces Diff	Paces PSD	ABS Paces PSD	Fractional Defect Minus Style	Fractional Defect Plus Style	Summary ABS Paces PSD	Comp Kms	Pedo Kms KMS	Kms Diff	Kms PSD	CALS	FAT BURN	GOAL	Goal Plus Paces	Notes
0	0	0.00	0.00	0:00:00	0.00	0.00	0	0	0					0	0	0	0	0	0	0	9000	
1	0	0.00	0.00	0:00:00	0.00	0.00	0	0	0					0	0	0	0	0	0	0	9000	
2	0	0.00	0.00	0:00:00	0.00	0.00	0	0	0					0	0	0	0	0	0	0	9000	
3	1	49.34	0.0307	0:00:45	0.75	64.6906	71	6.3094	9.7532	9.7532	0.09753152	2.09753152	6.539137	0.0427							9000	am 6 Aug 2016
4	1	49.34	0.0307	0:00:36	0.60	64.6906	62	-2.691	-4.159	4.1592	-0.041592194	1.958407806		0.0427	0.04	-0.003	-6.314	3	0.3	8938	9000	am 7 Aug 2016
5	1	49.34	0.0307	0:00:36	0.60	64.6906	61	-3.691	-5.705	5.705	-0.057050384	1.942949616		0.0427	0.04	-0.003	-6.314	3	0.3	8939	9000	pm 7 Aug 2016
6	2	98.68	0.0613	0:01:27	1.45	129.381	136	6.6188	5.1157	5.1157	0.051156949	2.051156949	4.478045	0.0854	0.08	-0.025	-29.74				9000	am 5 Aug 2016
7	2	98.68	0.0613	0:01:12	1.20	129.381	124	-5.381	-4.159	4.1592	-0.041592194	1.958407806		0.0854	0.08	-0.005	-6.314	7	0.76	8876	9000	am 7 Aug 2016
8	2	98.68	0.0613	0:01:14	1.23	129.381	124	-5.381	-4.159	4.1592	-0.041592194	1.958407806		0.0854				7	0.76	8876	9000	pm 7 Aug 2016
9	4	197.36	0.1226	0:02:54	2.90	258.763	273	14.238	5.5021	5.5021	0.055021497	2.055021497	4.349226	0.1708	0.17	-8E-04	-0.459	14	1.56		9000	am 5 Aug 2016
10	4	197.36	0.1226	0:02:22	2.37	258.763	248	-10.76	-4.159	4.159	-0.041592194	1.958407806		0.1708	0.33	0.1592	93.227	14	1.5	8752	9000	am 7 Aug 2016
11	4	197.36	0.1226	0:02:29	2.48	258.763	250	-8.762	-3.386	3.386	-0.033863098	1.966136902		0.1708							9000	pm 7 Aug 2016
12	8	394.72	0.2453	0:04:45	4.75	517.525	496	-21.53	-4.159	4.1592	-0.041592194	1.958407806	3.676151	0.3416	0.34	-0.002	-0.459	28	3.16	8504	9000	am 5 Aug 2016
13	8	394.72	0.2453	0:04:59	4.98	517.525	501	-16.53	-3.193	3.193	-0.031930825	1.968069175	3.054603	0.3416	0.33	-0.012	-3.386	28	3.16	8499	9000	am 7 Aug 2016
14	9	444.06	0.2759	0:06:23	6.38	582.216	600	17.784	3.0546	3.0546	0.030546028	2.030546028	2.898411	0.3843							9000	pm 7 Aug 2016
15	16	789.44	0.4905	0:11:09	11.15	1035.05	1050	14.95	1.4444	1.4444	0.014443747	2.014443747		0.6831	0.69	0.0069	1.0052	56	6.36		9000	am 5 Aug 2016
16	16	789.44	0.4905	0:09:32	9.53	1035.05	994	-41.05	-3.966	3.966	-0.039965992	1.96034008		0.6831	0.67	-0.013	-1.922	57	6.36	8006	9000	am 7 Aug 2016
17	16	789.44	0.4905	0:09:58	9.97	1035.05	1000	-35.05	-3.386	3.3863	-0.033863098	1.966136902		0.6831	0.75	0.0669	9.7883	60	6.76	8000	9000	pm 7 Aug 2016
18	32	1578.88	0.9811	0:11:03	11.05	2070.1	1064	28.95	2.797	2.797	0.027969663	2.027969663	2.536109	1.3663	1.37	0.0037	0.2733			7936	9000	am 5 Aug 2016
19	32	1578.88	0.9811	0:22:04	22.07	2070.1	2076	5.9	0.285	0.285	0.002850104	2.002850104		1.3663	1.33	-0.036	-2.654	112	12.45		9000	am 7 Aug 2016
20	32	1578.88	0.9811	0:19:24	19.40	2070.1	1964	-106.1	-5.125	5.1254	-0.051253563	1.948746437		1.3663	1.48	0.1137	8.3244	114	12.66	7036	9000	pm 7 Aug 2016
21	32	2318.98	1.4409	0:20:16	20.27	3040.46	1996	-74.1	-3.58	3.5795	-0.035795372	1.964204628		2.0067				109	13.26	7004	9000	am 6 Aug 2016
22	47	2368.32	1.4716	0:21:26	21.43	3105.15	2094	23.9	1.1545	1.1545	0.011545336	2.011545336		2.0494	2.08	0.0733	3.6526			6906	9000	am 7 Aug 2016
23	48	2368.32	1.4716	0:29:38	29.63	3105.15	2976	-64.46	-2.12	2.1201	-0.021200538	1.978799462	2.120054	2.0494	2.22	0.1706	8.3244	170	18.86	6024	9000	pm 7 Aug 2016
24	48	2368.32	1.4716	0:32:08	32.13	3105.15	3129	23.85	0.7681	0.7681	0.007680788	2.007680788	1.826535	2.0494	2.12	0.0706	3.445	178	19.86	5871	9000	am 6 Aug 2016
25	48	2368.32	1.4716	0:30:12	30.20	3105.15	3035	-70.15	-2.259	2.2592	-0.022591501	1.977408499		2.0494	2.02	-0.029	-1.435	173	19.26	5965	9000	am 7 Aug 2016
26	48	2368.32	1.4716	0:30:58	30.97	3105.15	3029	-76.15	-2.452	2.4524	-0.024523775	1.975476225		2.0494	2.82	0.0875	3.201	173	19.26	5971	9000	pm 7 Aug 2016
27	64	3157.76	1.9621	0:44:12	44.20	4140.2	4159	18.8	0.4541	0.4541	0.004540843	2.004540843	1.081107	2.7325	2.72	-0.013	-0.459	231	25.66		9000	am 7 Aug 2016
28	64	3157.76	1.9621	0:40:23	40.38	4140.2	4042	-98.2	-2.372	2.3719	-0.023718661	1.976281339		2.7325	2.94	0.2075	7.5925	232	25.86	4948	8990	am 5 Aug 2016
29	64	3157.76	1.9621	0:41:50	41.83	4140.2	4064	-76.2	-1.84	1.8405	-0.018404908	1.981595092		2.7325	2.95	0.2175	7.9985	237	26.36	4936	9000	am 7 Aug 2016
30	64	3157.76	1.9621	0:42:49	42.82	4140.2	4154	13.8	0.4058	0.3333	0.003333172	2.003333172		2.7325	3.7	0.2843	8.3244	237	26.36	4846	9000	pm 7 Aug 2016
31	64	3157.76	1.9621	0:42:41	42.68	4140.2	4157	16.8	0.7681	0.7681	0.004057775	2.004057775		2.7325	3.53	0.1143	3.3474	298	33.16	4843	9000	am 6 Aug 2016
32	80	3947.20	2.4527	0:53:43	53.72	5175.25	5215	39.75	0.7681	0.7681	0.007680788	2.007680788	1.521666	3.4157	3.42	0.0043	0.1269	288	32.06	3785	9000	am 6 Aug 2016
33	80	3947.20	2.4527	0:50:43	50.72	5175.25	5049	-126.3	-2.439	2.4395	-0.024394957	1.975605043		3.4157				291	32.46	3951	9000	am 7 Aug 2016
34	80	3947.20	2.4527	0:52:49	52.82	5175.25	5105	-70.25	-1.357	1.3574	-0.013574223	1.986425777		3.4157	4.44	0.3412	8.3244	357	39.76	3895	9000	pm 7 Aug 2016
35	96	4736.64	2.9432	1:04:19	64.32	6210.3	6253	42.7	1.3574	0.6876	0.006875674	2.006875674	1.3381	4.0988	4.42	0.3212	7.8365	346	38.56	2745	8998	am 6 Aug 2016
36	96	4736.64	2.9432	1:01:24	61.40	6210.3	6064	-146.3	-2.356	2.3558	-0.023557638	1.976442362		4.0988	4.12	0.0212	0.5173	351	39.06	2936	9000	am 7 Aug 2016
37	96	4736.64	2.9432	1:04:07	64.12	6210.3	6150	-60.3	-0.971	0.971	-0.009709676	1.990290324		4.0988						2850	9000	pm 7 Aug 2016

Summary statistics:

	Paces Diff	Paces PSD	Kms Diff	Kms PSD	CALS	FAT BURN	GOAL
Means	-22.23	-0.994	0.0685	3.7361	0.0465	8996.1	8998.4
CV	2.1294	3.3404	1.5569	4.781	0.1113	-157.3	0.0002

	Comp Paces	Pedo Paces STEPS	Comp Kms			
Intercept	-2.9928	-3E-04	-0.007	0.0465	8996.1	8998.4
Slope	0.99115	0.0007	1.0477	0.1113	-157.3	0.0002
RSQ	0.99955	0.9977	0.9969	0.9998	1	0.0423

Miles 2.94321
MPH 2.74568

The six columns with capitalised headings on blue fields indicate the pedometer data TMA, STEPS, KMS, CALS, FAT BURN and GOAL of which only STEPS is drawn from the environment. KMS, CALS, FAT BURN and GOAL are all (linear) functions of STEPS modified by pre-programmed Personal Metrics, i.e. KGS and HEIGHT.

As usual, Percentage Specific Defect is defined as:-

$$PSD = 100 \left(\frac{x_{emp} - x_{theo}}{x_{theo}} \right)$$
Equation 1

where Empirical Datum x_{emp} is identified with a pedometer reading; and Theoretical Datum x_{theo} is a value computed from mental pace counting and circuit ground measurement.

There were at least three re-zeroed records for each of the 1, 2, 4, 8, 16, 32, 48, 64, 80 and 96 lap experiments.

At the base of the column "Paces PSD" it is computed that there is essentially a one-percent shortfall in pedometer pace counts (-0.994) for a Coefficient of Variation of 3.3404: Reference to the "kms PSD" column shows a corresponding PSD of +3.7361 with a Variation of 4.781. Why these figures do not agree with the Paces PSD statistics I have no idea. Since only one paces-per-kilometer coefficient is programmed I would have expected a linear relationship between paces and distances paced.

Now the linear regression data at the foot of the "Pedo Paces STEPS" column does indeed infer strict linearity, suggesting with Slope = 0.007 a 0.7 meter pace length with a Coefficient of Determination r^2 = 0.9977.

Plot One shows the abundant linearity of pedometer KMS against pedometer STEPS clarifying what the pedometer "thinks" my Pace Length is.

The (linear) regression equation is:-

$$KMS = 0.000697417845 . STEPS - 0.000337736339$$
Equation 2

Discounting the intercept, this *regression plot equation* offers a Paces Length of 69.7417845 centimeters.

The four competing Pace Length estimates can now be summarised on Table Five.

But this is not the end of our pedometer perplexities. There are further known unknowns to be crystallised from our melt of unknown unknowns.

A linear regression of Computed Paces, P, versus STEPS gives:-

$$STEPS = 0.991148734185 . P - 2.992781374623$$
Equation 3

which shows something like a 0.89% disparity between the pedometer and the mental pace estimates and strongly suggests that the machine is programmed to ignore the first three steps.

STEPS versus KMS
According to the Pedometer

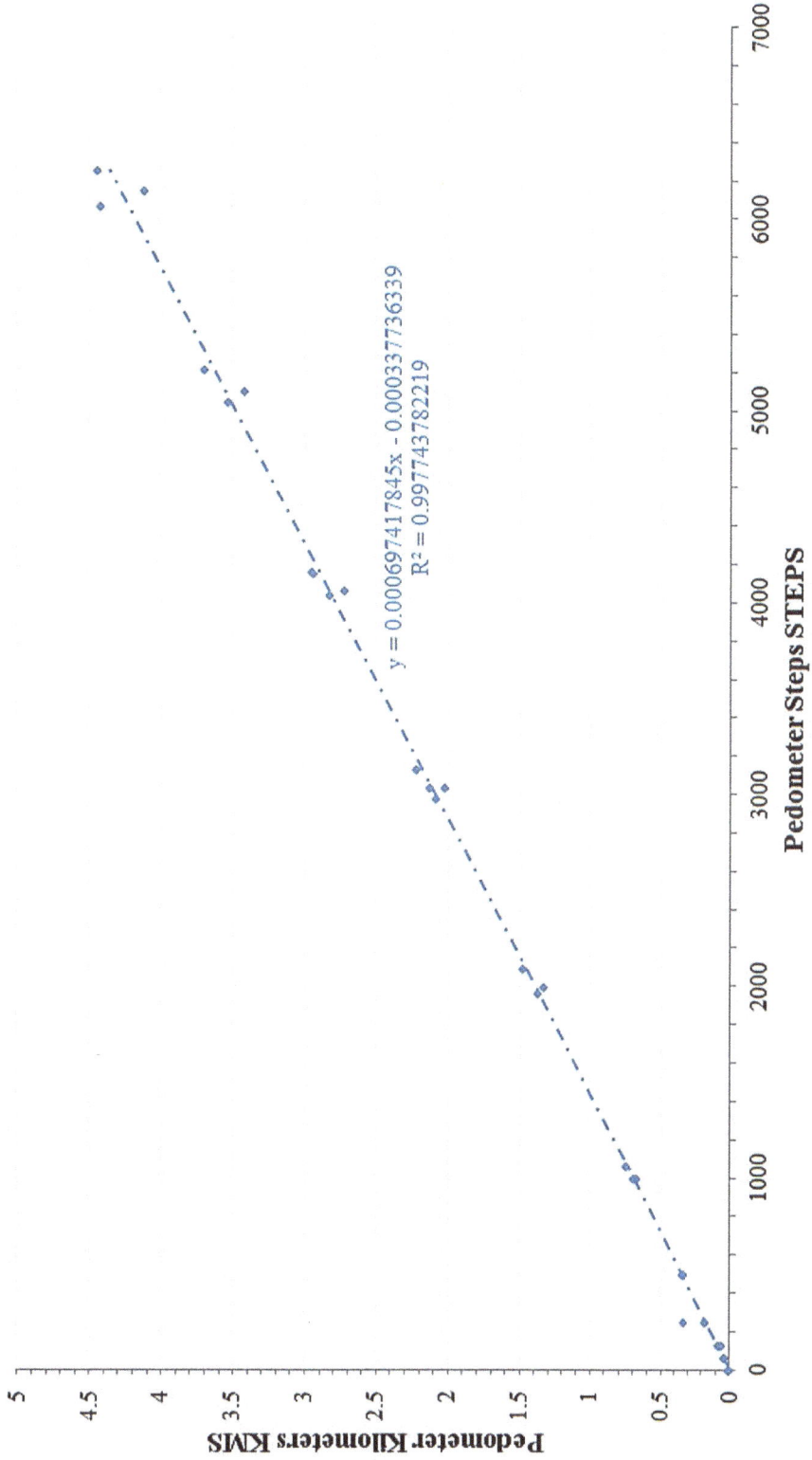

$y = 0.0006974178845x - 0.00033773363339$
$R^2 = 0.9977437882219$

Pedometer Steps STEPS

Pedometer Kilometers KMS

Plot One
Pedometer STEPS versus Pedometer KMS

Value	Unit	Source	Description
66	cm	Pedometer	The Default Stride that the Pedometer Persists in Quoting
69.74179	cm	BEST DATA 22 AUG 2016'!$H$2:$H$39,L2$L$2:$L$39	The Regression Grade of STEPS versus KMS according to EXCEL
76.2707	cm	PACE COUNT AVERAGE'!$D$24	The Mean Circuit Measured Stride Length
75	cm	PACE ESTIMATE (2)'!$E$10	The Stride I Thought I had Set

Table Five
Pace Estimate Anomalies

Notes upon Linear Regression Procedure

Linear Regression is the simplest of a family of algebraic polynomial regression fitments that attempt to correlate two variables by "best fitting" a mathematical lineament. In the context of linear, or first-degree, regression, that lineament, technically a "curve", is a straight line whose position is fully defined by the Intercept, α, at which it intersects the ordinate, and its Grade or Slope β.

The EXCEL® Form

Fitted Line Slope, β, is computed by MicroSoft Office 2010 EXCEL® using the formula:-

$$\beta = \frac{\sum (x - \bar{x})(y - \bar{y})}{\sum (x - \bar{x})^2}$$
Equation 4

where x are the Independent Feed Data and y the Dependent Feed Data, whilst $\bar{x}$ and $\bar{y}$ are their respective Sample Means, assumed equivalent to the Population Means.

The Intercept, α, is given by:-

$$\alpha = \bar{y} - \beta \bar{x}$$
Equation 5

Therefore the resulting best fit line describes:-

$$y = \alpha + \beta x$$
Equation 6

EXCEL® intrinsic functions exist for the means, and for intercept and slope and all three were employed by me and possibly by the EXCEL® trendline computation algorithms.

In terms of EXCEL® conventions the Coefficient of Determination, r^2, is given by:-

$$r^2 = \left[\frac{\sum (x - \bar{x})\,(y - \bar{y})}{\sqrt{\sum (x - \bar{x})^2 \sum (y - \bar{y})^2}} \right]^2$$
Equation 7

r^2 expresses the fraction of the data variation "accounted for" by the fitted model. If $r^2=0$, x and y are utterly unrelated. If $r^2=1$, the regression line is a perfect, deterministic fit. If $r^2=0.6$ then the fitted model "explains" 60% of the data variation.

There are many equivalent forms of these equations: Some are more aesthetic, some more mnemonic, some more economical, some less prone to numerical error.

The SSX, SXY, SYY Idiom

In certain circumstances brevity may be achieved by defining:-

$$SXX = \sum(x - \bar{x})^2 \qquad \textbf{Equation 8a}$$
$$SXY = \sum(x - \bar{x})(y - \bar{y}) \qquad \textbf{Equation 8b}$$
$$SYY = \sum(y - \bar{y})^2 \qquad \textbf{Equation 8c}$$

It then follows that:-

$$\alpha = \bar{y} - \beta\bar{x}$$
Equation 5

and:-

$$\beta = \frac{SXY}{SXX}$$
Equation 9

whilst:-

$$r^2 = \frac{SXY^2}{SXX.SYY}$$
Equation 10

Computed Paces versus Paces Percentage Specific Defect

This concerns the tendency of machine error to decrease as the number of paces increases during a particular measured walking episode. If practicable, we wish to quantify the error in order to plan a walk that achieves a particular accuracy of length estimation.

Computed Paces, P, is the product of Mean Paces per Lap, μ_{PL}; and Laps Unpaced, L, that is:-

$$P = L.\mu_{PL}$$
Equation 11

whilst Paces Percentage Specific Defect, PSD_P, is:-

$$PSD_P = 100\left(\frac{STEPS - P}{P}\right)$$
Equation 12

Plot Two displays the relation of P (x) to PSD_P (y) for the thirty-five non-zero data.

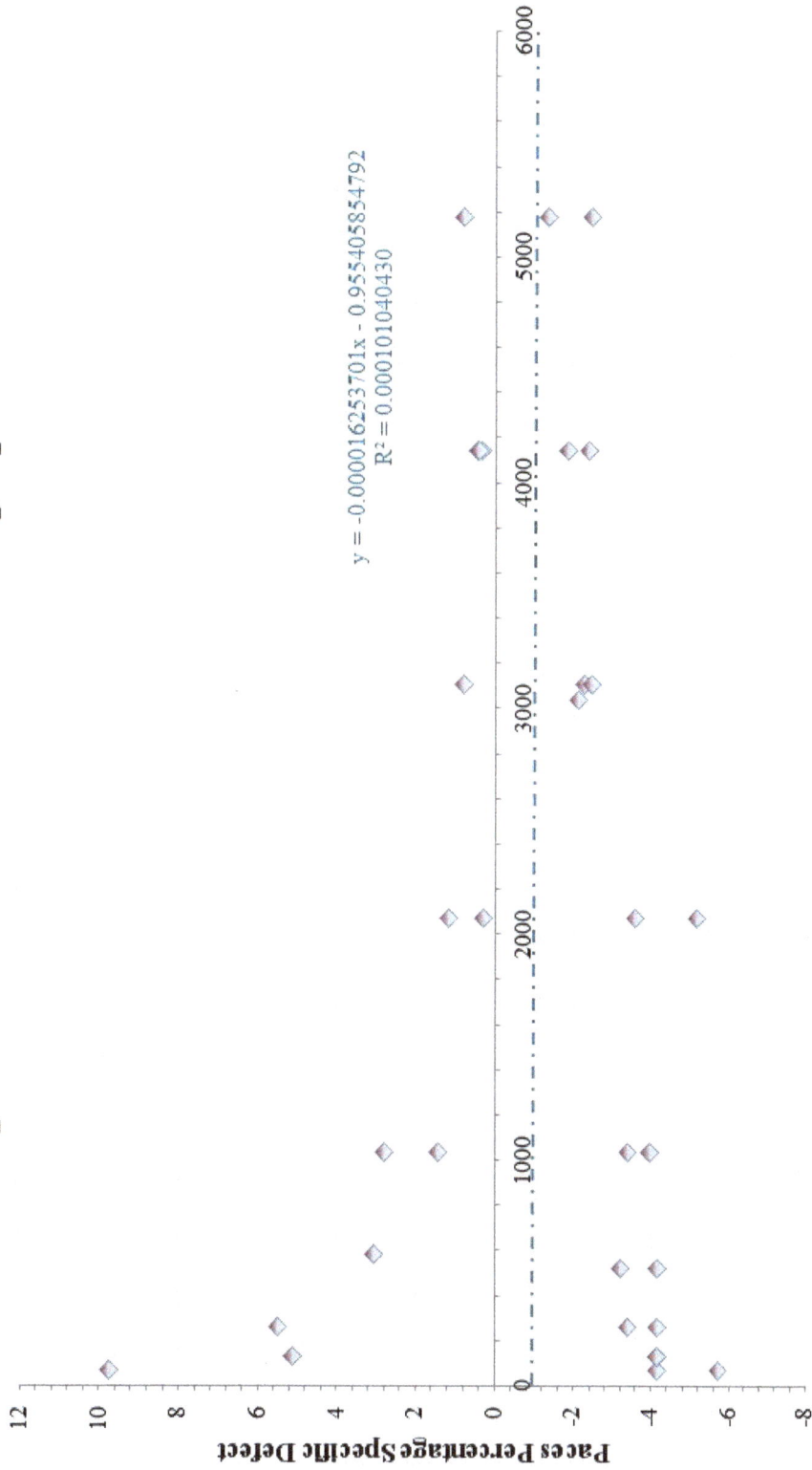

Computed Paces versus Paces Percentage Specific Defect

$y = -0.0000016253701x - 0.95540585479\overline{2}$
$R^2 = 0.000101040430$

Computed Paces

Paces Percentage Specific Defect

Plot Two
The Relation of PSDₚ to P

The plot and the data tabulation show that the Computed Paces (x) versus Paces Percentage Specific Defect (y) relation displays a systematic decrease in Paces PSD with increased Computed Paces; that is:-

$$PSD_P = f\left(\frac{1}{P}\right)$$
Equation 13

It is of course possible to re-arrange Equation Twelve the PSD equation in various ways with regard to function f(1/P), though this should be deferred at least until we have more information about the form of the function.

As you can see, I have fitted an EXCEL® linear regression trend to the Plot Two data.

This regression presents the equation:-

$$PSD_P = -0.995405854792 - 0.000016253701 \times P$$
Equation 14

or approximately:-

$$P \approx 61524(-PSD_P - 1)$$
Equation 15

another very weird "equation" that infers that the real paces unpaced may be positive or negative depending upon the local defect. It is of course a demonstration of the non-commutativity of actual and artificial statistics.

In order to frame a more useful P versus PSD_P relation we firstly need to compute the absolute value of PSD_P. This will in addition prevent the embarrassment of any subsequent logarithmic transformations.

If Δ is the Absolute Value of the Percentage Specific Defect of Pedometer and Mean Ground Paces we may define it as:-

$$\Delta = \left|100\left(\frac{STEPS - P}{P}\right)\right| \equiv \sqrt{\left[100\left(\frac{STEPS - P}{P}\right)\right]^2}$$
Equation 16

Readers who find the arbitrary multiplier tedious or prodigal may of course re-arrange fractional schemes.

All Δ are of course positive and this will facilitate more useful studies of pedometric error.

Exponential and Logarithmic Regressions

As noted, the Coefficient of Determination r^2 should be as close to unity as the data permit. Although we trespass dangerous ontological ground, we can say that the closer r^2 is to unity then the nearer our fitted model is to reality, even though we may remain wholly ignorant of the underlying science.

Exponential and Logarithmic Regressions are sophistications of Linear Regression which hinge upon the transformation of either environmental x or y data using an exponential or logarithmic function to provide the (linear) regression feed data X or Y.

Exponential Regression

Exponential Regression employs the transformations:-

$$X = P \qquad \textbf{Equation 17a}$$
$$Y = \ln \Delta \qquad \textbf{Equation 17b}$$

The (linear) regression output coefficients then express as:-

$$\alpha = e^{\bar{y} - \beta \bar{x}} \qquad \textbf{Equation 18a}$$
$$\beta = \frac{SXY}{SXX} \qquad \textbf{Equation 18b}$$
$$r^2 = \frac{SXY^2}{SXX.SYY} \qquad \textbf{Equation 18c}$$

Plot Three exhibits, for the thirty-five finite data: Linear, Exponential and Logarithmic Regressions of Absolute Paces Percentage Specific Defect, Δ, (y), versus Computed Paces, P, (x).

Also shown on Plot Three are the respective three EXCEL® fitted regression equations and their attendant Coefficients of Determination.

The Exponential Fitment is:-

$$\Delta = 4.028896921388e^{-0.000275156414P}$$
$$\textbf{Equation 19}$$

with:-

$$r^2 = 0.416309131712$$
$$\textbf{Equation 20}$$

The Exponential r^2 is inferior to the algebraic linear fitment which provides an $r^2 = 0.463536785948$ with equation:-

$$\Delta = 4.392491754196 - 0.000652566992P$$
$$\textbf{Equation 21}$$

In ordinary language the linear and exponential equations are as useless as one another.

It should be noted that arranging the data at each P (i.e. lap-count) level has no effect upon regression outcomes.

Logarithmic Regression

Logarithmic Regression employs the transformations:-

$$X = \ln P \qquad\qquad \textbf{Equation 22a}$$
$$Y = \Delta \qquad\qquad \textbf{Equation 22b}$$

The (linear) regression output coefficients then express as:-

$$\alpha = \bar{y} - \beta\bar{x} \qquad\qquad \textbf{Equation 23a}$$

$$\beta = \frac{SXY}{SXX} \qquad\qquad \textbf{Equation 23b}$$

$$r^2 = \frac{SXY^2}{SXX.SYY} \qquad\qquad \textbf{Equation 23c}$$

Now these facts lead directly to the third or fourth episode of unreasonable mathematics that we have encountered in the calibration of our pedometer.

Refer to Table Five where the discrepancies between the α, β and r^2 for Equations Twenty-Three; and the EXCEL® intrinsic values may be viewed.

Intercept and slope anomalies are minor though not negligible, but it is the determination coefficients that give real pause.

Equations Twenty-Three yield $r^2 = 0.620641301105$, whilst EXCEL® says $r^2 = 0.942022727844$, a lovely value that we earnestly hope is true.

I cannot account for this discrepancy.

Both correlations are markedly superior to either linear or exponential regressions. The EXCEL® form may be written:-

$$\Delta = 10.046859310454 - 1.020210957615 \times \ln P$$
$$\textbf{Equation 24}$$

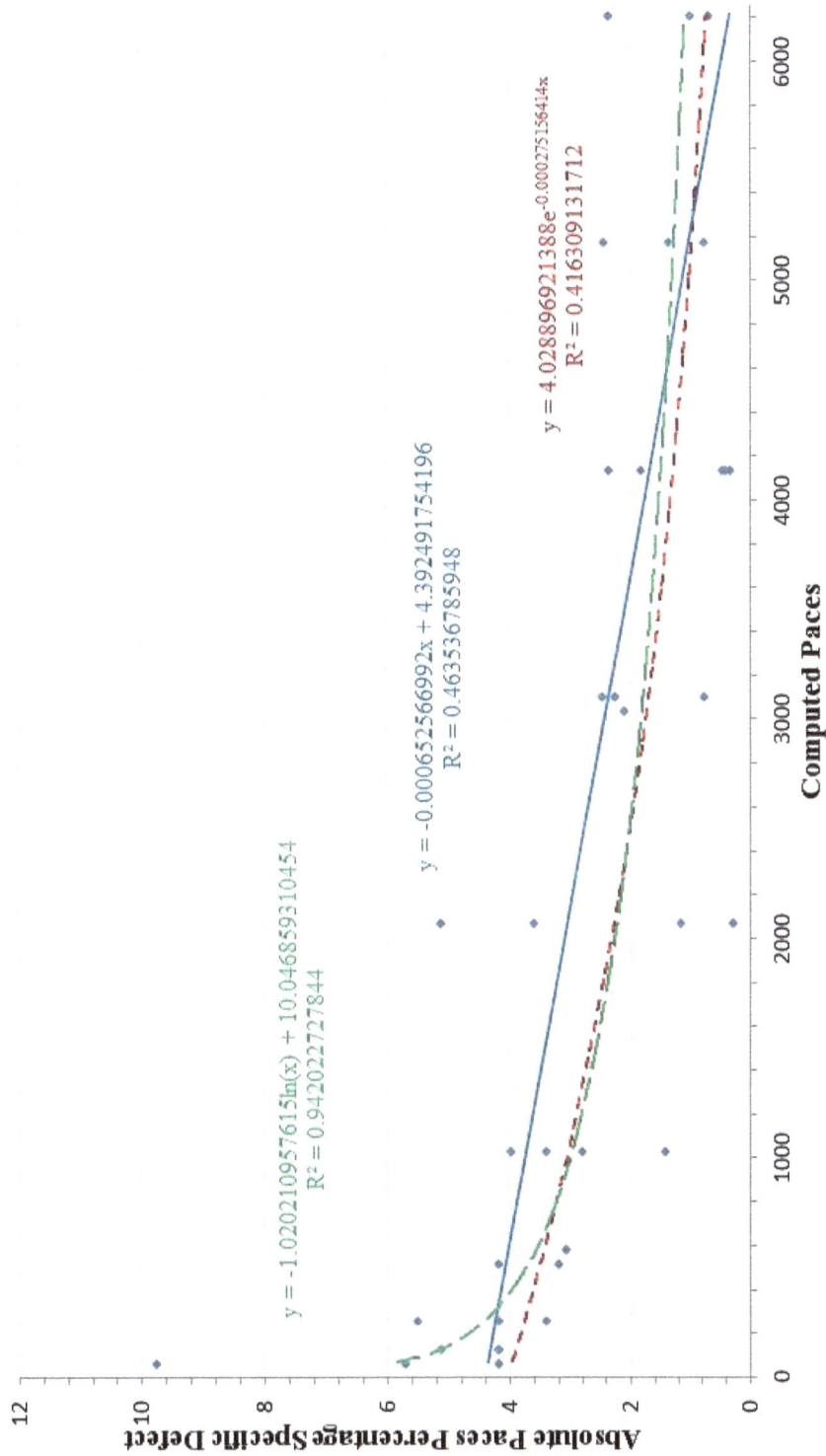

Plot Three
Computed Paces versus Absolute Paces
Percentage Specific Defect

	Linear Regression	Exponential Regression	Logarithmic Regression
Intercept	4.392491754196	4.028896921388	10.242559226167
Gradient	-0.000652566992	-0.000275156414	-1.048986958814
r^2	0.463536785948	0.416309131712	0.620641301105
Xbar	2360.283660714290	2360.283660714290	7.045188329059
Ybar	2.852248546598	0.744045434376	2.852248546598
SXX	147511651.953713000000	147511651.953713000000	76.435096948408
SXY	-96261.234929078100	-40588.777126899100	-80.179419894545
SYY	135.516546650896	26.826849326529	135.516546650896
α	4.392491754196	4.028896921388	10.242559226167
β	-0.000652566992	-0.000275156414	-1.048986958814
r^2	0.463536785948	0.416309131712	0.620641301105
EXCEL Plot α	4.392491754196	4.028896921388	10.046859310454
EXCEL Plot β	-0.000652566992	-0.000275156414	-1.020210957615
EXCEL Plot r^2	0.463536785948	0.416309131712	0.942022727844
EXCEL Model SSE	72.699642173554	71.777484649904	51.474404457040
EXCEL r^2	0.463536785948	-1.675583844239	0.620161480431

Table Five
P versus Δ Regressions Summary Data

	Linear Regression	Exponential Regression	Logarithmic Regression
Intercept	=INTERCEPT(B5:B39,A5:A39)	=EXP(INTERCEPT(D5:D39,A5:A39))	=INTERCEPT(B5:B39,C5:C39)
Gradient	=SLOPE(B5:B39,A5:A39)	=SLOPE(D5:D39,A5:A39)	=SLOPE(B5:B39,C5:C39)
r^2	=RSQ(B5:B39,A5:A39)	=RSQ(D5:D39,A5:A39)	=RSQ(B5:B39,C5:C39)
Xbar	=AVERAGE(A5:A39)	=AVERAGE(A5:A39)	=AVERAGE(C5:C39)
Ybar	=AVERAGE(B5:B39)	=AVERAGE(D5:D39)	=AVERAGE(B5:B39)
SXX	=SUM(F5:F39)	=SUM(K5:K39)	=SUM(P5:P39)
SXY	=SUM(G5:G39)	=SUM(L5:L39)	=SUM(Q5:Q39)
SYY	=SUM(H5:H39)	=SUM(M5:M39)	=SUM(R5:R39)
α	=W7-W14*W6	=EXP(X7-X14*X6)	=Y7-Y14*Y6
β	=W10/W9	=X10/X9	=Y10/Y9
r^2	=(W10^2)/(W9*W11)	=(X10^2)/(X9*X11)	=(Y10^2)/(Y9*Y11)
EXCEL Plot α	4.392491754196	4.028896921388	10.046859310454
EXCEL Plot β	-0.000652566992	-0.000275156414	-1.020210957615
EXCEL Plot r^2	0.463536785948	0.416309131712	0.942022727844
EXCEL Model SSE	=SUM(J5:J39)	=SUM(O5:O39)	=SUM(T5:T39)
EXCEL r^2	=1-W21/W11	=1-X21/X11	=1-Y21/Y11

Table Six
P versus Δ Regressions Summary Data
(formulae)

The Distance Unpaced for a Given Magnitude of Statistical Error

The Absolute Paces Percentage Specific Defect, Δ, is a measure of the likely error in the use of the pedometer over a given distance expressed in actual ground paces, P. (*Not* pedometer-logged paces, STEPS).

Specific Defect is a metric directly related to standard deviation or root mean square error, but in correctly-chosen contexts is simpler and more informative than either.

Therefore, for any required tolerance we may compute the number of actual paces, P, and thus assess the real ground distance that must be walked to achieve that accuracy.

Let the Mean Actual Paces per kilometer be π_{kms}, and the Mean Actual Paces per Mile be π_{miles}. Then the best estimates of these statistics for our pedometric experiments are:-

$$\pi_{kms} = 1311.119274$$
$$\pi_{miles} = 2110.041938$$

Suppose that we want to walk a distance whose pedometrically-assessed value is subject to only 1% deviation from the real ground value.

Then $\Delta = 1$ and the required actual ground paces are:-

$$P = e^{\frac{\Delta - \alpha}{\beta}}$$
Equation 25

Equation Twenty-Five yields P = 7098.47958536091, which value equates to 5.414061 kilometers or 3.364141 miles. The expected error in any further distance is less than one percent.

References

1   There are known knowns. (2016, September 16). In *Wikipedia, The Free Encyclopedia*. Retrieved 15:26, September 18, 2016, from https://en.wikipedia.org/w/index.php?
     title=There_are_known_knowns&oldid=739705802

2   Johari window. (2016, July 20). In *Wikipedia, The Free Encyclopedia*. Retrieved 15:28, September 18, 2016, from https://en.wikipedia.org/w/index.php?
     itle=Johari_window&oldid=730708814

CHAPTER TWO

Bhaskara's First Proof

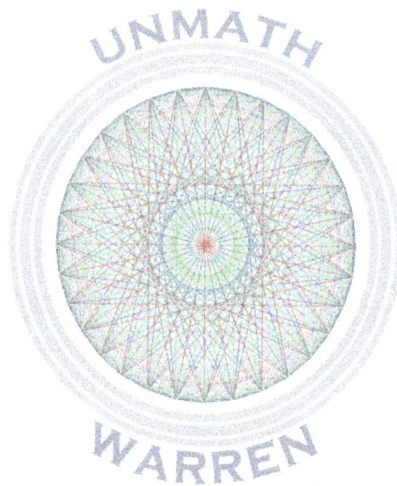

The Iteration of Bhaskara's First Proof of
The Theorem of Pythagoras

By
James R Warren BSc MSc PhD PGCE

PART ONE
INTRODUCTORY REMARKS

Sometime in remote antiquity, possibly in Neolithic prehistory, a forgotten genius made a startling but fecund observation with regard to any right angled triangle[1].

He or she declared that the sum of the squared lengths of the figure's opposite and adjacent sides was the same as the square of its hypotenuse.

Today we add that this condition is special to Euclidean plane right triangles, or make some other qualification we adjudge prudent or helpful.

The geometrical situation is illustrated by the sketch below:-

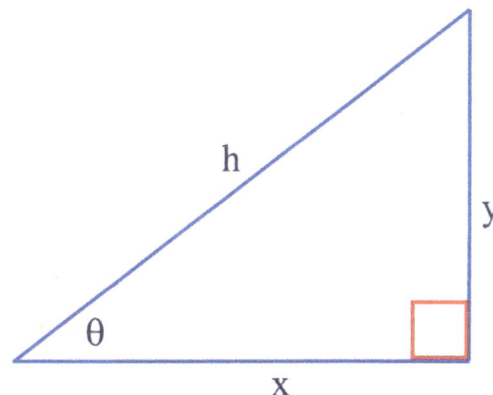

Figure One
Notation About a Right-Angled Triangle

So in algebraic terms this maxim, now conventionally called The Theorem of Pythagoras, may be couched as:-

$$h^2 = x^2 + y^2$$
Equation 1

Where h is the Hypotenuse; x is the Adjacent and y the Opposite. θ is the Angle Subtending the Opposite. The red square between the Adjacent and the Opposite only indicates that their interangle must be ninety degrees or $\pi/2$ radians as a condition of the rule.

We know that the Pythagorean Theorem was understood by the people of Ancient Mesopotamia who lived about four thousand years ago because they discuss it in their cuneiform literature which has come down to us preserved upon clay tablets.

Modern expressions such as the symbolism of Equation One are, on the other hand, no older than the sixteenth century *Anno Domini*.

Bhaskara's First Proof

There are tens of independent mathematical proofs of Equation One spanning the millennia separately developed by cultures as diverse as China, India, the Victorian USA, modern Iraq and Classical Greece. In many cases it was probably a natural fallout of cadastral survey or architectonic planning.

Our present discussion elaborates the First Proof of the Theorem of Pythagoras by the prolific Hindu mathematician Bhaskara who was active in the early twelfth century *Anno Domini*. I do not know if Bhaskara was influenced by earlier European or Arab work: Even Chinese antecedents should not be discounted.

Bhaskara's First Proof hinges upon visualising four identical right triangles inscribed in a (larger) square in the manner of Figure Two:-

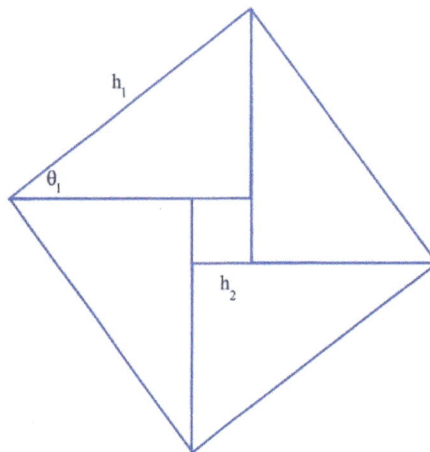

Figure Two
Bhaskara's Proof Scheme

It is evident that in addition to the four triangles there is a congeries of right angles, and that a smaller central square necessarily issues.

Bhaskara's Proof may be summarised by these steps[2]:-

(1) $A = h.h = h^2$

(2) $4t = 4 \times \left(\frac{1}{2}xy\right) = 2xy$

(3) $a = (x - y)(x - y) = x^2 - 2xy + y^2$

(4) $A = a + 4t$
$$\therefore h^2 = 2xy + (x^2 - 2xy + y^2)$$

$$\therefore h^2 = y^2 + x^2 \qquad \text{QED}$$

where A is the Area of the Large (Inclusive) Square; a is the Area of the Small Square; and t is the Area of a Triangle.

Bhaskara would not of course have used our modern notation, but his argument is the same.

A Trigonometric Form of Bhaskara's Proof

Note that for any one triangle:-

$$Opposite = y = h.\sin\theta \qquad \textbf{Equation 2}$$
$$Adjacent = x = h.\cos\theta \qquad \textbf{Equation 3}$$

So that:-

$$t = \frac{xy}{2} = \frac{h\sin\theta\, h\cos\theta}{2}$$

Equation 4

Or:-

$$t = \frac{h^2}{2}\sin\theta\cos\theta$$

Equation 5

From which:-

$$t = \frac{h^2}{4}\sin 2\theta$$

Equation 6

Accordingly, the Area of the Four Triangles, T, is given by:-

$$T = h^2 \sin 2\theta$$

Equation 7

Also:-

$$a = (h\cos\theta - h\sin\theta)(h\cos\theta - h\sin\theta)$$
$$= h^2(\cos\theta - \sin\theta)(\cos\theta - \sin\theta)$$
$$= h^2(\cos\theta - \sin\theta)^2$$
$$= h^2[(\cos\theta)^2 - 2\cos\theta\sin\theta + (\sin\theta)^2]$$
$$= h^2(1 - 2\cos\theta\sin\theta)$$

therefore:-

$$a = h^2[1 - \sin 2\theta]$$
Equation 8

We are now in a position trigonometrically to cast the steps of Bhaskara's First Proof as:-

(1) $A = h^2$
(2) $4t = T = h^2 \sin 2\theta$
(3) $a = h^2(1 - \sin 2\theta)$
(4) $A = h^2(1 - \sin 2\theta) + h^2 \sin 2\theta$
$$= h^2[(1 - \sin 2\theta) + \sin 2\theta]$$
$$= h^2$$
(5|) $A = h^2$
$$= h^2(\cos \theta)^2 + h^2(\sin \theta)^2$$
$$= h^2[(\cos \theta)^2 + (\sin \theta)^2]$$
$$= h^2 \times 1$$
$$= h^2$$
(6) $h^2 = y^2 + x^2$ QED

My trigonometric translation is strictly informal and of course probably includes an element of tautology.

An Iteration of Bhaskara's Proof Scheme

Our interest concerns the repetition of Bhaskara's four-triangle nests in the conception of Figure Three where the first, blue, iterative set of triangles is succeeded by the green set which is in turn superseded by the black set. At each stage the residual central square is of course smaller:-

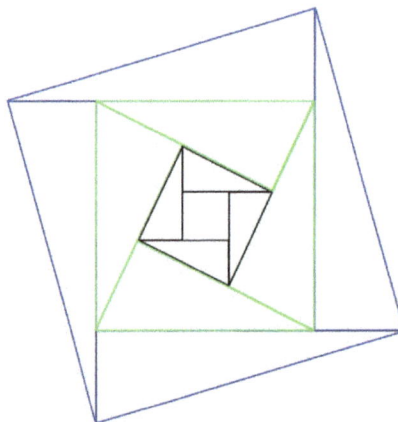

Figure Three
Three-Stage Iteration

In the light of the foregoing we may, for n iterations subscripted i for i=0..n, define convergent process for any series of positive θ_i *whatsoever*, for $0<\theta_i<\pi/4$.

If any $\theta_i=\pi/4$ then $a_{i+1}=0$ and the iterative series extinguishes.

As a matter of algebraic convenience we may assign θ_i's using a logistic model such as:-

$$z_i = z_S + i.z_C$$
Equation 9A

and:-

$$\theta_i = U\left[\frac{L}{1 + e^{-k(z_i-z_m)}}\right]$$
Equation 9B

where z_i is the Logistic Driver Value; z_S is the Driver Shift; and z_C is the Increment of z. Also, U is the Logistic Upper Bound; e is the Napierian Base ; k is the Logistic Steepness; and z_m is the Logistic Midpoint.

I emphasise that the generative angles sequences θ could literally be any set of values within the zero to ninety degree limit. This logistic model is only chosen as an illustration of a convergent continuous series yielding a convergence of h to zero. As we shall see later another function that converges to zero can generate h converging to a finite limit (that is, a residual square of finite area).

In my logistic convergence experiment the constants used were as follows:-

	Value
z_S	0
z_C	1/48
z_m	0.5
U	$\pi/4$
L	1
k	10

Table One
Logistic Curve Constants

I should say that this selection is arbitrary and is conditioned by the number of increments convenient using PTC Mathcad Express®, and by the need to prevent adventitious extinctions, which would of course vitiate the spirit of the Bhaskara method.

In practice, only thirty iterations exhausted both EXCEL® and Express® on my 64-bit HP Pavilion 500-400na.

The Local Check Sum, Z_i

At any iteration the Sum of the Areas of all the Triangles, W_i, with the Area of the Residual Square, a_i, must equal the Area of the Original Square, A_0.
Or equivalently Check Sum Z_i is given by:-

$$Z_i = 0 = A_0 - W_i - a_1$$
Equation 10

where:-

$$A_0 = h_0^{\,2}$$
Equation 11

$$W_i = \sum_{i=0}^{i} T_i$$
Equation 12

$$a_i = h_0^{\,2} = [h_i \cos \theta_i - h_i \sin \theta_i]^2$$
Equation 13

Therefore:-

$$Z_i = A_0 - \left(\sum_{i=0}^{i} T_i \right) - [h_i \cos \theta_i - h_i \sin \theta_i]^2$$

$$= A_0 - \left(\sum_{i=0}^{i} T_i \right) - [h_{i+1}]^2$$

$$= A_0 - \left(\sum_{i=0}^{i} h_i^{\,2} . \sin 2\theta_i \right) - [h_i \cos \theta_i - h_i \sin \theta_i]^2$$

$$= A_0 - \left(\sum_{i=0}^{i} h_i^{\,2} . \sin 2\theta_i \right) - h_i^{\,2} (\cos \theta_i - \sin \theta_i)^2$$

$$= A_0 - \left(\sum_{i=0}^{i} h_i^{\,2} . \sin 2\theta_i \right) - h_2^{\,2} (1 - 2 \cos \theta_i \sin \theta_i)$$

$$= A_0 - \left(\sum_{i=0}^{i} h_i^{\,2} . \sin 2\theta_i \right) - h_2^{\,2} (1 - \sin 2\theta_i)$$

Therefore:-

$$Z_i = A_0 - \left[\sum_{i=0}^{i} A_i \sin 2\theta_i\right] - A_i(1 - \sin 2\theta_i)$$

Equation 14

Allow that:-

$$S = \sin 2\theta_i$$

Equation 15

Then:-

$$Z_i = A_0 - \left[\sum_{i=0}^{i} A_i S_i\right] - A_i(1 - S_i)$$

Equation 16

Or:-

$$Z_m = [A_0 - A_m] - \sum_{i=0}^{m-1} A_i S_i$$

Equation 17

where 0<m<n.
All forms of Z_i are zero.

PART TWO
THE SUCCESSION OF HYPOTENUSES

If h_0 is the common Hypotenuse of the four first included triangles then it identifies with the Side of the primitive, inclusive square.

Accordingly:-

$$A = h_0{}^2$$
Equation 18

where A is the Area of the Primitive Square.

The geometry of the iterated triangles shows that the second triangles' hypotenuse, h_1, is given by:-

$$h_1 = x_0 - y_0 = h_0 \cos \theta_0 - h_0 \sin \theta_0$$
Equation 19

or:-

$$h_1 = h_0(\cos \theta_0 - \sin \theta_0)$$
Equation 20

More generally:-

$$h_{j+1} = h_0 \prod_{i=0}^{j-1}(\cos \theta_i - \sin \theta_i)$$
Equation 21

where i=0..n, and j=1..n.

Equation Twenty-One defines the successor hypotenuses from which it is manifest that a successor square's Area, a_j, is:-

$$a_j = h_j{}^2$$
Equation 22

An Identity of Cosθ-Sinθ

By the Sum Identity we may write:-

$$\cos(\alpha + \beta) = \cos \alpha \cos \beta - \sin \alpha \sin \beta$$
Equation 23

Setting:-

$$\beta = \theta$$
Equation 24

Equation Twenty-Three may be re-written as:-

$$\cos(\alpha + \beta) = q(\cos\theta - \sin\theta)$$
Equation 25

for some particular and unique value of q.
Now to eliminate α note that for Cosα=Sinα; α=π/4; and:-

$$q = \frac{\sqrt{2}}{2} = \frac{1}{\sqrt{2}}$$
Equation 26

Therefore:-

$$\cos\theta - \sin\theta \equiv \sqrt{2}.\cos\left(\frac{\pi}{4} + \theta\right) \equiv \sqrt{2}.\sin\left(\frac{\pi}{4} - \theta\right)$$
Equation 27

Some Approximations of Cosθ-Sinθ

Noting that:-

$$h_{i+1} = h_i.\sqrt{2}.\cos\left(\frac{\pi}{4} + \theta_i\right)$$
Equation 28

and that:-

$$h_j = h_0.2^{\frac{j}{2}}.\cos\left(\frac{\pi}{4} + \theta_i\right)$$
Equation 29

it is useful to consider the following:-

(1)   The Binomial Theoem[3]

$$_NC^K = \binom{N}{K} = \frac{N!}{(N-K)!\,K!}$$
Equation 30

where N is the Number of Entities and K is the Number of (non-empty) Subsets which may be drawn of N.

(2)   The Binomial Expansion[4]

$$BE = (\alpha + \beta)^N = \sum_{K=0}^{N} {}_N C^K . \alpha^{N-K} . \beta^N$$

Equation 31

(3)   The Cosine Series[5]

$$CS = \cos\theta = \frac{\theta^0}{0!} - \frac{\theta^2}{2!} + \frac{\theta^4}{4!} - \frac{\theta^6}{6!} + \cdots \pm \frac{\theta^{2i}}{(2i)!}$$

$$\therefore CS = \sum_{i=0}^{n} (-1)^i \frac{\theta^{2i}}{(2i)!}$$

Equation 32

(4)   The Sine Series[6]

$$SS = \sin\theta = \frac{\theta^1}{1!} - \frac{\theta^3}{3!} + \frac{\theta^5}{5!} - \frac{\theta^7}{7!} + \cdots \pm \frac{\theta^{(2i+1)}}{(2i+1)!}$$

$$\therefore CS = \sum_{i=0}^{n} (-1)^i \frac{\theta^{(2i+1)}}{(2i+1)!}$$

Equation 33

Conflation of CS and SS

Simple addition of the series terms of the Cosine and Sine Series permits us to approximate Cosθ –Sinθ in this manner:-

$$\cos\theta - \sin\theta = \sqrt{2}.\cos\left(\frac{\pi}{4} + \theta\right) = \sqrt{2}.\sin\left(\frac{\pi}{4} - \theta\right)$$

$$\therefore \cos\theta - \sin\theta \approx \sum_{i=1}^{i_{max}} sign\left[\tan\left\{\frac{(i+\varepsilon).\pi}{4}\right\}\right] . \frac{\theta^i}{i!}$$

Equation 34

Where i_{max} is the number of terms grazed from *both* series according to the power of θ.

Numerically, Equation Thirty-Four gave eleven-figure accuracy when i_{max} was nine.

ε is an arbitrary finitude-forcer which I chose to be 1+π/8, a figure which restricted the absolute size of the tangents computed.

This defines the jth. hypotenuse as:-

$$h_j = h_0 . \prod_{i=0}^{j-1} \sum_{l=1}^{l_{max}} sign\left[tan\left\{\frac{(l+\varepsilon).\pi}{4}\right\}\right] . \frac{\theta^l}{l!}$$

Equation 35

where l_{max} is chosen to return adequate accuracy of h_j (12-figure in my trials).

Conflation of BE and CS

The conflation of the Binomial Expansion and the Cosine Series may be expressed as:-

$$h_j = h_0 . 2^{\frac{j}{2}} . \prod_{i=0}^{j-1} \sum_{k=1}^{k_{max}} (-1)^k . \frac{\left(\frac{\pi}{4} + \theta_i\right)^{2k}}{(2k)!}$$

Equation 36

where k_{max} is chosen to return adequate accuracy of h_j (12-figure in my trials). In this context, Mathcad® trials ascertained that for a twelve-figure accuracy of Equation Thirty-Six h_j relative to Equation Twenty-Nine as fiducial, the value of k_{max} required is eight.

The structure of Equation Thirty-Six permits various algebraic approximations to be feasible. Fourth-order expansions are elaborated below to demonstrate a fourth-order polynomial estimator. This is offered as a *proof-of-principle only*.

Firstly, an expansion of Cos(α+β) may be expressed as:-

$$\cos(\alpha + \beta) = \frac{(\alpha + \beta)^0}{0!} - \frac{(\alpha + \beta)^2}{2!} + \frac{(\alpha + \beta)^4}{4!} - \cdots$$

Equation 37

The term $(\alpha+\beta)^0/0!$ Is of course unity whilst a quartic elaboration of the third term of Equation Thirty-Seven may be expressed as:-

$$(\alpha + \beta)^4 = \alpha^4 + 4\alpha^3\beta + 6\alpha^2\beta^2 + 4\alpha\beta^3 + \beta^4$$

Equation 38

Because in our application $\alpha=\pi/4$ we may recast Equation Thirty-Eight as:-

$$(\alpha + \beta)^4 = \frac{\pi^4}{256} + \frac{\pi^3}{16}\beta + \frac{6.\pi^2}{16}\beta^2 + \pi\beta^3 + \beta^4$$

Equation 39

Similarly, the simpler and shorter quadratic elaboration expands as:-

$$(\alpha + \beta)^2 = \alpha^2 + 2\alpha\beta + \beta^2$$
Equation 40

which by α substitution is:-

$$(\alpha + \beta)^2 = \frac{\pi^2}{16} + \frac{\pi}{2}\beta + \beta^2$$
Equation 41

Now by substitution of Equations Thirty-Nine and Forty-One in Equation Thirty-Seven we may write:-

$$\cos\left(\frac{\pi}{4} + \theta_i\right) \approx 1 - \frac{\frac{\pi^2}{16} + \frac{\pi}{2}\theta_i + \theta_i^2}{2} + \frac{\frac{\pi^4}{256} + \frac{\pi^3}{16}\theta_i + \frac{6.\pi^2}{16}\theta_i^2 + \pi\theta_i^3 + \theta_i^4}{24}$$
Equation 42

Simplification of this conflated partial series issues in the algebraic approximant:-

$$\cos\left(\frac{\pi}{4} + \theta_i\right) \approx \frac{1}{24}\theta_i^4 + \frac{\pi}{24}\theta_i^3 + \left(\frac{6\pi^2}{384} - \frac{1}{2}\right)\theta_i^2 + \left(\frac{\pi^2}{384} - \frac{\pi}{4}\right)\theta_i + \left(1 - \frac{\pi^2}{32} + \frac{\pi^4}{6144}\right)$$
Equation 43

To abbreviate Equation Forty-Three we may set:-

$$p = \frac{\pi}{4}$$
Equation 44a

and:-

$$r = \sqrt{\frac{\pi}{4}}$$
Equation 44b

which permit by substitution:-

$$\cos\left(\frac{\pi}{4} + \theta_i\right) \approx \frac{1}{24}\theta_i^4 + \frac{r}{6}\theta_i^3 + \left(\frac{p}{4} - \frac{1}{2}\right)\theta_i^2 + \left(\frac{pr}{6} - r\right)\theta_i + \left(1 - \frac{p}{2} + \frac{p^2}{24}\right)$$
Equation 45

Plots of Equation Forty-Three against $\cos(\pi/4+\theta)$ for sixty-four intervals of θ between 0 and $\pi/4$ demonstrated that $\cos(\pi/4+\theta)$ was significantly overestimated for larger θ.

Neglect of the high powers θ^3 and θ^4 led, on the other hand, to marked underestimation of larger θ.

Quadratic Regression as an Estimator of $\cos\theta$-$\sin\theta$

An EXCEL® trial of θ versus $\cos\theta$-$\sin\theta$ was made and plotted with an EXCEL® quadratic regression fitted according to:-

$$\cos\theta - \sin\theta = c_0 + c_1\theta + c_2\theta^2$$
Equation 46

For $0 \leq \theta \leq \pi/4$, in sixty-four intervals, where the coefficients are as listed in Table Two:-

Regression Coefficient	Value
c_0	1.004864660753
c_1	-1.075783034600
c_2	-0.267540162661

Table Two
θ versus $\cos\theta$-$\sin\theta$
Quadratic Regression Coefficients

These coefficients do not of course identify with those of the quadratic part of the Cosine-Binomial Expansion Conflation Model.

The Coefficient of Determination, R^2, proved to be 0.999951334620, indicating an excellent inverse fit, though there is slight discrepancy for larger values of angle.

There are of course many better ways of calculating trigonometric functions out there, but time and space preclude further digressions into computational efficacies and efficiencies.

This disquisition (like all my work) is merely indicative.

PART THREE
SOME TRIALS USING GENERATIVE ANGLE SEQUENCES

Generally:-

$$\cos\big(f(\theta)\big) - \sin\big(f(\theta)\big) \equiv \sqrt{2}.\cos\left[\frac{\pi}{4} + f(\theta)\right]$$
Equation 47

and correspondingly:-

$$h_j = h_0.\sqrt{2}.\prod_{j=0}^{j-1} \cos\left[\frac{\pi}{4} + f(\theta)\right]$$
Equation 48

when f(θ) issues in an angular value θ (or θ_i) which lies in the range 0≤θ≤π/4.

Case One: A Short Series of Arbitrary Angles

The set of θ_i for 0≤i≤n may be any without restriction, save that above.
To illustrate this we may consider the arbitrary four-number sequence:-

$$\theta_i = \{0.445637, 0.293668, 0.529421, 0.722979\}$$
Set 1

The computed results, including successive hypotenuses, are displayed in Table Three:-

Serial i	Data a	Data b	Data h	Check a²+b²	h²	Area ab/2	Acute Angle α Radians	Acute Angle α Radians/π	Acute Angle α Degrees	Trig Opposite	Trig Adjacent	Trig Triangle Area	Four Triangles' Area	Residual Square Area	Total Square Area =h²	Next Hypotenuse
0	66.5	123	140	19551	19551.3	4090	0.495637	0.157766	28.39789	66.5	123	4089.75	16359	3192.25	19551.25	56.5
1	16.3	53.9	56.3	3171	3170.9	439.3	0.293668	0.093477	16.82594	16.3	53.9	439.285	1757.14	1413.76	3170.9	37.6
2	18.9	32.3	37.4	1401	1400.5	305.2	0.529421	0.16852	30.33358	18.9	32.3	305.235	1220.94	179.56	1400.5	13.4
3	9	10.2	13.6	185	185.04	45.9	0.722979	0.230131	41.42367	9	10.2	45.9	183.6	1.44	185.04	1.2

Table Three
Arbitrary Angles' Bhaskara Iterates

Figures Four and Five respectively illustrate the successions of iterated Bhaskara triangles nested in the unit square:-

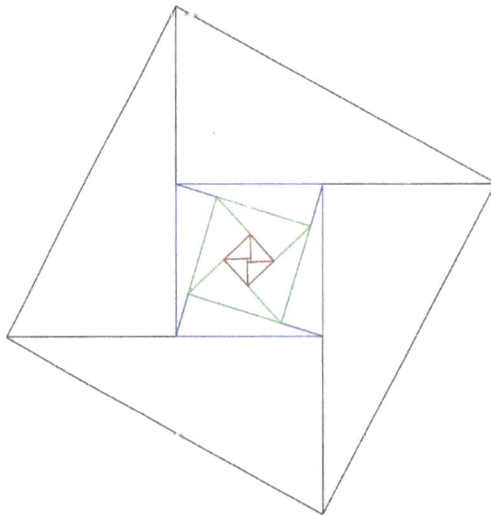

Figure Four
Arbitrary Angles' Iterated Bhaskara Triangles
As A Line Diagram

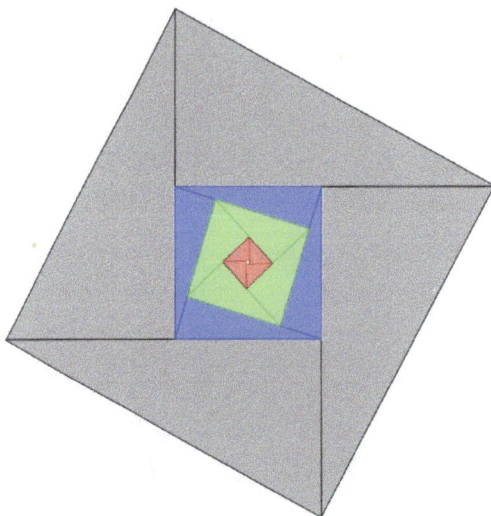

Figure Five
Arbitrary Angles' Iterated Bhaskara Triangles
As A Filled Diagram

Because all four angles are less than $\pi/4$ extinction never supervenes.

Case Two: A Logistic Model

A (Verhulstian) Logistic Model has the property of expanding and then diminishing the θ_i angle series as the values of θ_i converge upon, but never attain, zero.

Also, the succession of hypotenuses h_j (and residual squares h_j^2), continuously contract, but never extinguish.

The relevant equation is:-

$$f(\theta_i) = U.\frac{L}{[1 + e^{k(x-x_0)}]}$$

Equation 49

where U is the Limiting Upper Bound of $f(\theta_i)$; L is the Functional Maximum; e is the Napierian Base; k is the Steepness; x is the Driver Distribution; and x_0 is the Driver Distribution Midpoint.

Manipulation of the RHS parameters alters the conformation of the sigmoidal cumulate, and the equation is engineered so that $f(\theta_i)$ will itself lie on the interval $0 \leq f(\theta_i) \leq \pi/4$.

Table Four elaborates the computed results for a contrived angular succession illustrative of the logistical principle:

SCHEME B DIAGRAM III MODEL: LOGARITHMIC SPIRAL

Intervals	8
Initial Hypotenuse	1
Angular Intercept	0.392699
Angular Grade	0.098175
Diagram Scaling Constant	100
$\tan(\theta_i)/\sin(\theta_i)$	1.002442
$1/\cos(\theta_i)$	1.002442
SQRT(2)	1.414214

Serial i	θ_i (degrees)	Cumulate θ_i (degrees)	θ_i (radians)	$\cos(\theta)$	$\sin(\theta)$	Model h_i	θ_i (radians)	Computed h_i (mm)	$-\log_e(h_i)$
0	0	0	0	1	0	1	0	1.0	0
1	2.5	2.5	0.04363323	0.99904822	0.04361939	0.95542883	0.04363323	0.95542883	0.045595
2	5	7.5	0.08726646	0.9961947	0.08715574	0.86852203	0.08726646	0.86852203	0.14096233
3	7.5	15	0.13089969	0.99144486	0.13052619	0.74772683	0.13089969	0.74772683	0.29071757
4	10	25	0.17453293	0.98480775	0.17364818	0.60652578	0.17453293	0.60652578	0.50000805
5	7.5	32.5	0.13089969	0.99144486	0.13052619	0.52216937	0.13089969	0.52216937	0.64976329
6	5	37.5	0.08726646	0.9961947	0.08715574	0.47467229	0.08726646	0.47467229	0.74513062
7	2.5	40	0.04363323	0.99904822	0.04361939	0.4535156	0.04363323	0.4535156	0.79072562
8	4	44	0.06981317	0.99756405	0.06975647	0.42077521	0.06981317	0.42077521	0.86565654
44		204	0.76794487	8.95574737	0.7660073	6.04933593	0.76794487	6.04933593	4.02855901

Serial i	Hyp	Adj	Opp	$\tan(\theta)$	$\tan(\theta_i)/\sin(\theta_i)$	Area of Central Square	Azimuth Ex PhotoDraw (degrees)	Difference Ex PhotoDraw (degrees)	Measured from Paper	Specific Defect
0	100.0	100.0	0.0	0.0	1	1	0	0		0
1	95.5	95.5	4.2	0.043661	1.000953	0.912844	2.5	2.5	2.5	0
2	86.9	86.5	7.6	0.087489	1.00382	0.754331	8	5.5	5	-9.09090909091
3	74.8	74.1	9.8	0.131652	1.008629	0.559095	18	10	10	0
4	60.7	59.7	10.5	0.176327	1.015427	0.367824	33	15	15	0
5	52.2	51.8	6.8	0.131652	1.008629	0.272661	43.1	10.1	10	-0.99009901
6	47.5	47.3	4.1	0.087489	1.00382	0.225314	48.8	5.7	6.1	7.01754386
7	45.4	45.3	2.0	0.043661	1.000953	0.205676	51.3	2.5	3	20
8	42.1	42.0	2.9	0.069927	1.002442	0.177052	56	4.7	5	6.38297873
	604.9	602.2	47.9	0.771858	9.044671	4.474847	260.7	56	56.6	2.93493931
									Totals	Means

	R	G	B
Black	0	0	0
Maroon	135	0	32
Red	255	0	0
Orange	255	127	0
Yellow	255	255	0
Green	0	255	0
Blue	0	0	255
Indigo	181	0	252
Violet	233	0	254

Table Four
Logistic Models' Bhaskara Iterates

Figures Six and Seven respectively show the successions of iterated Bhaskara triangles for the illustrative Table Four:-

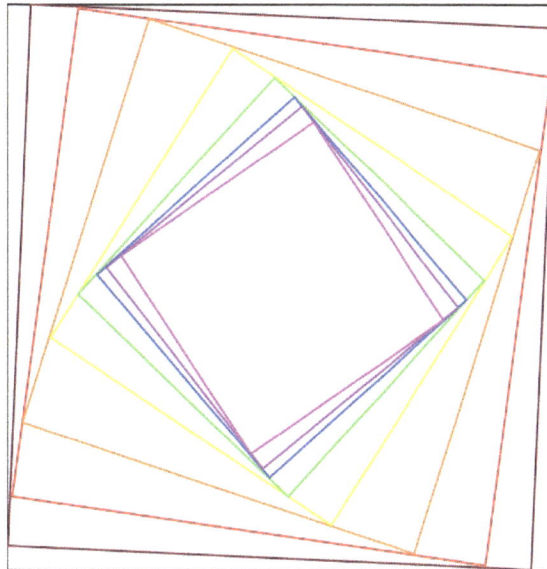

Figure Six
Hand-contrived Logistical Model Bhaskara Triangles
As a Line Diagram

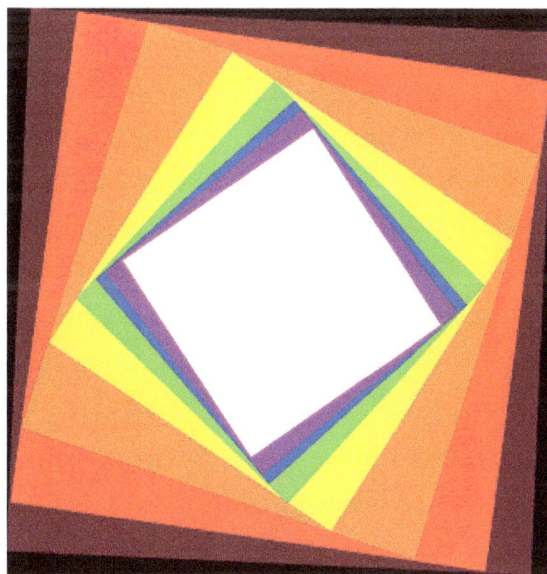

Figure Seven
Hand-contrived Logistical Model Bhaskara Triangles
As a Filled Diagram

A more realistic logistic scenario is given by the Logistic Driver Model of Table Five in which the Driver Function x conforms to a first-order polynomial advancement.

UPPER BOUND	UB	0.785398
LOWER BOUND	LB	0
ABSCISSAL SHIFT	X	0
MIDPOINT VALUE	x_0	0.5
FUNCTIONAL MAXIMUM	L	1
STEEPNESS	k	10
ABSCISSAL INCREMENT	x	0.033333

Serial			Acute Angle	
i	x	Radians	Radians /π	Degrees
0	0	0.005257	0.001673	0.301178
1	0.033333	0.007317	0.002329	0.419218
2	0.066667	0.010174	0.003238	0.582918
3	0.1	0.014126	0.004497	0.809379
4	0.133333	0.019576	0.006231	1.121599
5	0.166667	0.027053	0.008611	1.550034
6	0.2	0.037248	0.011856	2.134164
7	0.233333	0.051027	0.016242	2.923613
8	0.266667	0.069429	0.0221	3.977985
9	0.3	0.093622	0.029801	5.364131
10	0.333333	0.124776	0.039717	7.14911
11	0.366667	0.163841	0.052152	9.387384
12	0.4	0.211226	0.067235	12.10236
13	0.433333	0.266441	0.084811	15.26596
14	0.466667	0.327849	0.104357	18.78434
15	0.5	0.392699	0.125	22.5
16	0.533333	0.45755	0.145643	26.21566
17	0.566667	0.518957	0.165189	29.73404
18	0.6	0.574172	0.182765	32.89764
19	0.633333	0.621557	0.197848	35.61262
20	0.666667	0.660623	0.210283	37.85089
21	0.7	0.691776	0.220199	39.63587
22	0.733333	0.715969	0.2279	41.02201
23	0.766667	0.734371	0.233758	42.07639
24	0.8	0.74815	0.238144	42.86584
25	0.833333	0.758345	0.241389	43.44997
26	0.866667	0.765823	0.243769	43.8784
27	0.9	0.771272	0.245503	44.19062
28	0.933333	0.775224	0.246762	44.41708
29	0.966667	0.778081	0.247671	44.58078
30	1	0.780142	0.248327	44.69882
31	1.033333	0.781625	0.248799	44.78379
32	1.066667	0.782691	0.249138	44.84486
33	1.1	0.783456	0.249382	44.88873
34	1.133333	0.784006	0.249557	44.92022
35	1.166667	0.7844	0.249682	44.9428
36	1.2	0.784683	0.249772	44.959
37	1.233333	0.784885	0.249837	44.97062
38	1.266667	0.785031	0.249883	44.97894
39	1.3	0.785135	0.249916	44.98491
40	1.333333	0.785209	0.24994	44.98919
41	1.366667	0.785263	0.249957	44.99225
42	1.4	0.785301	0.249969	44.99445
43	1.433333	0.785329	0.249978	44.99602
44	1.466667	0.785348	0.249984	44.99715
45	1.5	0.785363	0.249989	44.99796
46	1.533333	0.785373	0.249992	44.99854
47	1.566667	0.78538	0.249994	44.99895
48	1.6	0.785385	0.249996	44.99925

Table Five
Logistic Driver Model

In any case, the infinitesimal limit of the residual square is a notable concomitant of cos($\pi/4+\theta$): A fact that is not general and is not a feature of our third illustrative model, which follows.

Case Three: An Exponential Decay Model

Our version of this model is one that halves $f(\theta_i)$ in every step of i. That is:-

$$f(\theta_i) = c.\,2^{(m_{max}-i)}$$
Equation 50

where Scaling Constant c is chosen to be unity and m_{max} is 5.

That this is indeed a special case of exponential decay is readily shown by the identity:-

$$f(\theta_i) \equiv c2^{(m_{max}-i)} \equiv ce^{log_n2(m_{max}-i)}$$
Equation 51

where log_n2 is 0.69314718055995.

The 16-interval series computed was commenced with $\theta_0=32°$ and ended with $\theta_{16}=1/2048°$ as tabulated in the EXCEL® worksheet of Table Six:-

SCHEME C — MODEL: POWER OF TWO ("P2")

m	5
Initial Hypotenuse	1
Square of Initial Hypotenuse	1
$Tan(\theta_n)/Sin(\theta_n)$	1
$1/Cos(\theta_n)$	1
h_n^2	0.02571497
h_n^2/h_0^2	0.02571497

	R	G	B
Black	0	0	0
Maroon	135	0	32
Red	255	0	0
Orange	255	127	0
Yellow	255	255	0
Green	0	255	0
Blue	0	0	255
Indigo	181	0	252
Violet	233	0	254

Serial i	θ_i (degrees)	Cumulate θ_i (degrees)	θ_i (radians)	$x_i = h_i \cdot Cos(\theta)$	$y_i = h_i \cdot Sin(\theta)$	Model h_i	$Tan(\theta_i)$	$Tan(\theta_i)/Sin(\theta_i)$	Area of Central Square
0	32	32	0.55850536	0.8480481	0.52991926	1	0.624869	1	1
1	16	48	0.27925268	0.30580506	0.08768819	0.31812883	0.286745	1.040299	0.101206
2	8	56	0.13962634	0.21599417	0.030356	0.21811687	0.140541	1.009828	0.047575
3	4	60	0.06981317	0.18518597	0.01294946	0.18563317	0.069927	1.002442	0.034462
4	2	62	0.03490659	0.17213158	0.00601097	0.1722365	0.034921	1.00061	0.029665
5	1	63	0.01745329	0.16609531	0.0028992	0.16612061	0.017455	1.000152	0.027596
6	0.5	63.5	0.00872665	0.16318989	0.00142414	0.16319611	0.008727	1.000038	0.026633
7	0.25	63.75	0.00436332	0.16176422	0.00070583	0.16176576	0.004363	1.00001	0.026168
8	0.125	63.875	0.00218166	0.161058	0.00035137	0.16105838	0.002182	1.000002	0.02594
9	0.0625	63.9375	0.00109083	0.16070653	0.0001753	0.16070662	0.001091	1.000001	0.025827
10	0.03125	63.96875	0.00054542	0.1605312	8.7556E-05	0.16053123	0.000545	1	0.02577
11	0.015625	63.984375	0.00027271	0.16044364	4.3754E-05	0.16044365	0.000273	1	0.025742
12	0.0078125	63.9921875	0.00013635	0.16039988	2.1871E-05	0.16039988	0.000136	1	0.025728
13	0.00390625	63.9960938	6.8177E-05	0.16037801	1.0934E-05	0.16037801	6.82E-05	1	0.025721
14	0.00195313	63.9980469	3.4088E-05	0.16036708	5.4667E-06	0.16036708	3.41E-05	1	0.025718
15	0.00097656	63.9990234	1.7044E-05	0.16036161	2.7332E-06	0.16036161	1.7E-05	1	0.025716
16	0.00048828	63.9995117	8.5221E-06	0.16035888	1.3666E-06	0.16035888	8.52E-06	1	0.025715
Totals	63.9995117	1024.00049	1.1170022	3.66281913	0.67265342	3.82980819	1.191903	17.05338	1.525182
Means	3.76467716	60.2353228	0.06570601	0.21545995	0.03956785	0.22528283	0.070112	1.00314	0.089717

Table Six
Decay Model's Bhaskara Iterates

Figures Eight and Nine respectively show the successions of iterated Bhaskara triangles for the illustrative Table Six:-

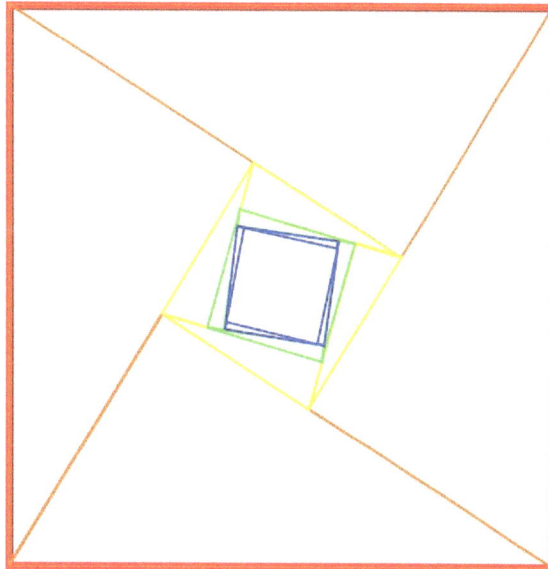

Figure Eight
Binary-Base Exponential Decay Model
As a Line Diagram

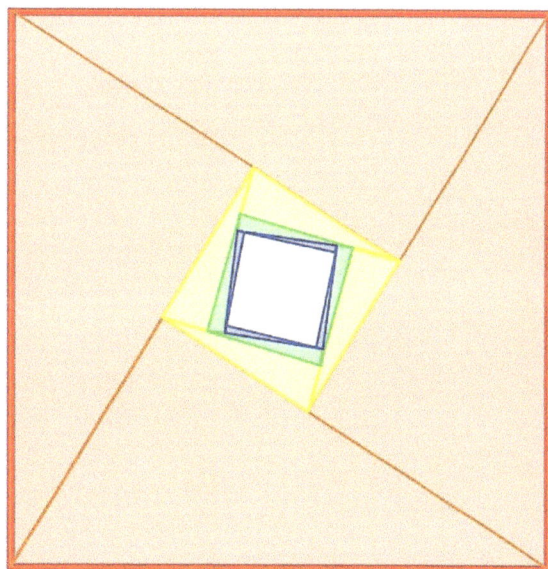

Figure Nine
Binary-Base Exponential Decay Model
As a Filled Diagram

The finite residual square of an exponential decay function is clear.

References

1   "Pythagorean Theorem"
         Angic Head
         http://jwilson.coe.uga.edu/emt668/
                emt668.student.folders/headangela/essay1/pythagor...

2   http://www.geom.uiue.edu/~demo5337/Group3/Bhaskara.html

3   "Handbook of Mathematical Functions
         with Formulas, Graphs and Mathematical Tables"
         Edited by Milton Abramowitz and Irene A Stegun
         Dover Publications of New York
         after The National Bureau of Standards
         December 1972
         ISBN 0-486-61272-4
         1046pp
         £18.95 paperback
         ("A&S Handbook")

         (1)   Binomial Coefficients 3.1.2 (p10)

4   A&S Handbook
         (2)   The Binomial Theorem 3.1.1 (p10)

5   A&S Handbook
         (3)   Cosine Series 4.3.66 (p74)

6   A&S Handbook
         (4)   Sine Series 4.3.65 (p74)

CHAPTER THREE

Smooth Transitions

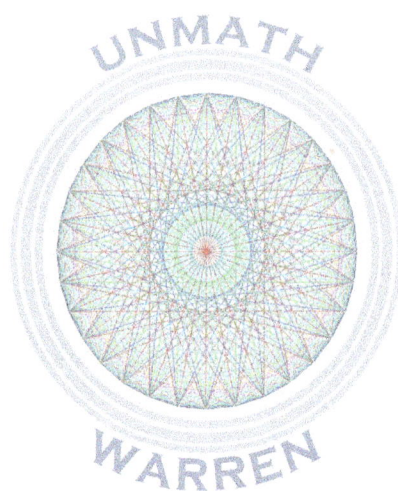

On Certain Smooth Transitions
Between
Disparate Functions

By
James R Warren BSc MSc PhD PGCE

PART ONE
INTRODUCTION

Sometimes, especially as an outcome of experiment, a quantitative binomial relationship appears to exist between two variables, other than as a smoothly-continuous function. In particular, should you draft a simple Cartesian x-y plot of the data involved you might discern a visible cusp or crick in any line interpolated between the plotted (x,y) points.

Notwithstanding that, it may be necessary or apt to relate the independent and dependent variables by some mathematical equation, if merely an empirical one for technical purposes.

Among the many models and contexts that apply to such problems, in both the natural and the social sciences, we may list the following:-

0   First-Order Polynomial to Third-Order Polynomial

1   First-Order Polynomial to Asymptotic
        (Elastico-Plastic Failure)

2   Asymptotic to First-Order Polynomial
        (Hardening of Concrete)

3   Three-Segment First-Order Polynomial
        (Regnault-Pfaundler Method)

4   Exponential Decay to Second-Order Polynomial
        (Heat of Fusion relations to the Mechanical Working of Polymers)
        (EOQ Bathtub Model)

As a primary step to the management of such data I propose the following simple crossover rule:-

$$Q(x) = (1 - H(x))F(x) + H(x)G(x)$$
Equation 1

which simplifies to:-

$$Q(x) = F(x) - H(x)(F(x) - G(x))$$
Equation 2

where $Q(x)$ is the Consolidated Descriptive Function of the Data; $F(x)$ is the Sinistral Component Function; $G(x)$ is the Dextral Component Function; and $H(x)$ is a suitable Heaviside Step Function. The Heaviside Step Function exhibits these properties:-

$$H(x) \begin{cases} 0 & x < x_{mid} \\ F(x_{mid}) \, or \, G(x_{mid}) & x = x_{mid} \\ 1 & x > x_{mid} \end{cases}$$
Equation 3

In contexts where both x and y values are normalised to the unitary interval we can take it that $H(x)=0.5$ at x_{mid}.

The Heaviside Step Function[1]

The Heaviside Step Function is formally defined as the integral of The Dirac Delta Function, which is a spike generator.
The relevant equation is:

$$H(x) = \frac{d}{dx} \max\{x,0\} = \int_{-\infty}^{x} \delta(s).ds$$
Equation 4

where $H(x)$ is the Heaviside step Function; $\max\{x,0\}$ is the Ramp Function; and $\delta(s)$ is the Dirac Delta Function.
Clearly Equation Four has no economically-interesting solution, but many *approximations* of $H(x)$ are possible and some are both simple and finely-controllable. In addition, whilst the ideal Heaviside Function is absolutely abrupt in transition from 0 to 1 at x_{mid} it is usually desirable, in practical curve-fitting, to have a smooth approximant with gradually-changing differentials along its path.

Some Heaviside Step Function Approximants

Among the interesting possibilities are:-

(a)   Logistic-Hyperbolic Distribution

$$H(x) \approx \frac{1}{2} + \frac{1}{2}\tanh(sx) \approx \frac{1}{1 + e^{-2sx}}$$ **Equation 5**

(b)   Cauchy Distribution

$$H(x) \approx \lim_{s \to \infty} \left(\frac{1}{2} + \frac{1}{\pi} \tan^{-1}(sx) \right)$$

Equation 6

(c)   Normal Distribution

$$H(x) \approx \lim_{s \to \infty} \left(\frac{1}{2} + \frac{1}{2} erf(sx) \right)$$

Equation 7

where:-

$$H(z) = \frac{2}{\sqrt{\pi}} \int_0^z e^{-t^2} .dt = \frac{2}{\sqrt{\pi}} \sum_{n=0}^{\infty} \frac{z}{2n+1} \prod_{k=1}^{n} \frac{-z^2}{k}$$

Equation 8

or for $z \geq 0$ we may use Craig's Formula:-

$$erf(z) = 1 - \frac{2}{\pi} \int_0^{\pi/2} \exp\left(-\frac{z^2}{\sin^2(\theta)} \right) .d\theta$$

Equation 9

where s is the Steepness of the functional transition; x is the Ordinate and z is equivalent to x.

The Transition Point (x_{mid}, y_{mid})

In order to control the location of the transition we also need to specify the Ordinal Transition Location x_{mid}, the critical point at which G(x) supersedes F(x). This is effectively where the sign of the second differential of H(x) changes, the mid-point of H(x), the dyadic symmetry-center of its sigmoid.

y_{mid} is of course readily soluble from either function F(x) or G(x), and the co-ordinate (x_{mid}, y_{mid}) marks the triple chiasm at which F(x), G(x) and H(x) all intersect: The cusp whose discontinuity is to be smoothed.

In principle, x_{mid} is the solution of the simultaneous equation F(x) versus G(x) and in many cases may be analytically-determinable. Notwithstanding that, numerical determinations may be more convenient, though like all such resorts they should be treated with intelligent care, not to say circumspection.

In my trials I use The Method of Regula Falsi (False Position) to establish x_{mid}.

Regula Falsi[2]

The Method of Regula Falsi is a robust and efficient, but not infallible, method of solving simultaneous equations. Firstly, it is required to define the *range* of the independent variable as:-

where x_1 is an iterate (*not* in a direct way a model independent variable).

The first approximation of x_{mid} is then the first iterate (i.e. x_3) which may be expressed as:-

$$x_3 = \frac{x_1.D(x_2) - x_2.D(x_1)}{D(x_2) - D(x_1)}$$

Equation 11

where:-

$$D(x_i) = F(x_i) - G(x_i)$$

Equation 12

Equation Eleven generalises to:-

$$x_{i+2} = \frac{x_{i-2}.D(x_{i-1}) - x_{i-1}.D(x_{i-2})}{D(x_{i-1}) - D(x_{i-2})}$$

Equation 13

The Model Zero Trial Functions F(x) and G(x)

For the Trial Model Zero I elected to define two simple algebraic polynomials which more or less simulate the elastico-plastic failure of a ductile material under stress where Q(x) surrogates the Applied Stress and x is the Strain.

Firstly, the Sinistral Function F(x) is a First-Degree (Linear) Polynomial which simulates the Hooke's Law elastic response of a completely-restorable mass under restricted stress, that is:-

$$F(x) = c_0 x^0 + c_1 x^1 + \varepsilon = \varepsilon + \sum_{j=0}^{1} c_j x^j$$

Equation 14

by regressing the F(x) data of Table One this became:-

$$F(x) = 0.007510548523 x^0 + 1.565400843882 x^1$$

Equation 15

Secondly, the Dextral Function G(x) is a Third-Degree (Cubic) Polynomial which simulates plastic failure of the mass, that is:-

$$G(x) = c_0 x^0 + c_1 x^1 + c_2 x^2 + c_3 x^3 + \varepsilon = \varepsilon + \sum_{j=0}^{3} c_j x^j$$

Equation 16

by regressing the G(x) data of Table One this became:-

$$G(x) = 0.174088926995 x^0 + 1.493260549484 x^1 - 1.393861059214 x^2 + 0.427232423753 x^3$$

Equation 17

where c_j are all Empirical Term Coefficients and ε is a Residual Error.

Coupling Policy

It is notable that many of the available approximants of H(x) are fundamentally Napierian in the sense that they intimately involve the exponent base e in their definitions. This is true even of the Cauchy Model which appeals to complex exponents of the Napierian base. In many instances we will adjudge it expedient also to model F(x) and G(x) as e-based curves, and this might facilitate Q(x) simplifications due to the co-ordination of factors.

It will often be wise to attempt the mathematical co-ordination of F(x), G(x) and H(x) sub-models in this way.

But the present discussion involves of course the marriage of algebraic polynomials and whilst this obviously does not prohibit a Napierian H(x) it might be useful slightly to digress to examine possible algebraic or pseudo-algebraic approximations of the Error Function such as those developed by Hastings[3] during the middle of the last century.

We will further pause to quote some of the formal summation approximations of erf also given by Abramowitz and Stegun[4].

My choice of a Napierio-algebraic Hastings formula for you to consider for your own fitments is:-

$$|\varepsilon(x)| = 1.5 \times 10^{-7} \qquad \text{and:-} \qquad t = \frac{1}{1+px} \qquad \text{where:-} \qquad p = 0.3275911$$

$$erf(x) = 1 - (a_1 t + a_2 t^2 + a_3 t^3 + a_4 t^4 + a_5 t^5) e^{-(x^2)} + \varepsilon(x)$$

Equation 18

or equivalently:-

$$erf(x) = 1 - \left(\sum_{i=1}^{5} a_i t^i \right) e^{-(x^2)} + \varepsilon(x)$$

Equation 19

where:-

$a_1 = 0.254829592$ $\qquad a_2 = -0.284496736$ $\qquad a_3 = 1.421413741$

$$a_4 = -1.453152027 \qquad a_5 = 1.061405429$$

I have not adequately trialled this myself but Hastings Polynomials are usually reliable and this one might be convenient, especially where proprietary intrinsic functions are expensive or unavailable.

I happen to be a MicroSoft EXCEL® customer and I used the company's ERF() intrinsic during the current work.

Selected Summations

Alternatively, researchers may prefer one or more of these Abramowitz and Stegun offerings, all of which are of course easily programmable:-

(A&S 7.1.5 p 297)

$$erf(z) = \frac{2}{\sqrt{2}} \sum_{n=0}^{\infty} \frac{(-1)^n z^{2n+1}}{n!(2n+1)}$$

Equation 20

(A&S 7.1.6 p297)

$$erf(z) = \frac{2}{\sqrt{\pi}} e^{-z^2} \sum_{n=0}^{\infty} \frac{2^n}{1 \times 3 \times 5 \times 7 \times ... \times (2n+1)} . z^{(2n+1)}$$

Equation 21

or:-

$$erf(z) = \frac{2}{\sqrt{\pi}} e^{-z^2} \sum_{n=0}^{\infty} \frac{2^n}{\prod_{i=1}^{n}(2i+1)} . z^{(2n+1)}$$

Equation 22

Equation Twenty is a purely-algebraic Maclaurin Expansion which could itself readily be truncated to any desired accuracy.

The Selection and Adjustment of a Heaviside Approximant

It remains for us to define an amenable and convenient form of H(x).

As adumbrated above, we need to incorporate x_{mid} into all the candidate models in order to control the position of H(x).

This is done by subtracting x_{mid} from x in any H(x) equation selected.

For example, were we to decide to use the Logistic-Hyperbolic Model of Equation Five that expression would be modified to:-

$$H(x) \approx \frac{1}{2} + \frac{1}{2} \tanh(s[x - x_{mid}]) \approx \frac{1}{1 + e^{-2s(x - x_{mid})}}$$

Equation 5(a)

Plot One shows the relative abruptitude of the Logistic, Cauchy and Error Function Models with the common Steepness value s = 50.

Heaviside Models Response Comparisons

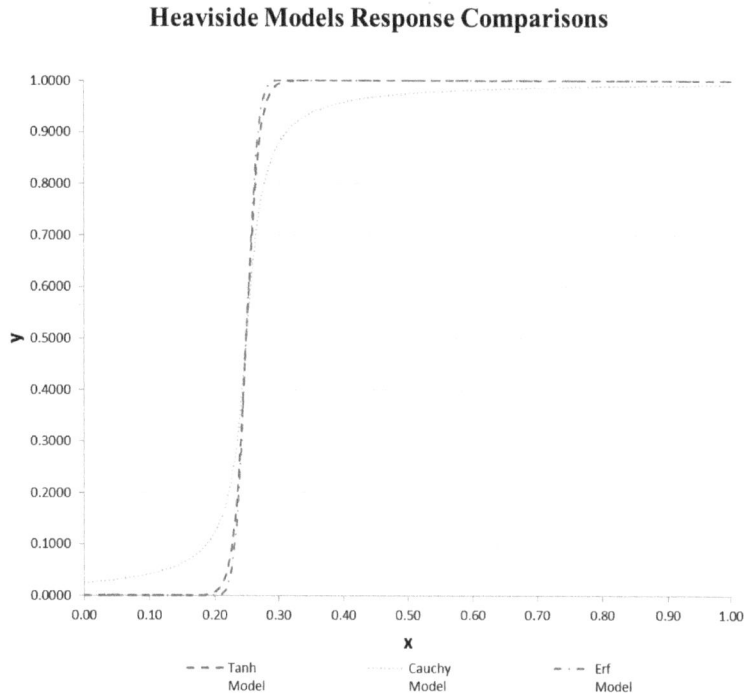

**Plot One
The Cut-off Efficiency of
Heaviside Step Function Approximants
Compared**

Plot One shows that the Erf Model (i.e. the Error Function Approximant) gives the best abruptitude of transition. If, however, we find it *too* precipitous we can always change the steepness parameter. In this example, note that the x_{mid} is artificially set to 0.25.
(The Logistic Model is not plotted because it is identical to the Tanh Model).
Therefore, we choose that our H(x) model shall be:-

$$H(x) = \frac{1}{2} + \frac{1}{2} erf\left[s\left(x - x_{mid}\right)\right]$$

Equation 23

Model Zero F(x), G(x), H(x) and Q(x) Functional Presentations

We shall examine the behaviours of these four functions and their curves; and also the establishment of the appropriate x_{mid} value via Regula Falsi.

I normalised the demonstration data to the intervals x={0,1) and H(x)={0,1), and computed the independent and dependent variables for (x_i,function(x_i)) where i=0…100.

Steepness, s, was adjudged to be optimal at 12.

Table One the functional components' values:-

x

Lower Bound: 0
Upper Bound: 1
Intervals: 100
Increment: 0.01

F(x) a+bx
a   0.007510548523
b   1.565400843882

G(x) c+dx+fx²+gx³
c   0.174088926995
d   1.493260549484
f   -1.393861059214
g   0.427232423758

H(x) x_{mod} 0.337368254

Serial	x	F(x)= data	G(x)= data	F(x)= a+bx	G(x)= c+dx+fx²+gx³	H(x) (ERF model)	Q(x) (ERF model) (simplified)	Q(x)= (1-H(x))*F(x) +H(x)*G(x)
0	0.000000	0		0.00751055	0.17408893	0.00000001	0.00751055	0.00751055
1	0.010000			0.02316456	0.18888257	0.00000001	0.02316456	0.02316456
2	0.020000			0.03881857	0.20340001	0.00000004	0.03881857	0.03881857
3	0.030000			0.05447257	0.21764880	0.00000009	0.05447259	0.05447259
4	0.040000			0.07012658	0.23161651	0.00000022	0.07012662	0.07012662
5	0.050000			0.08578059	0.24532071	0.00000054	0.08578068	0.08578068
6	0.060000	0.13		0.10143460	0.25875894	0.00000126	0.10143480	0.10143480
7	0.070000			0.11708861	0.27193379	0.00000285	0.11708905	0.11708905
8	0.080000			0.13274262	0.28484780	0.00000628	0.13274357	0.13274357
9	0.090000			0.14839662	0.29750358	0.00001347	0.14839863	0.14839863
10	0.100000			0.16405063	0.30990860	0.00002809	0.16405473	0.16405473
11	0.110000			0.17970464	0.32205051	0.00005703	0.17971276	0.17971276
12	0.120000			0.19535865	0.33394685	0.00011263	0.19537426	0.19537426
13	0.130000			0.21101266	0.34559518	0.00021646	0.21104179	0.21104179
14	0.140000	0.17		0.22666667	0.35699805	0.00040486	0.22671943	0.22671943
15	0.150000			0.24232068	0.36815805	0.00073702	0.24241342	0.24241342
16	0.160000			0.25797468	0.37907772	0.00180607	0.25813285	0.25813285
17	0.170000	0.28		0.27362869	0.38975963	0.00225332	0.27389037	0.27389037
18	0.180000			0.28928270	0.40020635	0.00378548	0.28970260	0.28970260
19	0.190000			0.30493671	0.41042044	0.00619352	0.30559002	0.30559002
20	0.200000	0.36		0.32059072	0.42040445	0.00987097	0.32157598	0.32157598
21	0.210000			0.33624473	0.43016097	0.01532795	0.33768427	0.33768427
22	0.220000			0.35180873	0.43969254	0.02319621	0.35393522	0.35393522
23	0.230000			0.36755274	0.44900174	0.03421993	0.37033992	0.37033992
24	0.240000			0.38320675	0.45809112	0.04922712	0.38689309	0.38689309
25	0.250000			0.39886076	0.46696825	0.06907860	0.40356518	0.40356518
26	0.260000			0.41451477	0.47562070	0.09459437	0.42029505	0.42029505
27	0.270000	0.42		0.43016878	0.48406602	0.12646176	0.43698472	0.43698472
28	0.280000			0.44582278	0.49230178	0.16513466	0.45349808	0.45349808
29	0.290000			0.46147679	0.50033054	0.21073721	0.46966472	0.46966472
30	0.300000	0.5		0.47713080	0.50815487	0.26298811	0.48528976	0.48528976
31	0.310000			0.49278481	0.51577733	0.32116101	0.50016911	0.50016911
32	0.320000			0.50843882	0.52320048	0.38409280	0.51410867	0.51410867
33	0.330000	0.54		0.52409283	0.53042689	0.45024459	0.52694470	0.52694470
34	0.340000			0.53974684	0.53745912	0.51781172	0.53856223	0.53856223
35	0.350000			0.55540084	0.54429973	0.58486989	0.54890814	0.54890814
36	0.360000			0.57105485	0.55095129	0.64953800	0.55799682	0.55799682
37	0.370000			0.58670886	0.55741636	0.71013498	0.56590723	0.56590723
38	0.380000			0.60236287	0.56369750	0.76530893	0.57277191	0.57277191
39	0.390000			0.61801688	0.56979727	0.81412238	0.57876022	0.57876022
40	0.400000			0.63367089	0.57571825	0.85608544	0.58405848	0.58405848
41	0.410000			0.64932489	0.58146299	0.89119778	0.58885059	0.58885059
42	0.420000			0.66497890	0.58703406	0.91958824	0.59330174	0.59330174
43	0.430000			0.68063291	0.59243402	0.94202622	0.59754725	0.59754725
44	0.440000	0.6		0.69628692	0.59766543	0.95922114	0.60168711	0.60168711
45	0.450000			0.71194093	0.60273086	0.97202495	0.60578602	0.60578602
46	0.460000			0.72759494	0.60763287	0.98128898	0.60987749	0.60987749
47	0.470000			0.74324895	0.61237403	0.98780202	0.61397044	0.61397044
48	0.480000			0.75890295	0.61695689	0.99225128	0.61805679	0.61805679
49	0.490000			0.77455696	0.62138402	0.99520464	0.62211854	0.62211854
50	0.500000			0.79021097	0.62565799	0.99710951	0.62613363	0.62613363
51	0.510000			0.80586498	0.62978135	0.99830312	0.63008011	0.63008011
52	0.520000	0.65		0.82151899	0.63375668	0.99903032	0.63393875	0.63393875
53	0.530000			0.83717300	0.63758653	0.99946049	0.63769421	0.63769421
54	0.540000			0.85282700	0.64127347	0.99970782	0.64133528	0.64133528
55	0.550000			0.86848101	0.64482005	0.99984600	0.64485450	0.64485450
56	0.560000			0.88413502	0.64822886	0.99992101	0.64824749	0.64824749
57	0.570000			0.89978903	0.65150244	0.99996058	0.65151222	0.65151222
58	0.580000			0.91544304	0.65464336	0.99998086	0.65464835	0.65464835
59	0.590000			0.93109705	0.65765418	0.99999096	0.65765666	0.65765666
60	0.600000	0.66		0.94675105	0.66053748	0.99999584	0.66053867	0.66053867
61	0.610000			0.96240506	0.66329580	0.99999814	0.66329636	0.66329636
62	0.620000			0.97805907	0.66593173	0.99999919	0.66593198	0.66593198
63	0.630000			0.99371308	0.66844780	0.99999966	0.66844792	0.66844792
64	0.640000			1.00936709	0.67084661	0.99999986	0.67084665	0.67084665
65	0.650000			1.02502110	0.67313069	0.99999994	0.67313071	0.67313071
66	0.660000			1.04067511	0.67530263	0.99999998	0.67530263	0.67530263
67	0.670000			1.05632911	0.67736497	0.99999999	0.67736497	0.67736497
68	0.680000			1.07198312	0.67932029	1.00000000	0.67932029	0.67932029
69	0.690000			1.08763713	0.68117115	1.00000000	0.68117115	0.68117115
70	0.700000	0.68		1.10329114	0.68292011	1.00000000	0.68292011	0.68292011
71	0.710000			1.11894515	0.68456974	1.00000000	0.68456974	0.68456974
72	0.720000			1.13459916	0.68612260	1.00000000	0.68612260	0.68612260
73	0.730000			1.15025316	0.68758125	1.00000000	0.68758125	0.68758125
74	0.740000			1.16590717	0.68894825	1.00000000	0.68894825	0.68894825
75	0.750000			1.18156118	0.69022617	1.00000000	0.69022617	0.69022617
76	0.760000			1.19721519	0.69141758	1.00000000	0.69141758	0.69141758
77	0.770000			1.21286920	0.69252503	1.00000000	0.69252503	0.69252503
78	0.780000			1.22852321	0.69355109	1.00000000	0.69355109	0.69355109
79	0.790000			1.24417722	0.69449832	1.00000000	0.69449832	0.69449832
80	0.800000	0.7		1.25983122	0.69536929	1.00000000	0.69536929	0.69536929
81	0.810000			1.27548523	0.69616656	1.00000000	0.69616656	0.69616656
82	0.820000			1.29113924	0.69689269	1.00000000	0.69689269	0.69689269
83	0.830000			1.30679325	0.69755025	1.00000000	0.69755025	0.69755025
84	0.840000			1.32244726	0.69814179	1.00000000	0.69814179	0.69814179
85	0.850000			1.33810127	0.69866989	1.00000000	0.69866989	0.69866989
86	0.860000			1.35375527	0.69913711	1.00000000	0.69913711	0.69913711
87	0.870000			1.36940928	0.69954600	1.00000000	0.69954600	0.69954600
88	0.880000			1.38506329	0.69989914	1.00000000	0.69989914	0.69989914
89	0.890000			1.40071730	0.70019909	1.00000000	0.70019909	0.70019909
90	0.900000	0.7		1.41637131	0.70044840	1.00000000	0.70044840	0.70044840
91	0.910000			1.43202532	0.70064965	1.00000000	0.70064965	0.70064965
92	0.920000			1.44767932	0.70080539	1.00000000	0.70080539	0.70080539
93	0.930000			1.46333333	0.70091820	1.00000000	0.70091820	0.70091820
94	0.940000			1.47898734	0.70099063	1.00000000	0.70099063	0.70099063
95	0.950000			1.49464135	0.70102524	1.00000000	0.70102524	0.70102524
96	0.960000			1.51029536	0.70102461	1.00000000	0.70102461	0.70102461
97	0.970000			1.52594937	0.70099129	1.00000000	0.70099129	0.70099129
98	0.980000			1.54160338	0.70092784	1.00000000	0.70092784	0.70092784
99	0.990000			1.55725738	0.70083684	1.00000000	0.70083684	0.70083684
100	1.000000	0.7		1.57291139	0.70072084	1.00000000	0.70072084	0.70072084

Table One
Model Zero Function Components Table

The six stated regression coefficients were as computed for the original six $F(x)$ points and nine $G(x)$ points, as computed by EXCEL® to twelve places of decimals, *assuming the least squares fitments of algebraic polynomials.*

The development of the Table One functions was of course predicated upon accurate knowledge of x_{mid} as determined by Regula Falsi.

The relevant regula falsi tableau is presented in Table Two:-

x_1	0.000000000000
x_2	1.000000000000

Power	Fc_{power}	Gc_{power}
0	0.007510548523	0.174088926995
1	1.565400843882	1.493260549484
2		-1.393861059214
3		0.427232423753

Serial, i	x_i	$F(x_i)$	$G(x_i)$	$D(X_i)=F(x_i)-G(x_i)$
1	0.000000000000000	0.000000000000000	0.174088926995000	-0.174088926995000
2	1.000000000000000	1.572911392405000	0.700720841018000	0.872190551387000
3	0.166388551617410	0.267975327637197	0.385929177109281	-0.117953849472084
4	0.265694955022219	0.423429655329972	0.480456110291428	-0.057026454961456
5	0.358643163075999	0.568930858654679	0.550059771660667	0.018871086994012
6	0.335532615949854	0.532753588680836	0.534341385331723	-0.001587796650887
7	0.337326205960138	0.535561275996513	0.535597719383886	-0.000036443387372
8	0.337368339816474	0.535627232370778	0.535627157943753	0.000000074427026
9	0.337368253943383	0.535627097944969	0.535627097948440	-0.000000000003472
10	0.337368253947388	0.535627097951239	0.535627097951239	0.000000000000000
11	0.337368253947388	0.535627097951239	0.535627097951239	0.000000000000000

Table Two
Regula Falsi Tableau for the Determination of x_{mid}
For the Demonstrational Data of Model Zero

In our context the regula falsi iterate converges to fifteen-figure accuracy at x_{10}, i.e. at the eighth iteration.

The resulting x_{mid} proved to be 0.337368253947388.

I especially draw your attention to Plot Two which depicts the behaviour of the four functions in the relevant space. The blue triangles plot the available first-order data $F(x)$ assumed to be linear in an "elastic" deformational zone (for sake of argument). The red squares mark the available nine "plastic" readings which we have modelled with a cubic equation. The blue dashed line is the appropriate EXCEL regression curve and the red solid one the regression line (obscured) for the cubic data. The thick purple solid line is the trajectory of $Q(x)$, and as we would anticipate, or at least hope, it virtually obliterates the underlying plots: This $Q(x)$ is our desired outcome.

The green dashed sigmoid is the course of $H(x)$, our steepness-12 Heaviside Step Function.

Note that the $F(x)$, $G(x)$, $H(x)$ and $Q(x)$ curves all converge at (x_{mid}, y_{mid}), the grand chiasm that marks the cusp of $F(x)=G(x)$.

Blue and red color-coded regression equations and their *polynomial algebraic* Determination Coefficients are also presented to twelve decimal positions, and you may discern pecked black lines that extrapolate the F(x) and G(x) regression curves.

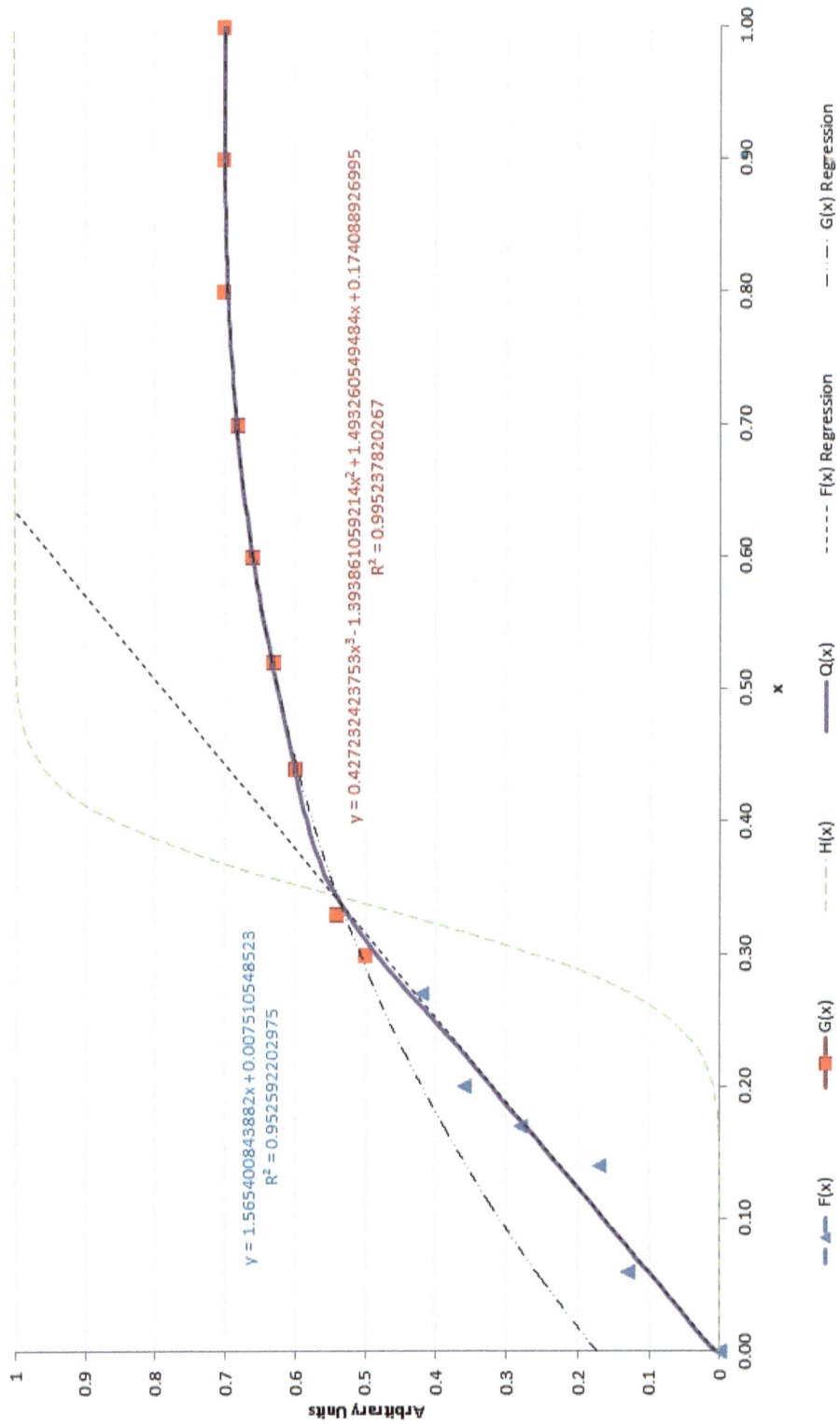

Model Zero: First-Order Polynomial F(x) to Asymptotic Data G(x)

$y = 0.42723423753x^3 - 1.3938610592142x^2 + 1.4932605494484x + 0.174088926995$
$R^2 = 0.995237820267$

$y = 1.565400843882x + 0.007510548523$
$R^2 = 0.952592202975$

x

F(x) G(x) H(x) Q(x) G(x) Regression F(x) Regression G(x) Regression

Arbitrary Units

Plot Two
Functional Data and their Fitted Curves

Coefficients of Correlation and Determination

The statistical Coefficient of Correlation (r or R) and its square the Coefficient of Determination (r^2 or R^2) measure the quality of fitments to empirical data, but both need to be applied with extreme circumspection.

Both are defined on the range zero (for complete non-concordance between the model and the data) to unity (for perfect fit of the data upon the curve of the equation, i.e. determinacy). In the case of the Coefficient of Determination defined upon the range $-1 \leq r \leq +1$, negative values indicate inverse correlation. In both cases, out-of-range coefficients are absurd and are a handy indicator of some misprision in the model.

Thoroughgoing treatments of correlation in the scientific and engineering literature tend implicitly to assume fitments to *continuous, algebraic polynomial* functions. Ironically, renditions of correlation algebra in social science texts, deliberately simplified for the benefit of less numerate scholars, often offer more robust mathematics. Note that our Q(x) fitments are *by definition* neither continuous nor algebraic.

I may illustrate this difficulty firstly by reference to the rigorously linear propositions of Davis and Sampson[5]:-

$$Total\,Sum\,of\,Squares\left(SS_T\right) = \sum Y^2 - \frac{\left(\sum Y\right)^2}{n} = \sum \left(Y - \bar{Y}\right)^2$$

Equation 24

$$Sum\,of\,Squares\,Due\,To\,Regression\left(SS_R\right) = \sum \hat{Y}^2 - \frac{\left(\sum \hat{Y}\right)^2}{n} = \sum \left(\hat{Y} - \bar{Y}\right)^2$$

Equation 25

$$Sum\,of\,Squares\,Due\,To\,Deviations\left(SS_D\right) = SS_T - SS_R = \sum \left(\hat{Y} - Y\right)^2$$

Equation 26

$$R^2 = Goodness\,Of\,Fit = \frac{SS_R}{SS_T}$$

Equation 27

Applied to our Model 0 data the system of Equations Twenty-Four to Twenty-Seven issues in $R^2 \approx 1.02$ which is absurd.

An alternative and more general (PRE) treatment due to Sonia R Wright[6] is:-

PRE (Proportional Reduction in Error) Measure r^2
(Coefficient of Determination)

$$E_1 = Total\,Sum\,of\,Squares = \sum \left(Y - \overline{Y}\right)^2$$

Equation 28

$$E_2 = Error\,Sum\,of\,Squares = \sum \left(Y - \hat{Y}\right)^2$$

Equation 29

$$Total\,Sum\,Of\,Squares = Error\,Sum\,Of\,Square + Explained\,Sum\,Of\,Squares$$

Equation 30

$$\begin{aligned}
r^2 &= \frac{Total\,Sum\,Of\,Squares - Error\,Sum\,Of\,Squares}{Total\,Sum\,Of\,Squares} \\
&= \frac{Explained\,Variation}{Total\,Variation} \\
&= \frac{E_1 - E_2}{E_1} \\
&= \frac{\sum \left(Y - \overline{Y}\right)^2 - \sum \left(Y - \hat{Y}\right)^2}{\sum \left(Y - \overline{Y}\right)^2}
\end{aligned}$$

Equation 31

where:-

$\hat{Y}$ is a Model (Postdicted) Ordinate
$\overline{Y}$ is the Mean Observed Ordinate
Y is an Observed Ordinate

Application of Equation Thirty-One to the Model 0 data gives an R^2 value within bounds (less than unity).

To be specific, I elaborated an EXCEL® tabulation (presented as Table Four) to compute F(x), G(x) and Q(x) coefficients of determination. These are rubricated in Table Four but are presented in Table Three below for convenience:-

Function Fitted to the Data	Coefficient of Determination R^2	
	From EXCEL Worksheet	From EXCEL Regression Plot
F(x)	0.952592202975	0.952592202975
G(x)	0.995237820323	0.995237820267
Q(x)	0.998728284312	not computed

Table Three
Model Zero Coefficients of Determination

Clearly, we are working close to the arithmetic limits of our equipment (I used Microsoft EXCEL® upon Windows 7 mediated by a 64-bit Hewlett-Packard 500-400na).

Notwithstanding that, we can examine the apparent paradox that the determination of our fitted Q(x) is superior to either of those of F(x) or G(x): But this is only to be expected of a correlation that suppresses the cusp of contact and includes many more data points than either component function.

s	Steepness		12
x			
	Lower Bound:		0
	Upper Bound:		1
	Intervals:		100
	Increment:		0.01
F(x)	a+bx		
	a		0.007510548523
	b		1.565400843882
G(x)	c+dx+fx²+gx³		
	c		0.174088926995
	d		1.493260549484
	f		-1.393861059214
	g		0.427232423753
H(x)	x_{mid}		0.337368254

Count	101	6	101	9	101	101	6	6	9	9	101	15	15	15	15 Count
Mean	0.500000	0.226667	0.500000	0.634444	0.790211	0.561652	0.020322	0.019359	0.004869	0.004846	0.661022	0.470398	0.011050	0.011303	0.000014 Mean
Sum	50.500000	1.360000	50.500000	5.710000	79.811308	56.726845	0.121933	0.116153	0.043822	0.043614	66.763175	7.055973	0.165756	0.169552	0.000211 Sum
R²					0.992592	0.998238						0.9987283			R²

Serial	x_d	$F(x_d)$= data	x_g	$G(x_g)$= data	$F(x_d)$= a+bx	$G(x_g)$= c+dx+fx²+gx³	F(x) Observed Squares SS_{yy}	F(x) Modelled Squares $SS_{\hat y\hat y}$	G(x) Observed Squares SS_{yy}	G(x) Modelled Squares $SS_{\hat y\hat y}$	H(x) (ERF model)	Q(x)= (1-H(x))*F(x)+H(x)*G(x)	Q(x) Observed Squares SS_{yy}	Q(x) Modelled Squares $SS_{\hat y\hat y}$	Q(x) Error Squares $SS_{\varepsilon\varepsilon}$
0	0.000000	0	0.000000		0.00751055	0.17408893	0.05137778	0.04802940			0.00000001	0.00751055	0.05137778	0.04802940	0.00001121
1	0.010000		0.010000		0.02316456	0.18888257					0.00000001				
2	0.020000		0.020000		0.03881857	0.20340001					0.00000004				
3	0.030000		0.030000		0.05447257	0.21764380					0.00000009				
4	0.040000		0.040000		0.07012658	0.23161651					0.00000022				
5	0.050000		0.050000		0.08578059	0.24532071					0.00000054				
6	0.060000	0.13	0.060000		0.10143460	0.25875894	0.00934444	0.01568907			0.00000126	0.10143480	0.00934444	0.01568302	0.00004018
7	0.070000		0.070000		0.11708861	0.27193379					0.00000285				
8	0.080000		0.080000		0.13274262	0.28484780					0.00000628				
9	0.090000		0.090000		0.14839662	0.29750355					0.00001347				
10	0.100000		0.100000		0.16405063	0.30990360					0.00002809				
11	0.110000		0.110000		0.17970464	0.32206051					0.00005703				
12	0.120000		0.120000		0.19535865	0.33394685					0.00011263				
13	0.130000		0.130000		0.21101266	0.34559518					0.00021646				
14	0.140000	0.17	0.140000		0.22666667	0.35699805	0.00321111	0.00000000			0.00040486	0.22671943	0.00321111	0.00000000	0.00001031
15	0.150000		0.150000		0.24232068	0.36815805					0.00073760				
16	0.160000		0.160000		0.25797469	0.37907772					0.00130607				
17	0.170000	0.28	0.170000		0.27362869	0.38975963	0.00284444	0.00220543			0.00228892	0.27369037	0.00284444	0.00223008	0.00000038
18	0.180000		0.180000		0.28928270	0.40020635					0.00378548				
19	0.190000		0.190000		0.30493671	0.41042043					0.00619352				
20	0.200000	0.36	0.200000		0.32059072	0.42040445	0.01777778	0.00882173			0.00987097	0.32157598	0.01777778	0.00900778	0.00007691
21	0.210000		0.210000		0.33624473	0.43016097					0.01532795				
22	0.220000		0.220000		0.35189873	0.43969254					0.02319621				
23	0.230000		0.230000		0.36755274	0.44900174					0.03421993				
24	0.240000		0.240000		0.38320675	0.45809112					0.04922712				
25	0.250000		0.250000		0.39886076	0.46696325					0.06907860				
26	0.260000		0.260000		0.41451477	0.47562070					0.09459437				
27	0.270000	0.42	0.270000		0.43016878	0.48406602	0.03737778	0.04141311			0.12646176	0.43698472	0.03737778	0.04423368	0.00004700
28	0.280000		0.280000		0.44582278	0.49230178					0.16513466				
29	0.290000		0.290000		0.46147679	0.50033054					0.21073721				
30	0.300000		0.300000	0.5	0.47713080	0.50815487			0.01807531	0.01594906	0.26298811	0.48528976	0.01807531	0.02224712	0.00001740
31	0.310000		0.310000		0.49278481	0.51577733					0.32116101				
32	0.320000		0.320000		0.50843882	0.52320048					0.38409280				
33	0.330000		0.330000	0.54	0.52409283	0.53042689			0.00891975	0.01081965	0.45024459	0.52694470	0.00891975	0.01155619	0.00000695
34	0.340000		0.340000		0.53974684	0.53745912					0.51781172				
35	0.350000		0.350000		0.55540084	0.54429973					0.58480989				
36	0.360000		0.360000		0.57105485	0.55095129					0.64853800				
37	0.370000		0.370000		0.58670886	0.55741636					0.71013498				
38	0.380000		0.380000		0.60236287	0.56369750					0.76530893				
39	0.390000		0.390000		0.61801688	0.56979227					0.81412238				
40	0.400000		0.400000		0.63367089	0.57571825					0.85608544				
41	0.410000		0.410000		0.64932489	0.58146299					0.89113778				
42	0.420000		0.420000		0.66497890	0.58703406					0.91958824				
43	0.430000		0.430000		0.68063291	0.59243402					0.94202622				
44	0.440000		0.440000	0.6	0.69628692	0.59766543			0.00118642	0.00135270	0.95922114	0.60168711	0.00118642	0.00107304	0.00000001
45	0.450000		0.450000		0.71194093	0.60273086					0.97202495				
46	0.460000		0.460000		0.72759494	0.60763287					0.98128898				
47	0.470000		0.470000		0.74324895	0.61237403					0.98780202				
48	0.480000		0.480000		0.75890295	0.61695689					0.99225128				
49	0.490000		0.490000		0.77455696	0.62138402					0.99520464				
50	0.500000		0.500000		0.79021097	0.62565799					0.99710951				
51	0.510000		0.510000		0.80586498	0.62978135					0.99830332				
52	0.520000		0.520000	0.63	0.82151899	0.63375668			0.00001975	0.00000047	0.99903032	0.63393875	0.00001975	0.00000026	0.00000000
53	0.530000		0.530000		0.83717300	0.63758653					0.99946049				
54	0.540000		0.540000		0.85282700	0.64127347					0.99970782				
55	0.550000		0.550000		0.86848101	0.64482005					0.99984600				
56	0.560000		0.560000		0.88413502	0.64822886					0.99992101				
57	0.570000		0.570000		0.89978903	0.65150244					0.99996058				
58	0.580000		0.580000		0.91544304	0.65464336					0.99998086				
59	0.590000		0.590000		0.93109705	0.65765418					0.99999096				
60	0.600000		0.600000	0.66	0.94675105	0.66053748			0.00065309	0.00068085	0.99999584	0.66053867	0.00065309	0.00068091	0.00000000
61	0.610000		0.610000		0.96240506	0.66329580					0.99999814				
62	0.620000		0.620000		0.97805907	0.66593173					0.99999919				
63	0.630000		0.630000		0.99371308	0.66844780					0.99999866				
64	0.640000		0.640000		1.00936709	0.67084661					0.99999986				
65	0.650000		0.650000		1.02502110	0.67313069					0.99999994				
66	0.660000		0.660000		1.04067511	0.67530263					0.99999998				
67	0.670000		0.670000		1.05632911	0.67736497					0.99999999				
68	0.680000		0.680000		1.07198312	0.67932029					1.00000000				
69	0.690000		0.690000		1.08763713	0.68117115					1.00000000				
70	0.700000		0.700000	0.68	1.10329114	0.68292011			0.00207531	0.00234989	1.00000000	0.68292011	0.00207531	0.00234989	0.00000008
71	0.710000		0.710000		1.11894515	0.68456974					1.00000000				
72	0.720000		0.720000		1.13459916	0.68612260					1.00000000				
73	0.730000		0.730000		1.15025316	0.68758125					1.00000000				
74	0.740000		0.740000		1.16590717	0.68894825					1.00000000				
75	0.750000		0.750000		1.18156118	0.69022617					1.00000000				
76	0.760000		0.760000		1.19721519	0.69141758					1.00000000				
77	0.770000		0.770000		1.21286920	0.69252503					1.00000000				
78	0.780000		0.780000		1.22852321	0.69355109					1.00000000				
79	0.790000		0.790000		1.24417722	0.69449832					1.00000000				
80	0.800000		0.800000	0.7	1.25983122	0.69536979			0.00429753	0.00371184	1.00000000	0.69536929	0.00429753	0.00371184	0.00000034
81	0.810000		0.810000		1.27548523	0.69616656					1.00000000				
82	0.820000		0.820000		1.29113924	0.69689269					1.00000000				
83	0.830000		0.830000		1.30679325	0.69755025					1.00000000				
84	0.840000		0.840000		1.32244726	0.69814179					1.00000000				
85	0.850000		0.850000		1.33810127	0.69866989					1.00000000				
86	0.860000		0.860000		1.35375527	0.69913711					1.00000000				
87	0.870000		0.870000		1.36940928	0.69954600					1.00000000				
88	0.880000		0.880000		1.38506329	0.69989914					1.00000000				
89	0.890000		0.890000		1.40071730	0.70019909					1.00000000				
90	0.900000		0.900000	0.7	1.41637131	0.70044840			0.00429753	0.00435652	1.00000000	0.70044840	0.00429753	0.00435652	0.00000000
91	0.910000		0.910000		1.43202532	0.70064865					1.00000000				
92	0.920000		0.920000		1.44767932	0.70080539					1.00000000				
93	0.930000		0.930000		1.46333333	0.70091820					1.00000000				
94	0.940000		0.940000		1.47898734	0.70099063					1.00000000				
95	0.950000		0.950000		1.49464135	0.70102524					1.00000000				
96	0.960000		0.960000		1.51029536	0.70102461					1.00000000				
97	0.970000		0.970000		1.52594937	0.70099129					1.00000000				
98	0.980000		0.980000		1.54160338	0.70092784					1.00000000				
99	0.990000		0.990000		1.55725738	0.70083684					1.00000000				
100	1.000000		1.000000	0.7	1.57291139	0.70072084			0.00429753	0.00439256	1.00000000	0.70072084	0.00429753	0.00439256	0.00000001

Table Four
Model Zero Data Correlations Tableau

PART TWO
SOME CONVENIENT MODELS
FOR THE COMPONENT FUNCTIONS F(x) AND G(x)

Algebraic Polynomials

In Part One we explored the use of particular algebraic polynomials (a linear and a cubic) for the approximation of F(x) and G(x):-

$$E(x) = c_0 x^0 + c_1 x^1 + c_2 x^2 + c_3 x^3 + ... c_n x^n + \varepsilon = \varepsilon + \sum_{j=0}^{n} c_j x^j$$

Equation 32

where E(x) is a General Fitted Function (in the present context either F(x) or G(x)); c_j is some Coefficient; and n is the Number of Terms in the Polynomial.

Polynomials are very versatile but they lack truly transcendental characteristics and in particular they inadequately simulate convergent conditions including asymptotic approach. They also lack "physical realism".

The Exponential Function

The Exponential Function is defined by:-

$$E(x) = y = c_0 e^{c_1 x} \equiv c_0 \exp(c_1 x)$$

Equation 33

where e is the Napierian Base and c_0 and c_1 are Coefficients.

Hyperbolic and Reciprocal Functions[7]

(a) The Hyperbolic Function

This is given by:-

$$y = \frac{x}{c_0 + c_1 x}$$

Equation 34

whose Coefficients via simultaneous equations are given by:-

$$c_1 = \frac{1}{(x_1 - x_2)}\left(\frac{x_1}{y_1} - \frac{x_2}{y_2}\right)$$ **Equation 35a**

$$c_0 = \frac{x_1}{y_1} - c_1 x_1 \qquad \textbf{Equation 35b}$$

c_0 and c_1 are coefficients. y, the Dependent Variable E(x), may of course be construed as either F(x) or G(x).

(b)   The Reciprocal Function

$$y = c_0 + c_1\left(\frac{1}{x}\right) \quad \text{study } c_1 < 0 \ (\ c_0 \text{ is } y_{limit} \text{ i.e. the ceiling })$$

Equation 36

The Coefficient equations are:-

$$c_1 = \frac{(y_1 - y_2)}{\left(\dfrac{1}{x_1} - \dfrac{1}{x_2}\right)} \qquad \textbf{Equation 37a}$$

$$c_0 = y_1 - c_1\left(\frac{1}{x_1}\right) \qquad \textbf{Equation 37b}$$

Napierian Functions[8]

(a)   Logarithmic Functions[9]

$$y = \frac{\log_n x}{x}$$

Equation 38

(b)   Exponential Functions[10]
        Increasing to a Limit

$$y = y_\infty (1 - \exp(-ax)) \qquad y_\infty \text{ is the ceiling}$$

Equation 39

a is a Coefficient.
Exponential Rise Function[11]

$$u = u_\infty + (u_0 - u_\infty)e^{-x/\tau} \qquad [12]\text{Eqn.3.4}$$

Equation 40

$$u_0 = \text{Initial State} \qquad \tau = \text{Time or its surrogate}$$
$$u_\infty = \text{Final State} \qquad (\text{ the independent variable })$$

Hot Filament Model[12]

$$T^* = T_\infty \left(1 - e^{-x/\lambda}\right) \qquad\qquad {}^{[13]}\text{Eqn(4.13)}$$

Equation 41

$T_\infty = \text{Plateau Temperature}$
$x = \text{Distance along Filament}$
$\lambda = \text{Length Constant}$

Linear Regression[14]

For the statistical regression of a *first-degree algebraic polynomial* we may define the Coefficient of Determination as:-

$$r^2 = \frac{S_{xy}^{\ 2}}{S_{xx}S_{yy}} = \frac{SSR}{S_{yy}}$$

Equation 42

where r^2 is the Sample Coefficient of Determination and the implicated Sums of Squares are:-

$$S_{xx} = \sum \left(x - \bar{x}\right)^2 \qquad\qquad \textbf{Equation 43}$$
$$S_{yy} = \sum \left(y - \bar{y}\right)^2 \qquad\qquad \textbf{Equation 44}$$
$$S_{xy} = \sum \left(x - \bar{x}\right)\left(y - \bar{y}\right) \qquad\qquad \textbf{Equation 45}$$

Also, the Coefficients of the First Degree Polynomial:-

$$f(x) = c_0 + c_1.x$$

Equation 46

are yielded by:-

$$b = \frac{n\sum xy - \sum x \sum y}{n\sum x^2 - \left(\sum x\right)^2}$$

Equation 47

$$a = \frac{\sum y - b \sum x}{n}$$

Equation 48

In the context of regression both Equation Forty-Seven and Equation Forty-Eight may be generalised as:-

$$c_1 = \frac{n \sum s(x)t(x) - \sum s(x) \sum t(x)}{n \sum [s(x)]^2 - \left(\sum s(x)\right)^2}$$

Equation 49

$$c_0 = \frac{\sum t(x) - c_1 \sum s(x)}{n}$$

Equation 50

where $c_0 \equiv a$, $c_1 \equiv b$; and $s(x)$ and $t(x)$ are (transcendental) transforms respectively of x and y data such that $t(x)$ is a function of x. It is of course possible, according to context, that $s(x)$ and $t(x)$ are the native data, i.e. x and y.

These transformations can be used to "linearise" the data so that a simple linear model may furnish the coefficients c_0 and c_1. But such mathematical wizardry must be handled with extreme caution because any non-linear model so treated is *by definition* not appropriate to linear regression, and accordingly the coefficients arrived at may not resemble true and optimal variance-minimising parameters native to the curve analysed. As with so much empirical craftwork we are still working well within the doctrine of "Goodenough Theory".

Back-transformation of the coefficients found may or may not be needed.

It is possible to construct theoretically-respectable (but much more intricate and expensive) solutions by appeal to Gauss-Jordan Elimination and other advanced methods such as discussed in my PhD thesis[15] and numerous other sources in the literature.

The Linear Regression Treatment of Selected Functions

The Exponential Function

Recall that:-

$$E(x) = y = c_0\, e^{c_1 x} \equiv c_0 \exp(c_1 x)$$

Equation 51

The linear transform of Equation Fifty-One is:-

$$\log_n(y) = \log_n(c_0) + c_1(x)$$

Equation 52

For this transformation, the relevant data transforms are:-

$$s(x) = x \qquad \textbf{Equation 53a} \qquad t(x) = \log_n(y) \qquad \textbf{Equation 53b}$$

Regression by Equation Forty-Nine and Fifty establishes the Coefficients c_0 and c_1
as:-

$$c_1 = b \qquad \textbf{Equation 54a} \qquad c_0 = \exp(a) \qquad \textbf{Equation 54b}$$

The Exponential Limit Function

Recall that:-

$$y = y_\infty(1 - \exp(-ax)) \qquad y_\infty \text{ is the ceiling}$$
$$\textbf{Equation 55}$$

By equivalating $c_0 \equiv y_\infty$ the linear transform of Equation Fifty-Five is:-

$$\log_n(c_0 - y) = \log_n(c_0) + c_1(x)$$
$$\textbf{Equation 56}$$

For this transformation, the relevant data transforms are:-

$$s(x) = x \qquad \textbf{Equation 57a} \qquad t(x) = \log_n(c_0 - y) \qquad \textbf{Equation 57b}$$

Regression by Equation Forty-Nine and Fifty establishes the Coefficients c_0 and c_1
as:-

$$c_1 = -b \qquad \textbf{Equation 58a} \qquad c_0 = \exp(a) \qquad \textbf{Equation 58b}$$

The Hyperbolic Function

Recall that:-

$$y = \frac{x}{c_0 + c_1 x}$$
$$\textbf{Equation 59}$$

No logarithmic transformation of Equation Fifty-Nine is required.
For this transformation, the relevant data transforms are:-

$$s(x) = \frac{1}{x} \qquad \textbf{Equation 60a} \qquad t(x) = \frac{1}{y} \qquad \textbf{Equation 60b}$$

Regression by Equation Forty-Nine and Fifty establishes the Coefficients c_0 and c_1 as:-

$$c_1 = a \qquad \textbf{Equation 61a} \qquad c_0 = b \qquad \textbf{Equation 61b}$$

Note the unexpected switch of coefficients in Equations Sixty-One.

The Reciprocal Function

Recall that:-

$$y = c_0 + c_1 \left(\frac{1}{x} \right) \qquad \text{study } c_1 < 0 \ (\ c_0 \text{ is } y_{limit} \text{ i.e. the ceiling })$$

$$\textbf{Equation 62}$$

No logarithmic transformation of Equation Sixty-Two is required.
For this transformation, the relevant data transforms are:-

$$s(x) = \frac{1}{x} \qquad \textbf{Equation 63a} \qquad t(x) = y \qquad \textbf{Equation 63b}$$

Regression by Equation Forty-Nine and Fifty establishes the Coefficients c_0 and c_1 as:-

$$c_1 = b \qquad \textbf{Equation 64a} \qquad c_0 = a \qquad \textbf{Equation 64b}$$

Note that in this case switching of the regression coefficients is not required.

An Illustrative Plot of Selected F(x) and G(x) Component Functions

Plot Three shows specimen trajectories of selected functions on the interval $0 \leq E(x) \leq 1$ together with their equations.

Exponential Limit Coefficient Equations

For the pure Exponential Limit Equation Thirty-Nine we may define:-

$$c_1 = -\frac{1}{x} . \log_n \left(1 - \frac{y}{c_0} \right) \qquad \textbf{Equation 65a}$$

$$c_0 = \frac{y}{1 - e^{-c_1 x}}$$ **Equation 65b**

These equations imply circularity of definition, but may have virtue in the stochastic estimation of c_0 and c_1.

Some Non-Linear Part Two Models

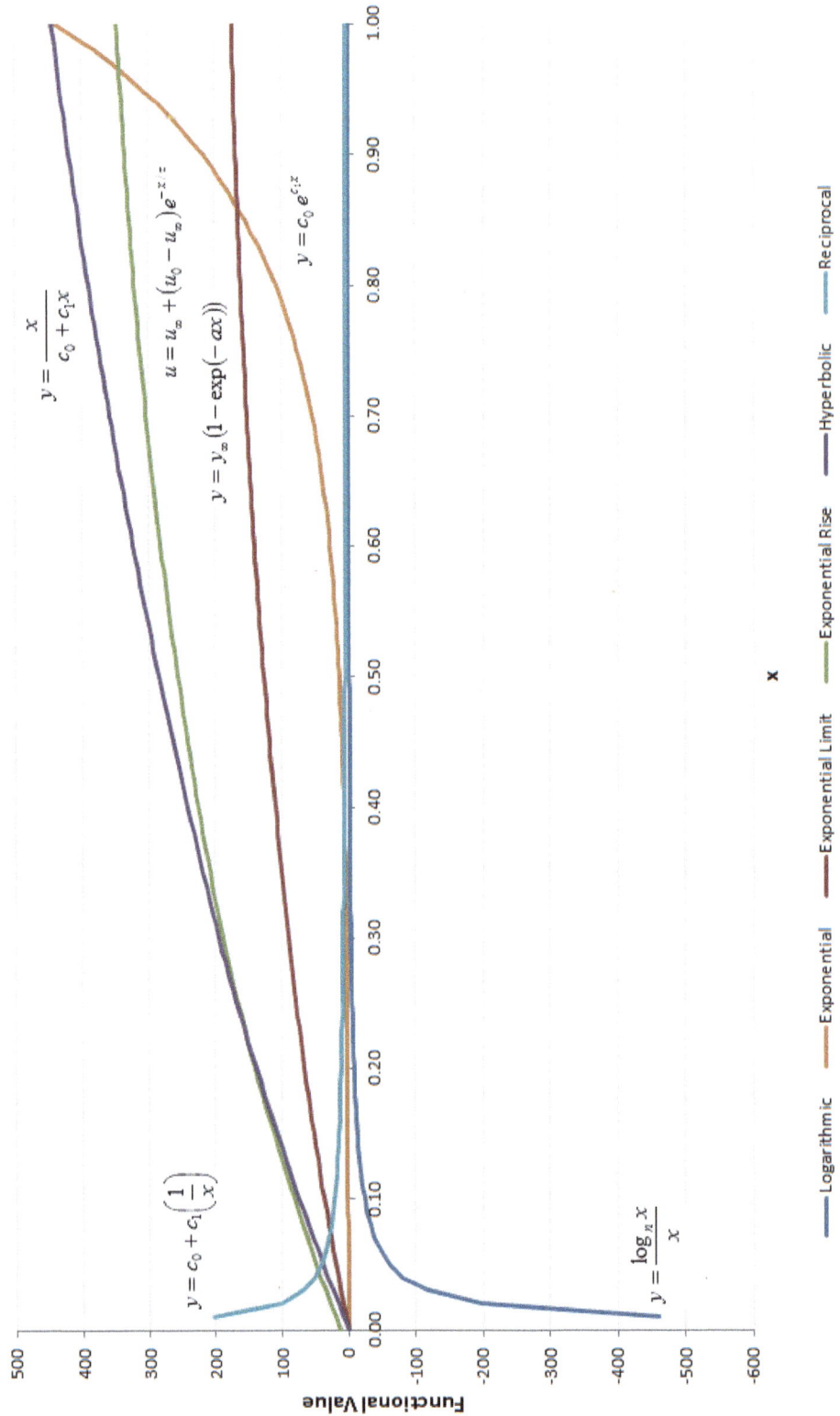

$$y = \frac{x}{c_0 + c_1 x}$$

$$u = u_\infty + (u_0 - u_\infty)e^{-x\tau}$$

$$y = y_\infty (1 - \exp(-\alpha x))$$

$$y = c_0 e^{c_1 x}$$

$$y = c_0 + c_1 \left(\frac{1}{x}\right)$$

$$y = \frac{\log_n x}{x}$$

Functional Value

x

— Logarithmic   — Exponential   — Exponential Limit   — Exponential Rise   — Hyperbolic   — Reciprocal

Plot Three
Selected Specimen Function Plot

PART THREE
TRANSITION MODELS IN
THE STRENGTH OF CONCRETE

      Traditionally, the strength of concrete (or indeed competent natural rock) was tested with a uniaxial rig, a crushing device that gradually and destructively applied stress (which is physically a pressure $ML^{-1}T^{-2}$) to a shaped specimen, whilst the deformationary response of the workpiece was measured in terms of dimensionless strain ($M^0L^0T^0$).   At first, the specimen would respond elastically according to Hooke's Law in that any strain was strictly proportional to stress and could wholly be recovered when the applied stress was lightened. But with increased pressure a point would be reached where the specimen was permanently deformed and no longer recovered upon release. This Yield Point onset suddenly; mathematically-discontinuously. The specimen was said to transit from its Elastic Zone to its Plastic Zone. Eventually, as pressure mounted, the specimen would reach a Failure Point at which spallation and crumbling would reduce the stress borne but the specimen would lose all structural integrity.

      The outcome for several identical samples would be a tabulation of comparative readings of Stress and Strain which could be plotted as a discontinuous empirical curve.

      The typical experimental history can be envisioned with the help of Figure One:-

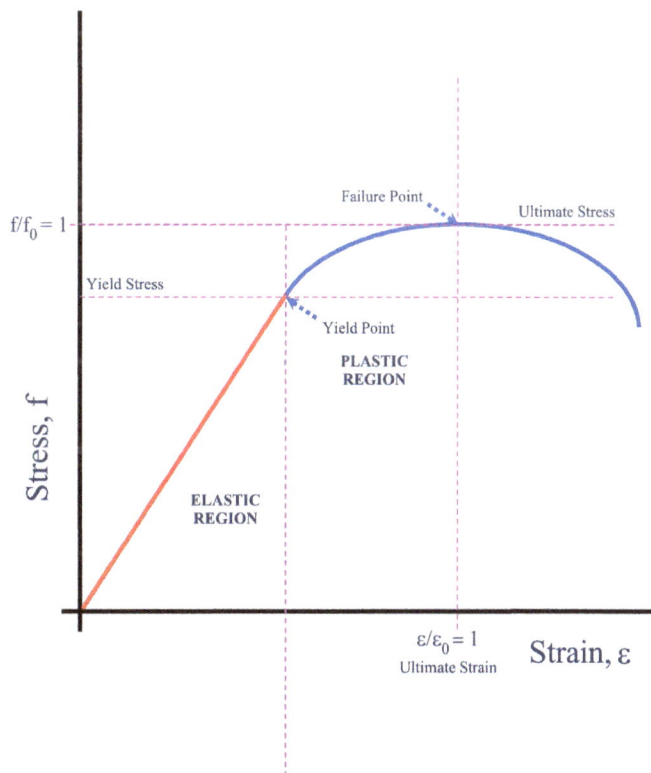

Figure One
Typical Stress and Strain Behaviour of a Brittle Member

The red line denotes the elastic response of the specimen and the blue its plastic response. The yield point marks the mathematical change of regime, which in terms of a Heaviside Step Function we may identify with x_{mid}. The red line is obviously a First-Degree Polynomial whilst the blue is some non-linear function not of course the arc of an ideal ellipse. It is the treatment of this highly-variable plastic curve which will account for almost all the intellectual effort, and almost all of the error, in anyone's analyses.

Stress is noted as f, and Strain as ε.

In order to impose some degree of generality and comparability it is interesting to treat stress and strain not as their raw physical values but as fractions of their respective maxima: f_0 the Ultimate Stress where complete breakdown occurs at maximum stress; and ε_0, the strain where that total failure supervenes. Such normalisation allows us to relate a Normalised Stress, f/f_0, to a Normalised Strain, $\varepsilon/\varepsilon_0$, given that the regime of total destruction beyond the failure point is of only limited engineering interest.

Some Models of Concrete Deformation

The deformation of concrete is a very complex outcome of age, composition, fabric and reinforcement (if any). Accordingly, there are myriads of test findings in the literature, and models to fit them.

We shall discuss a handful, and in particular those recently offered[16].

The Hognestad (1951) Relations[17]

$$f_c = f_c' \left[2\left(\frac{\varepsilon}{\varepsilon_0} \right) - \left(\frac{\varepsilon}{\varepsilon_0} \right)^2 \right] \qquad 0 < \varepsilon < \varepsilon_0$$

Equation 66

$$f_c = f_c' - 0.15 f_c' \left[\frac{\varepsilon - \varepsilon_0}{\varepsilon_\infty - \varepsilon_0} \right] \qquad \varepsilon_0 < \varepsilon < \varepsilon_\infty$$

Equation 67

where f_c is Stress in Concrete; f_c' is the Peak Concrete Stress; ε is the Strain in Concrete; ε_0 is the Strain at Peak Concrete Stress; and ε_∞ is the Ultimate Concrete Strain.

Desayi-Krishnan (1964) Relations[18]

$$f_c = \frac{E_0 \varepsilon}{1 + \left(\dfrac{\varepsilon}{\varepsilon_0} \right)^2}$$

Equation 68

$$E_0 = 2 f_c' / \varepsilon_0$$

Equation 69

where E_0 is the Initial Tangent Modulus.

The Ali-Farid-Al-Janabi Relations[19]

$$\frac{f}{f_0} = \frac{R\left(\dfrac{\varepsilon}{\varepsilon_0}\right)}{1 + (R-1)\left(\dfrac{\varepsilon}{\varepsilon_0}\right)^{\beta}}$$

Equation 70

$$\beta = \frac{R}{R-1}$$

Equation 71

$$R = \frac{E_c}{E_0}$$

Equation 72

$$E_0 = \frac{f_0}{\varepsilon_0}$$

Equation 73

where R is a Material Parameter Depending Upon the Shape of the Stress-Strain Curve; E_c is the Elastic Modulus of Concrete; E_0 is the Dividend of the Maximum Stress over the Maximum Strain.

The Development of Models

For this Model One illustration we will choose the Primary Model, F(x), to be a simple first-order polynomial (i.e. straight line) indicative of elastic response:-

$$F(x) = c_1 + c_2 x$$

Equation 74

It is obvious from the physical predicament that c_1 is theoretically zero at the origin of the line where zero stress engenders zero strain. Furthermore, it is clear that the gradient c_2 identifies with λ, Young's Modulus, a property of the given material, which is defined by:-

$$\lambda \equiv E_c = \frac{Elastic\ Stress}{Elastic\ Strain}$$

Equation 75

and accordingly *at the analytical level* that:-

$$F(x) \equiv \lambda x$$

Equation 76

Turning now to the Secondary Model, $G(x)$, we shall choose the Ali-Farid-Al-Janabi equation for f/f_0 (Equation Seventy) and specify:-

$$G(x) \equiv \frac{f}{f_0} = \frac{c_3 x}{1 + (c_3 - 1)x^{c_4}}$$

Equation 77

In my opinion, Equation Seventy-Seven is also capable of appropriate simplification, though this was not done for the current exercise.

Having these facts to hand we may now specify the Composite Function, $Q(x)$, as:-

$$Q(x) = [1 - H(x)]F(x) + H(x)G(x)$$

Equation 78

or:-

$$Q(x) = [1 - H(x)]\lambda x + H(x)\frac{c_3 x}{1 + (c_3 - 1)x^{c_4}}$$

Equation 79

We may also examine a second composite, $P(x)$, defined by:-

$$P(x) = \frac{F(x) + G(x)}{2}$$

Equation 80

Whatever the virtues or vices of $P(x)$ it is demonstrably the same as $Q(x)$ at x_{mid}, and also it is very simple.

<u>Procedure</u>

To illustrate the technique I randomly choose seven F(x) data pairs and fourteen G(x) data pairs from Figure 2 in the Ali-Farid-Al-Janabi paper "Stress-Strain Relationship for Concrete in Compression Made of Local Materials". I did this by inspection, attempting to include pairs that would regress to representative curves. The relevant stress-strain plot includes about a hundred data pairs from numerous concrete strength experiments.

There issues, of course, a certain subjective imprecision but this need not detract from the generally of our arguments.

Simple linear regression using the MicroSoft EXCEL® 2010 worksheet intrinsic functions gave $c_1 = -0.012566753840$ and $c_2 = 1.839449282765$. The former figure is not of course a perfect zero because of experimental, procedural and regression errors. The latter result is the average Young's Modulus of the seven relevant specimens.

To regress the non-linear G(x) I resorted to the EXCEL® Data Solver utility applied to the fourteen-pair (x,y) tableau shown below:-

Error Sum	0.174007390251097
a	8100.424593001720000
b	0.486332295163104

x	y_{data}	y_{est}	Error Term
0.164	0.36	0.395016	0.007858
0.309	0.42	0.546974	0.053889
0.229	0.49	0.468934	0.002018
0.403	0.525	0.626943	0.02644
0.336	0.57	0.571029	3.25E-06
0.321	0.61	0.557787	0.008763
0.436	0.735	0.652814	0.01585
0.603	0.736	0.771154	0.002078
0.586	0.87	0.759907	0.020989
0.729	0.872	0.850117	0.000663
0.671	0.888	0.81467	0.008102
0.893	0.95	0.94352	4.72E-05
0.721	0.985	0.845311	0.027308
1	1	1	0

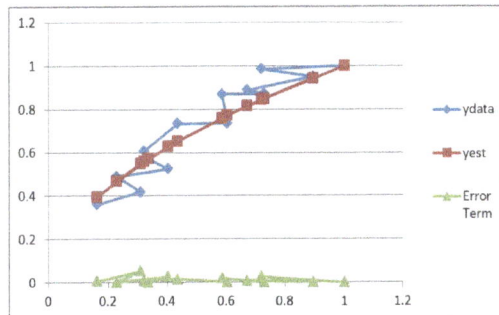

Table Five
Non-Linear Solutions Tableau

In this context, $a \equiv c_3$ and $b \equiv c_4$ whilst the error sum to be minimised is defined by:-

$$Error\ Sum = \sum \left(\frac{y_{data} - y_{est}}{y_{est}} \right)^2$$

Equation 81

where y_{est} is given by:-

$$y_{est} = \frac{ax}{1 + (a-1)x^b}$$

Equation 82

The outcome of this routine was that $c_3 = 8100.424593001720000$ and $c_4 = 0.486332295163104$.

The Determination of x_{mid}

The midpoint of the Step Function $H(x)$ was determined by Regula Falsi iteration using the normalisation end-points $x_1 = 0$ and $x_2 = 1$ for which the iteration Start Point was $x_0 = (x_2-x_1)/2$, i.e. 0.5.

Successor points were computed by:-

$$x_i = \frac{x_{i-2}D(x_{i-1}) - x_{i-1}D(x_{i-2})}{D(x_{i-1}) - D(x_{i-2})}$$

Equation 83

for $x_i = 3\ldots$, where $D(x_i)=F(x_i)-G(x_i)$.
x converged to fifteen decimal positions after six iterations so that:-

$$x_8 \equiv x_{mid} = 0.299413207852393$$

Equation 84

The system also converged using Newton-Raphson Iteration for the root $D(x)=0$ but convergence took fifteen iterations. Halley Iteration was not attempted.
The relevant Regula Falsi tableau may be inspected as Table Six.

| x_1 | 0.000000000000000 |
| x_2 | 1.000000000000000 |

Power	Fc_{power}	Gc_{power}
0	-0.012566753840380	8100.424593001720000
1	1.839449282765390	0.486332295163104

For Row 18:-

F	0.538188656594188		Serial, i	x_i	$F(x_i)$	$G(x_i)$	$D(X_i)=F(x_i)-G(x_i)$
G		0.538188656594188	1	0.500000000000000	0.907157887542313	0.700404811630494	0.206753075911819
F-G	0.000000000000000		2	1.000000000000000	1.826882528925010	1.000000000000000	0.826882528925005
			3	0.333298454937745	0.600518850041666	0.568665668137460	0.031853181904205
			4	0.306586779865293	0.551384078488183	0.544775309152857	0.006608769335326
			5	0.299593893869933	0.538521019359558	0.538355495721841	0.000165523637717
			6	0.299414250258249	0.538190574046891	0.538189619255578	0.000000954791313
			7	0.299413208006549	0.538188656877750	0.538188656736551	0.000000000141199
			8	0.299413207852393	0.538188656594188	0.538188656594188	0.000000000000000
			9	0.299413207852393	0.538188656594188	0.538188656594188	0.000000000000000

Table Six
Model One Regula Falsi Tableau

Selected Results

Table Seven presents the functional top matter for the Model One trial fitments. Please note that the Exponential Limit $F(x)$ was *not* employed in this instance: The Ali-Farid $G(x) = \alpha x/(1+(\alpha-1)x^\beta)$ model was run instead.

Ali-Farid Eqn.2 adverts to a cubic algebraic polynomial that Ali, Farid and Al-Janabi offer in their paper as an alternative Stress-Strain fitment to Equation Seventy.

To be clear that fitted equation is:-

$$\frac{f}{f_0} = 2.1\left(\frac{\varepsilon}{\varepsilon_0}\right) - 1.33\left(\frac{\varepsilon}{\varepsilon_0}\right)^2 + 0.2\left(\frac{\varepsilon}{\varepsilon_0}\right)^3$$

Equation 85

Equation Eighty-Five is computed in my Data Correlations Tableau and is plotted with the functional curves for comparison. Equation Eighty-Five was of course regressed upon the whole collection of these workers' data, not just my selected twenty-one points, so there would have been little purpose in a formal corroborative regression, or indeed the pursuit of a determination coefficient.

It seems to me that Ali, Farid and Al-Janabi suppressed the zeroth coefficient in their routine because it is theoretically zero, given that Stress-Strain relations theoretically intersect the origin. I am not sure how this might affect the correlation of the measured points.

s	Steepness	12
x		
	Lower Bound:	0
	Upper Bound:	1
	Intervals:	15
	Increment:	0.066666667
F(x)	$a_1 + b_1.x$	
	a_1	-0.012566753840
	b_1	1.839449282765
G(x)	$1 - a_2 \exp(-b_2.x)$	
	a_2 (known limit y_{inf})	1.626533775993920
	b_2	1.856504274198260
G(x) Error Term		0.051022896139
H(x)	x_{mid}	0.299413207852
Σ(data)/n(data)		0.524809523810
Ali-Farid Eqn.2 c_0		0.000000000000
Ali-Farid Eqn.2 c_1		2.100000000000
Ali-Farid Eqn.2 c_2		-1.330000000000
Ali-Farid Eqn.2 c_3		0.200000000000
G(x)	$\alpha x/(1+(\alpha-1)x^{\beta})$	
Ali-Farid Eqn.3 α		8100.424593001720000
Ali-Farid Eqn.3 β		0.486332295163104

Table Seven
Model One Functional Tabulation Top Matter

Table Eight
Model One Functional Tabulation Bottom Matter

Summary block (per column):

	x_o (F)	F(x_o)=data	x_o (G)	G(x_o)=data	F(x_o)=a₁+b₁x	G(x_o)=αx/(1+(α−1)x²)	SqErr F	SqErr G	SS_Fo	SS_Fw	SS_Go	SS_Gw	H(x)	P(x)	SS_Po	SS_Pw	SS_Pe	Q(x)	SS_Qo	SS_Qw	SS_Qe	AF Eqn.2	AF Eqn.3	AF Eqn.2
Count	7	7	14	14	21	21	7	14	7	21	14	21	21	21	21	21	21	21	21	21	21	21	21	21
Mean	0.085271	0.144286	0.715071	0.528643	0.687992	0.557340	0.002673	0.005460	0.008432	0.580164	0.044330	0.088283	0.525017	0.622666	0.104763	0.162482	0.027685	0.508701	0.032364	0.022494	0.000461	0.528271	0.557340	0.528271
Sum	0.596900	1.010000	10.011000	7.401000	14.447830	11.704149	0.018710	0.076434	0.059021	12.183437	0.620619	1.853953	11.025365	13.075990	2.200023	3.412129	0.581394	10.682712	0.679640	0.472364	0.009691	11.089481	11.704149	11.089481
a	−0.012567	−0.012567	0.290610	0.290610	−0.012567	0.324264	0.000428	0.006219														−0.150014	0.534692	−0.150014
b	1.839449	1.839449	0.802927	0.802927	1.839449	0.711321	0.015560	−0.001436														1.216623	49.076948	1.216623
R²	0.682994	0.682994	0.869208	0.869208	1.000000	0.876842229	0.391459	0.003062							0.735732861			0.985740865				0.988041	0.033896	0.988041

Main data:

Serial	x_Row	x_o (F)	F(x_o)=data	x_o (G)	G(x_o)=data	F(x_o)=a₁+b₁x	G(x_o)=αx/(1+(α−1)x²)	SqErr (F(data)−a₁+b₁x)	SqErr (G(data)−αx/(1+(α−1)x²))	SS_Fo	SS_Fw	SS_Go	SS_Gw	H(x) [ERF model]	P(x)=F(x)+G(x)	SS_Po	SS_Pw	SS_Pe	Q(x)=(1+H(x))·F(x)+H(x)·G(x)	SS_Qo	SS_Qw	SS_Qe	AF Eqn.2 f/f_E	AF Eqn.3 f/f_E	AF Eqn.2 f/f_E
0	0.014	0.014	0.04			0.01318554	0.11115985	0.00071902		0.01087551	0.01718726		0.36427451	0.00000064	0.06235259	0.23504027	0.21386632	0.00044834	0.01318560	0.01087551	0.01718728	0.00003984	0.02913987	0.11151985	0.02913987
1	0.069	0.0686	0.065			0.11361947	0.25214082	0.00236385		0.00628622	0.00094042		0.21405182	0.00004482	0.18301677	0.21142480	0.11682228	0.00094964	0.11362569	0.10287551	0.00094042	0.00002858	0.13786564	0.25241408	0.13786564
2	0.071	0.071	0.1			0.11803415	0.25691485	0.00032523		0.00196122	0.00068914		0.20990806	0.00005303	0.18747417	0.18046313	0.11379514	0.00444462	0.11804151	0.00196122	0.00065106	0.00000162	0.14246205	0.25691419	0.14246205
3	0.071	0.0714	0.07			0.11876992	0.25765968	0.00237851		0.00551837	0.00065106		0.20922799	0.00005453	0.18821345	0.20685170	0.11329692	0.00875250	0.11877750	0.00551837	0.00065106	0.00002369	0.14323251	0.25765697	0.14323251
4	0.093	0.0929	0.23			0.15831808	0.29495774	0.00513830		0.00734694	0.00019691		0.17648036	0.00002861	0.22664693	0.08691266	0.08890093	0.00000395	0.15834933	0.00734694	0.00019691	0.00005112	0.18377391	0.29497577	0.18377391
5	0.129	0.129	0.295			0.22472220	0.34917820	0.00493897		0.02271480	0.00647003		0.13387785	0.00191394	0.28695020	0.05281242	0.05657706	0.00001417	0.22496041	0.02271480	0.00647003	0.00026389	0.24919981	0.34917821	0.24919981
6	0.15	0.15	0.21			0.26335064	0.37731437	0.00284629		0.00431837	0.01417646		0.11407979	0.00561232	0.32033254	0.09910504	0.04181084	0.00328263	0.26399024	0.00431837	0.01417646	0.00009718	0.28575000	0.37731444	0.28575000
7	0.164			0.164	0.36	0.28910293	0.39501629		0.00122615		0.02097203	0.12607572	0.10243523	0.01077980	0.34205965	0.10243523	0.03339752	0.00003888	0.29024465	0.12607572	0.10243523	0.00055687	0.30951051	0.39501638	0.30951051
8	0.229			0.229	0.49	0.40866713	0.46893428		0.00044376		0.06989753	0.05065715	0.06058350	0.11605394	0.43880070	0.06058350	0.00739752	0.00003826	0.41566131	0.05065715	0.06058350	0.00009653	0.41355527	0.46893428	0.41355527
9	0.309			0.309	0.42	0.55582307	0.54697205		0.01612245		0.16936300	0.08706715	0.02825668	0.55139864	0.56446002	0.02825668	0.00070698	0.00010564	0.55082683	0.08706715	0.02825668	0.00345867	0.52781300	0.54697420	0.52781300
10	0.321			0.321	0.61	0.57786647	0.55778524		0.00277625		0.18801828	0.11104001	0.02473854	0.64294431	0.56784149	0.00725742	0.00185175	0.00002922	0.55496689	0.01104001	0.02473854	0.00018765	0.54367670	0.55778652	0.54367670
11	0.336			0.336	0.57	0.60548821	0.57102915		0.00000106		0.21270774	0.02104572	0.02074818	0.73266678	0.67783722	0.00204218	0.00402579	0.00000393	0.58024120	0.02104572	0.02074818	0.00000009	0.56303493	0.57102915	0.56303493
12	0.403			0.403	0.525	0.72873131	0.62694314		0.01099240		0.34157665	0.03651715	0.00776660	0.96061996	0.91262889	0.00000004	0.02341748	0.00054438	0.63095156	0.03651715	0.00776660	0.00008032	0.64338420	0.62694314	0.64338420
13	0.436			0.436	0.735	0.78943313	0.65281591		0.00675461		0.41621519	0.00039715	0.00387604	0.98977411	0.93388761	0.04418004	0.03853912	0.00003182	0.75990743	0.00039715	0.03387604	0.00001110	0.67934469	0.65281359	0.67934469
14	0.586			0.586	0.87	1.06535053	0.75990725		0.01212041		0.84836039	0.02400286	0.00020102	0.99999942	0.91262889	0.11915646	0.15040386	0.00097640	0.75990743	0.02400286	0.00201025	0.00004367	0.81412533	0.75990725	0.81412533
15	0.603			0.603	0.736	1.09662116	0.77115405		0.00123581		0.90694281	0.00043801	0.00043801	0.99999987	0.93388761	0.00460142	0.16734488	0.01506596	0.77115409	0.00043801	0.00314526	0.00007733	0.82655528	0.77115405	0.82655528
16	0.671			0.671	0.888	1.22170371	0.81466953		0.00537736		1.16082955	0.02990429	0.02462658	1.00000000	1.01818662	0.13190732	0.24342096	0.01243529	0.81466953	0.02990429	0.00919378	0.00039938	0.87070181	0.81466093	0.87070181
17	0.721			0.721	0.985	1.31328177	0.84531133		0.01951293		1.36747406	0.07286143	0.07286143	1.00000000	1.07949375	0.21177527	0.30767459	0.00919668	0.84531133	0.07286143	0.03312470	0.03312470	0.89767254	0.84531133	0.89767254
18	0.729			0.729	0.872	1.32839177	0.85011691		0.00047087		1.40210716	0.02462658	0.01823728	1.00000000	1.08925434	0.12054123	0.31859795	0.03922047	0.85011691	0.02462658	0.01823728	0.00004082	0.90156757	0.85011691	0.90156757
19	0.893			0.893	0.95	1.63006146	0.94351965		0.00004199		2.20752955	0.05519143	0.05218859	1.00000000	1.26679055	0.18078694	0.58061509	0.15986255	0.94351965	0.05519143	0.05218859	0.00000902	0.95711222	0.94351965	0.95711222
20	1			1	1	1.82688253	1.00000000		0.00000000		2.83113204	0.08118429	0.08118429	1.00000000	1.41344126	0.25680599	0.78966637	0.31793853	1.00000000	0.08118429	0.08118429	0.00000000	0.97000000	1.00000000	0.97000000

In regard to Table Eight we may segregate the Coefficients of Determination of the key fitted functions in Table Nine below:-

Function Fitted to the Data	Coefficient of Determination R^2	
	From EXCEL Worksheet	From EXCEL Regression Plot
F(x)	0.682993774110	not computed
G(x)	0.876842229183	not computed
Q(x)	0.985740864552	not computed
P(x)	0.735732860805	not computed
$ARIT(R^2_F;R^2_G)$	0.779918001646	
$GEOM(R^2_F;R^2_G)$	0.773871942513	
$HARM(R^2_F;R^2_G)$	0.767872753474	

Table Nine
Model One Key Fitted Function
Coefficients of Determination

Table Nine advertises the very marked superiority of our Composite Function Q(x). This Q(x) is a fit to discontinuous experimental data, exceeding the separate merits of F(x) and G(x). But caution is necessary as part of this precision may be an artefact of a larger data population or something.

There is no simple relation of the Coefficients of Correlation R_F^2 and R_G^2 to Pythagorean means.

It is clear that P(x) is markedly worse than any of the other fitments.

The Functional Comparison Plot

The blue triangles mark the chosen elastic-response data to constitute F(x) whilst the red squares represent the non-linear plastic-response G(x) data.

As noted above, Ali et al provided "Eqn.3", a hyperbolic-style non-linear formula; and also a cubic polynomial "Eqn.2". These are respectively Equations Seventy and Eighty-Five of the current discussion. "Eqn.3" forms a bold magenta long-dot-short line seen to depart the data field significantly in the elastic region, and clearly optimised for the plastic response. The polynomial "Eqn.2" on the other hand shows its greatest weakness in the plastic region, but continues to intersect the data field. "Eqn.2" is denoted by a navy blue bold dash-dot line.

The feint green dashed line is the trajectory of the Step Function H(x). I have not attempted to suppress the segmentality of this or any trace, especially noticeable in this graph of only twenty intervals.

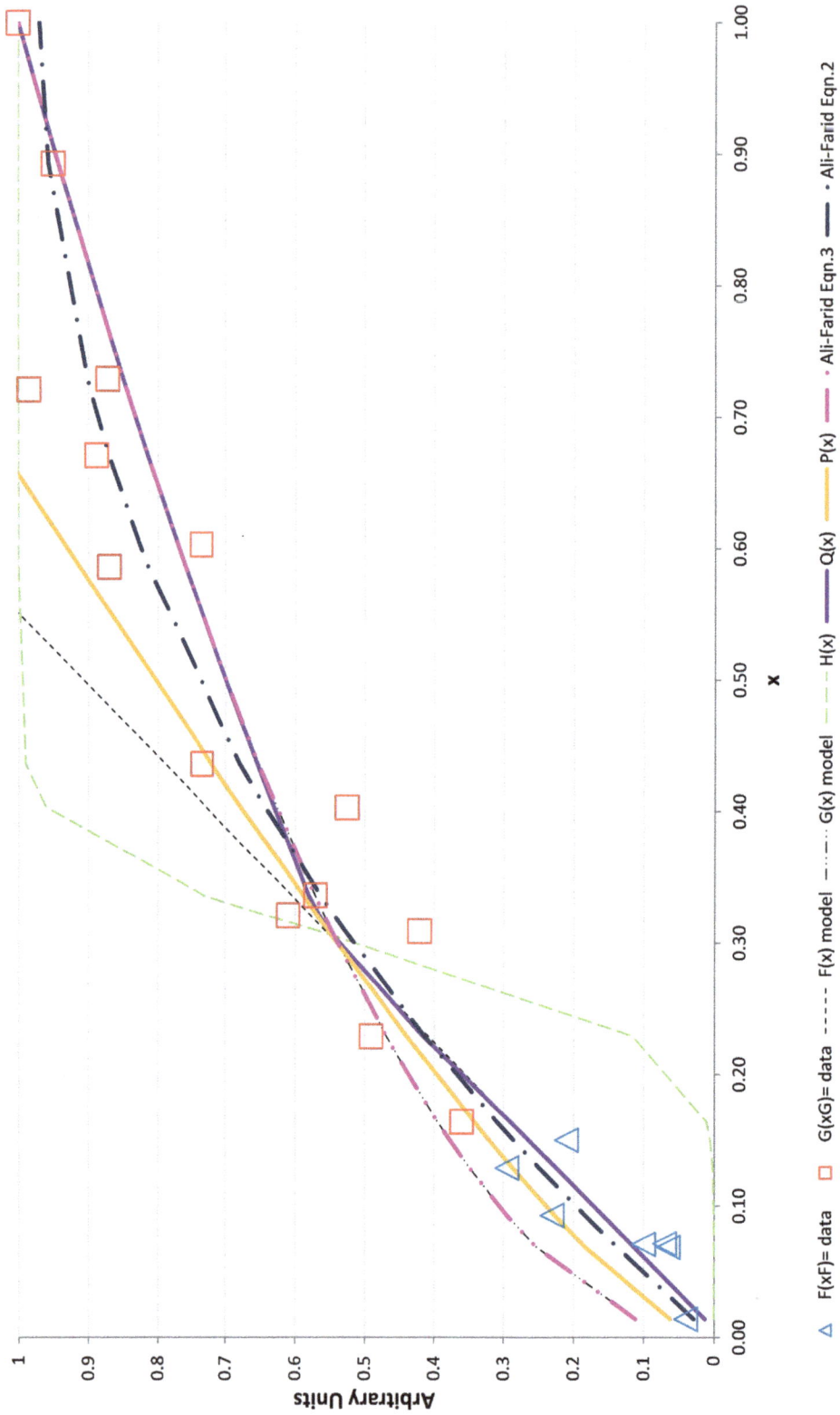

Model One: First-Order Polynomial to Exponential Limit Formula

Plot legend: △ F(xF)= data □ G(xG)= data ---- F(x) model ---·--- G(x) model —— H(x) □ F(xF)= data —— Q(x) —— P(x) ---·--- Ali-Farid Eqn.3 ···· Ali-Farid Eqn.2

x

Arbitrary Units

Plot Four
The Fitment of Functional Models to Concrete Strength Data

The feint black dashed trace of the $F(x)$ regression may be discerned in the region of 0.3<x<0.5, but this and the trace of similar trace of $G(x)$ are generally obliterated by the fitments, as of course we should wish.

It can be seen that the bold yellow course of $P(x)$ is a poor fit, overestimating the best course in the elastic left and wildly overestimating in the plastic right.

The best composite function is $Q(x)$ represented by a bold continuous purple line that overlays both $F(x)$ and $G(x)$ models in their respective regions and has a pronounced crick or discontinuity very near to $x = 0.3$.

x_{mid} is 0.2994 and at that cuspate point $F(x)$, $G(x)$, $H(x)$, $P(x)$ and $Q(x)$ all converge. Only the Ali-Farid "Eqn.2" cubic polynomial misses the point (x_{mid}, y_{mid}) by any visible margin.

PART FOUR
TRANSITION MODELS IN THE CURING OF CONCRETE

The curing of concrete is a process of setting and hardening of laid concrete over time. Engineers make a distinction between curing and drying. For sure, curing involves a decrease in fluidity, but as well as simple drying to the atmosphere it involves complex chemical and crystallisation changes that depend upon the composition and fabric of the mix, as well as temperature, atmospheric humidity and time. We do not concern ourselves here with any structural reinforcement that may or may not be incorporated.

In formal terms, curing involves the progressive gain of Concrete Compressive Strength, f_c, with Age, typically logged in Days, because most of the gain of strength occurs in the first few weeks. Concrete Compressive Strength is physically a pressure ($ML^{-1}T^{-2}$) whilst Age is of course a time (T^1).

In a sense of speaking, curing is a temporal converse of the elastico-plastic stress-distance relationship, and we may model it as beginning with a steep non-linear temporal strength gain ($G(x)$) followed by a quasi-linear, essentially asymptotic model of long term hardening ($F(x)$).

The general conformation of the strength gain history is illustrated in Figure Two:-

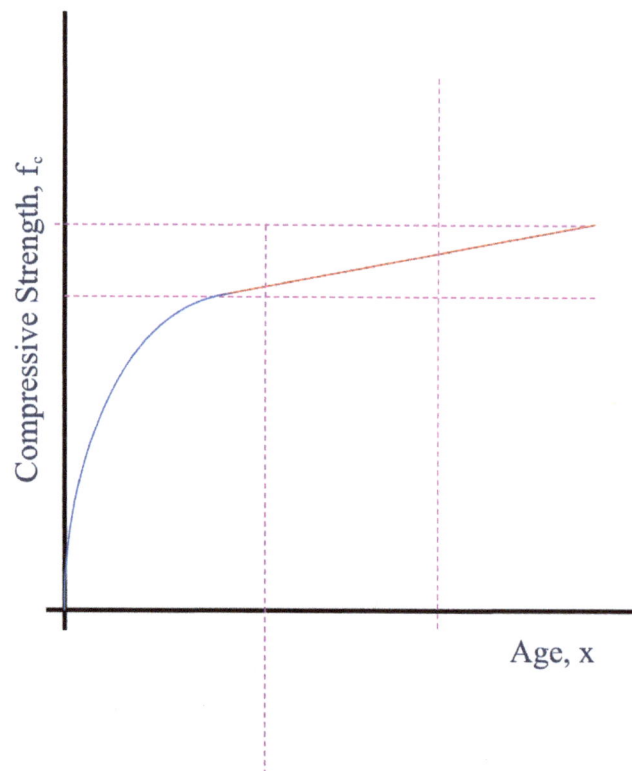

Figure Two
Typical Curing (Strength gain with Age) Behaviour of Concrete

As with all my sketch graphs there is an element of caricature. The initial $G(x)$ phase is not necessarily the arc of an ellipse, and both the rectitude and gradient of $F(x)$ are grossly exaggerated. I write this for the benefit of students and laymen who may naturally over-interpret what they see.

Concrete laid by the Romans over two thousand years ago is still curing. Therefore we cannot claim that the strength of concrete reaches some absolute limit. But we can say that, *in approximate terms*, the strength reaches a practical limit, and its continued history becomes a constant.

Some Models of Concrete Curing

The literature abounds with empirical studies of concrete curing and often enough this or that algebraic model has been fitted to the data.

It would be tedious for us to survey more than enough of these studies but for the purposes of comparison we will inspect the excellent review thesis of Carmichael[20], and in particular the computer-fitted equation that he plots with his Strength-Age data in Figure 7 of his disquisition:-

$$f_c' = 3500 - 2300e^{-0.18t}$$

Equation 86

which may be generalised as:-

$$f_c' = c_5 + c_6 e^{c_7 t}$$

Equation 87

where f_c' is the Compressive Strength of the Concrete (psi); t is the Age of the Mix in Days; and c_5, c_6 and c_7 are all Coefficients of Best Fit. Carmichael quotes the appropriate Coefficient of Determination as 0.9494. Equation Eighty-Six has c_5=3500, c_6=-2300 and c_7=-0.18.

The Development of Models

We will apply an Exponential Limit model to the rapid-curing phase denoted by the blue line:-

$$G(x) = a_2(1 - \exp(-b_2 x))$$
Equation 88

and a First-Order Polynomial (straight line equation) to the red line of dry aging:-

$$F(x) = a_1 + b_1 x$$
Equation 89

The appropriate formulation of the Composite Model Q(x) is:-

$$Q(x) = H(x)\,F(x) + (1 - H(x))\,G(x)$$
Equation 90

which we may elaborate as:-

$$Q(x) = \frac{1}{2} + \frac{1}{2}erf\{k[x - x_{mid}]\}\,F(x) + \left[1 - \left(\frac{1}{2} + \frac{1}{2}erf\{k[x - x_{mid}]\}\right)\right]G(x)$$
Equation 91

giving:-

$$Q(x) = \frac{1}{2} + \frac{1}{2}erf\{k[x - x_{mid}]\}(a_1 + b_1 x) + \left[1 - \left(\frac{1}{2} + \frac{1}{2}erf\{k[x - x_{mid}]\}\right)\right](a_2(1 - \exp(-b_2 x)))$$
Equation 92

Equation Ninety-Two is algebraically co-ordinate and obviously capable of simplification, but in EXCEL® and other elaborations we will utilise Equations Eighty-Eight through Ninety.

To contextualise Equation Ninety-Two we may write:-

$$f_c' = \frac{1}{2} + \frac{1}{2}erf\{k[t - x_{mid}]\}(a_1 + b_1 t) + \left[1 - \left(\frac{1}{2} + \frac{1}{2}erf\{k[t - x_{mid}]\}\right)\right](a_2(1 - \exp(-b_2 t)))$$
Equation 93

where f_c' is Compressive Strength and t is Time. x_{mid} is of course a time value. The simple combined function P(x) is yielded by:-

$$P(x) = \frac{G(x) + F(x)}{2}$$
Equation 94

Procedure

I utilised the curing history data tabulated by Ryan P Carmichael and shown *inter alia* in Table Ten.

I divided the sixteen Average f_c versus Day data into an initial eight G(x) points and a final eight F(x) points.

I then used EXCEL® to regress or otherwise optimise for the five equation coefficients.

The following coefficients and tableau are computed for normalised Carmichael data, (i.e. $0 < x < 1$; $0 < f(x) < 1$).

In terms of the definition of F(x) I relied upon EXCEL® to furnish a_1 and b_1 via its intrinsic regression functions INTERCEPT() and SLOPE(), together with RSQ() to give the *linear* regression Coefficient of Determination R^2.

EXCEL® yielded a_1=0.547044, b_1=0.428081 and R^2=0.976025378.

Day UB	25
f_c UB	4007

Day	Specimen 1		Specimen 2		Average					Normalisations		
	Compressive Strength, f_c psi	Unit Weight lb/f^2	Compressive Strength, f_c psi	Unit Weight lb/f^2	Compressive Strength, f_c psi	Unit Weight lb/f^2	E psi	ν in/in	Day	Mean f_c psi	Day	Mean f_c psi
1	1519	153.1	1611	155.1	1565	154.1	3079082	na	1	1565	0.04	0.390567
2	1733	155.3	1949	154.1	1841	154.7	3490590	na	2	1841	0.08	0.459446
3	2032	154.4	2518	153.6	2275	154	3226864	0.186	3	2275	0.12	0.567756
4	2050	154.1	2824	155.3	2437	154.7	3641757	0.184	4	2437	0.16	0.608186
5	2715	153	2742	155	2728.5	154	3880445	0.183	5	2728.5	0.2	0.680933
6	3185	154.5	2559	155.2	2872	154.85	3898696	0.178	6	2872	0.24	0.716746
7	2759	153.8	3243	153.6	3001	153.7	4012590	0.203	7	3001	0.28	0.748939
9	3392	153.5	3340	154.6	3366	154.05	3935557	0.174	9	3366	0.36	0.84003
11	3240	154.5	2783	153.9	3011.5	154.2	3891686	0.248	11	3011.5	0.44	0.75156
13	3244	154	3022	154.9	3133	154.45	3980031	0.184	13	3133	0.52	0.781882
14	3117	154.1	3359	154.6	3238	154.35	4122259	0.194	14	3238	0.56	0.808086
15	3136	154.1	3162	154.8	3149	154.45	4198048	0.179	15	3149	0.6	0.785875
17	3725	154.2	2805	155.6	3265	154.9	4314568	0.184	17	3265	0.68	0.814824
19	3211	152.4	3431	151.9	3321	152.15	4590814	0.188	19	3321	0.76	0.8288
23	3619	154.3	4004	153.4	3811.5	153.85	5264952	0.188	23	3811.5	0.92	0.95121
25	3944	154.2	4070	153.4	4007	153.8	5359575	0.179	25	4007	1	1

Table Ten
Ryan Carmichael's Concrete Curing Data

To define G(x) I exploited the EXCEL® Solver utility to identify the optimal non-linear best fit coefficients as a_2=0.750277465918 and b_2=11.646735891812.

The relevant G(x) Exponential Limit solutions tableau is reproduced in Table Eleven:-

Error Sum	0.002111946755585

a	0.750277465918433
b	11.646735891812300

x	y_{data}	y_{est}	Error Term
0.04	0.390567	0.279411	0.158261
0.08	0.459446	0.454767	0.000106
0.12	0.567756	0.564818	2.71E-05
0.16	0.608186	0.633885	0.001644
0.2	0.680933	0.677231	2.99E-05
0.24	0.716746	0.704434	0.000305
0.28	0.748939	0.721507	0.001446
0.36	0.84003	0.738946	0.018713

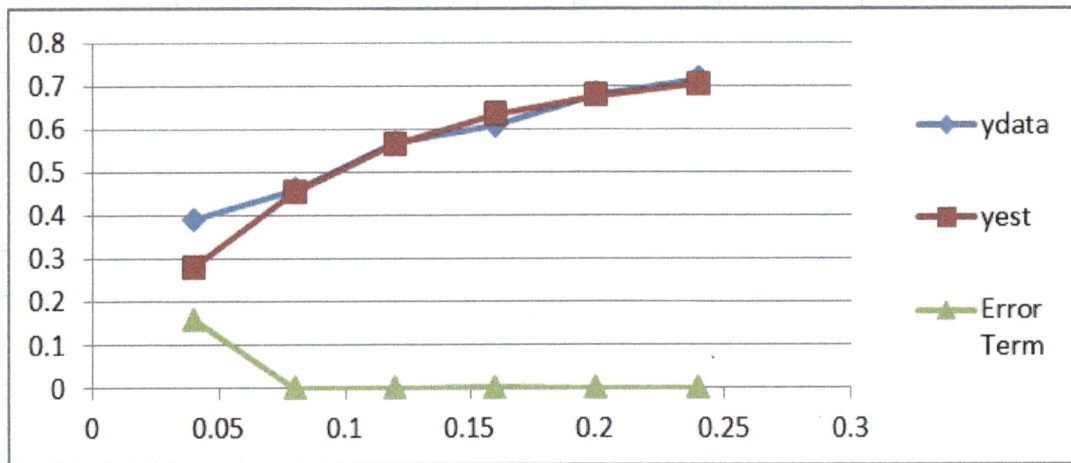

Table Eleven
The G(x) Solutions Tableau

The Determination of x_{mid}

The midpoint of the Step Function $H(x)$ was determined by Regula Falsi iteration using the normalisation end-points $x_1 = 0$ and $x_2 = 1$ for which the iteration Start Point was $x_0 = (x_2-x_1)/2$, i.e. 0.5.

Successor points were computed by:-

$$x_i = \frac{x_{i-2}D(x_{i-1}) - x_{i-1}D(x_{i-2})}{D(x_{i-1}) - D(x_{i-2})}$$

Equation 95

for $x_i = 3\ldots$, where $D(x_i) = F(x_i)-G(x_i)$.

x converged to fifteen decimal positions after five iterations so that:-

$$x_8 \equiv x_{mid} = 0.467155577071259$$

Equation 96

The system also converged using Newton-Raphson Iteration but very slowly and there was only six figure accuracy after eighteen iterations. Halley Iteration was not attempted.

The relevant Regula Falsi tableau may be inspected as Table Twelve.

| x_1 | 0.000000000000 |
| x_2 | 1.000000000000 |

Power	Fc_{power}	Gc_{power}
0	0.547043783667	0.750277465918
1	0.428081353560	11.646735891812

For Row 19:-

F	0.747024375423	
G		0.747024375423
F-G	0.000000000000	

Serial, I	x_i	$F(x_i)$	$G(x_i)$	$D(X_i)=F(x_i)-G(x_i)$
1	0.500000000000000	0.761084460447363	0.748058424860306	0.013026035587057
2	1.000000000000000	0.975125137227469	0.750270902822190	0.224854234405278
3	0.469253301355229	0.747922372074001	0.747102890726976	0.000819481347025
4	0.467311919433016	0.747091302672963	0.747030293514979	0.000061009157985
5	0.467155760673194	0.747024454019688	0.747024382379405	0.000000071640283
6	0.467155577087479	0.747024375430067	0.747024375423738	0.000000000006329
7	0.467155577071259	0.747024375423123	0.747024375423123	0.000000000000000
8	0.467155577071259	0.747024375423123	0.747024375423123	0.000000000000000

Table Twelve
Model Two Regula Falsi Tableau

Selected Results

Table Thirteen offers the controlling top matter of the fitted models whilst Table Fourteen shows the resulting point-wise elaborations.

It was found that where the Heaviside Steepness, k, was set to twelve an unrealistic hump developed near the x_{mid} triple point. I found that by setting k equal to unity I was able to model a more realistic monotonic curve with a smooth and simple transition from the non-linear to the linear aging regimes. Such a value of k, whatever its correlative merits, further facilitates any algebraic simplifications one might adjudge desirable.

Table Fifteen exhibits the Coefficients of Determination and their Pythagorean means for the various functions. These need not be taken too seriously because for one thing none of the functions G(x), Q(x) or P(x) are linear. They are only useful in assessing the comparative merits of the fitments. In particular, it is not realistic to think that P(x) has greater fidelity to the data than the smoothly-discontinuous function Q(x).

s	Steepness	1
x		
	Lower Bound:	0
	Upper Bound:	1
	Intervals:	15
	Increment:	0.066666667
F(x)	$a_1+b_1.x$	
	a_1	0.547043783667
	b_1	0.428081353560
G(x)	$a_2(1-exp(-b_2.x))$	
	a_2 (known limit y_{inf})	0.750277465918
	b_2	11.646735891812
G(x)	Error Term	1.402175342489
H(x)	x_{mid}	0.467155577071
	$\Sigma(data)/n(data)$	0.733427439481

Table Thirteen
Model Two Functional Tabulation Top Matter

Summary statistics (top matter)

Stat	X_B	F(x_B)=data	X_B	G(x_B)=data	F(x)=a₁+b₁x	G(x)=a₂(1-exp(-b₂x))
Count	8	8	8	8	16	16
Mean	0.685000	0.840280	0.626575	0.732259	0.673050	
Sum	5.480000	6.722236	5.012603	11.732147	10.768799	
a		0.547044	0.369693	0.547044	0.351285	
b		0.428081	1.388555	0.428081	1.327512	
R²		0.919094	0.967367	0.976029538	0.869797786	

Far-right column summary: R^2 values 0.985768841, 0.974604223; final error column — Count 16, Mean 0.009338, Sum 0.004407.

Data tabulation

Serial	X_{BOTH}	X_B	$F(x_B)$=data	X_B	$G(x_B)$=data	$F(x)=a_1+b_1x$	$G(x)=a_2(1-e^{-b_2x})$	Sq Err F	Sq Err G	SS_{FO}	SS_{FP}	SS_{GO}	SS_{GG}	H(x) (ERF)	$P(x)=F(x)+G(x)$	SS_{PO}	SS_{PP}	$SS_{P err}$	Q(x)	SS_{QO}	SS_{QQ}	SS_{QC}
0	0.04			0.04	0.39056651	0.56416704	0.27941142		0.05731783			0.05570018	0.12052281	0.25441591	0.42178923	0.11755362	0.09711838	0.00041760	0.35185778	0.05570018	0.12052281	0.00421197
1	0.08			0.08	0.45944597	0.58129029	0.45476701		0.14045031			0.02793224	0.02951811	0.28556881	0.51802865	0.07506585	0.04639664	0.00082192	0.49089811	0.02793224	0.02951811	0.00000252
2	0.12			0.12	0.56775643	0.59841355	0.56481829		0.19786331			0.00045967	0.00381394	0.31842915	0.58161592	0.02744688	0.02304674	0.00001936	0.57551600	0.00345967	0.00381394	0.00000013
3	0.16			0.16	0.60818568	0.61553680	0.63388529		0.22456726			0.00033818	0.00005343	0.35278400	0.62471104	0.01568550	0.01181925	0.00001495	0.62741223	0.00033818	0.00005343	0.00000008
4	0.2			0.2	0.68093337	0.63266005	0.67723099		0.22774942			0.00295479	0.00256599	0.38838371	0.65494552	0.00275563	0.00615941	0.00001159	0.65992036	0.00295479	0.00256599	0.00000015
5	0.24			0.24	0.7167457	0.64978331	0.70443428		0.21569920			0.00813069	0.00606201	0.42494717	0.67710880	0.00027828	0.00317179	0.00000837	0.68121051	0.00813069	0.00606201	0.00000428
6	0.28			0.28	0.74893936	0.66690656	0.72156678		0.19492824			0.01497295	0.00901197	0.46216840	0.69420667	0.00024062	0.00153827	0.00000168	0.69627228	0.01497295	0.00901197	0.00002553
7	0.36			0.36	0.84002995	0.70115307	0.73894561		0.14359977			0.04556286	0.01262707	0.49972416	0.72004934	0.01136409	0.00017897	0.00012511	0.72005976	0.04556286	0.01262707	0.00108477
8	0.44	0.44	0.75156			0.73539958	0.74581421	0.08726091		0.00787119	0.01099980			0.53728237	0.75908281	0.00032878	0.00005154	0.00000008	0.74021861	0.00787119	0.01099980	0.00000379
9	0.52	0.52	0.781882			0.76964609	0.74851953	0.06232317		0.00341030	0.00498908			0.57451087	0.76797177	0.00234782	0.00032820	0.00000285	0.76065697	0.00341030	0.00498908	0.00000249
10	0.56	0.56	0.808086			0.78676934	0.74917420	0.05142433		0.00103643	0.00286334			0.61108625	0.77673883	0.00557388	0.00119331	0.00001919	0.77214808	0.00103643	0.00286334	0.00000334
11	0.6	0.6	0.785875			0.80389260	0.74958507	0.04157219		0.00295988	0.00132401			0.64670219	0.79407193	0.00275072	0.00187588	0.00000077	0.78470587	0.00295988	0.00132401	0.00000268
12	0.68	0.68	0.814824			0.83813910	0.75000475	0.02500798		0.00064798	0.00000458			0.68107718	0.81127783	0.00662541	0.00367775	0.00000869	0.81003105	0.00064798	0.00000458	0.00000341
13	0.76	0.76	0.8288			0.87238561	0.75017005	0.01263053		0.00013179	0.00103080			0.71396108	0.84556972	0.09909585	0.06606068	0.00000921	0.83742721	0.00013179	0.00103080	0.00003080
14	0.92	0.92	0.95121			0.94067863	0.75026080	0.00043592		0.01230566	0.01012018			0.74514037	0.86258802	0.04742941	0.01257589	0.00121477	0.89229784	0.01230566	0.01012018	0.00000878
15	1	1	1			0.97512514	0.75027090	0.00061876		0.02551063	0.01818334			0.77444201		0.07106093	0.01671088	0.00295393	0.92440747	0.02551063	0.01818334	0.00005369

Table Fourteen
Model Two Functional Tabulation Bottom Matter

Function Fitted to the Data	Coefficient of Determination R^2	
	From EXCEL Worksheet	From EXCEL Regression Plot
F(x)	0.976025378033	0.952592202975
G(x)	0.869792785755	0.995237820267
Q(x)	0.974604222621	not computed
P(x)	0.985768409928	not computed
ARIT(R^2_F;R^2_G)	0.973915011621	
GEOM(R^2_F;R^2_G)	0.973681563804	
HARM(R^2_F;R^2_G)	0.973448171945	

Table Fifteen
Model Two Key Fitted Function
Coefficients of Determination

The Functional Comparison Plot

The most striking feature of the Functional Comparison Plot, Plot Five, is the behaviour of the Heaviside Step Function H(x) trace denoted by the green dashed line that makes its unsteady linear course in the lower part of the diagram. H(x) is here much less abrupt than heretofor in this disquisition. As aforementioned, more sudden rises in H(x) tended to engender unsatisfactory peak-and-trough formations that unrealistically implied greater model complexities. We prefer simple monotonic rises because we conceive that they better reflect the experimental data and afford, we believe, greater physical fidelity.

The non-linear G(x) part of that data is represented by the blue triangles and the linear part F(x) by the red squares.

The linear regression line for the latter is represented by the feint dashed black straight line and the trace of the exponential limit model G(x) by a similar feint black line which is patterned dash-dot-dot.

P(x) is the bold yellow line which overweights the Normalised Strength dependent values for the non-linear region: Underweights them in the linear.

Q(x) is a bold purple line of the Composite Model that is almost coincident with P(x) in its central reaches, but better intersects the data of both fields.

P(x), Q(x) and the F(x) and G(x) optimisation traces all intersect at x_{mid}, but H(x) does not meet them: It only controls their crossing at a distance.

Lastly, I have incorporated a normalised form of Carmichaels Model based upon his stated coefficients. I have not attempted to metricate his model, to "non-dimensionalise" it or otherwise to tamper with the original. The trajectory of the Carmichael fitment is given by the bold dashed magenta line which my fitments closely follow, the Carmichael line re-crossing them in several places.

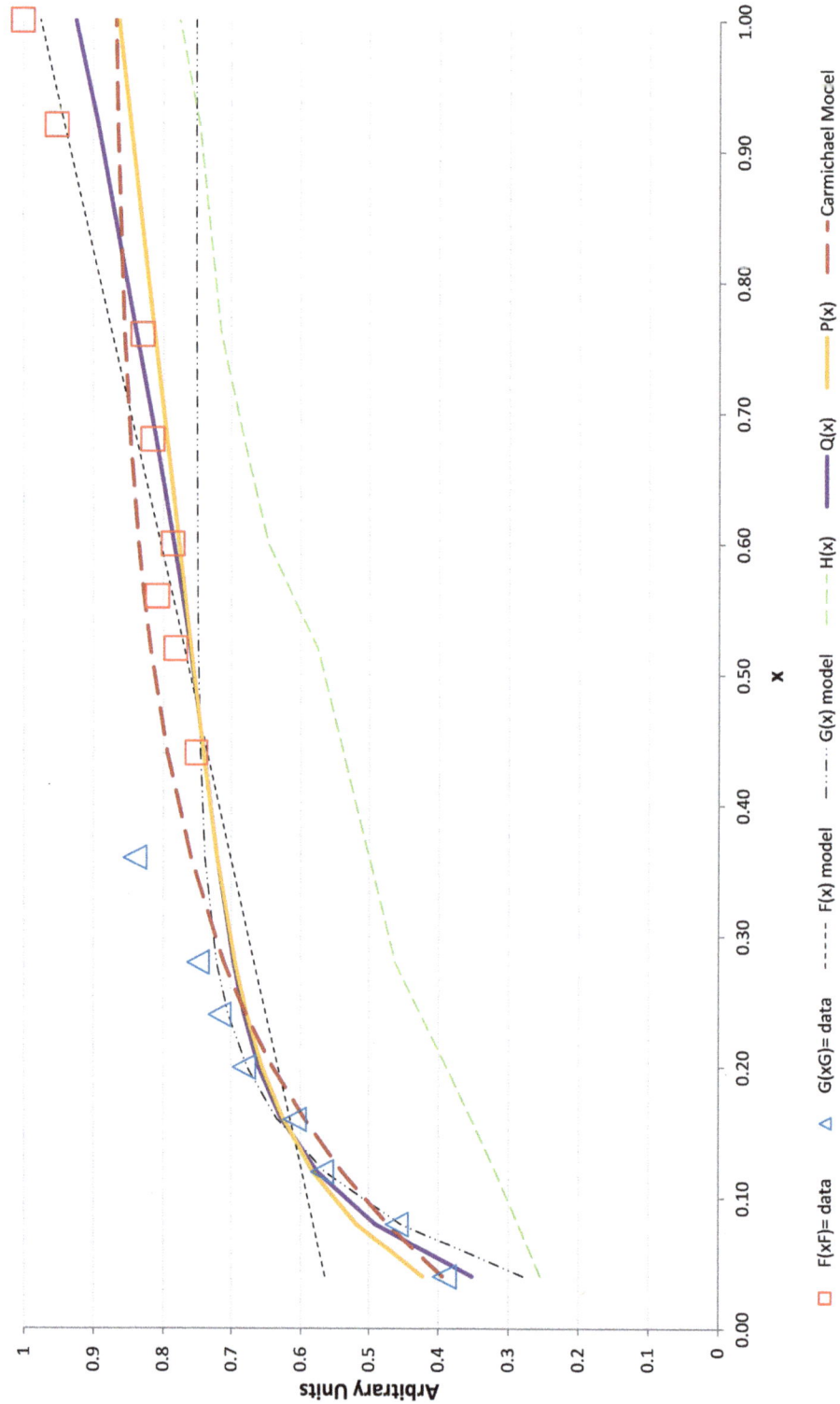

Model Two: First-Order Polynomial to Exponential Limit Formula

Plot Five axis labels: x (horizontal), Arbitrary Units (vertical)

Legend: F(xF)= data · G(xG)= data · F(x) model · G(x) model · H(x) · Q(x) · P(x) · Carmichael Mocel

Plot Five
The Fitment of Functional Models to Concrete Curing Data

PART FIVE
SIMPLE TRIPARTITE TRANSITIONS MODELS

In this Part we will discuss composite functions Q(x) that devolve to three separate and successive component regimes, say E(x), F(x) and G(x).

As part of our explication of the principle we will study E(x), F(x) and G(x) all as First-Degree Polynomials (FDPs), that is to say straight line functions susceptible of linear regression.

In the development of our program we will utilise the Boxcar Function variant of the coupled Heaviside Step Function defined by:-

$$f(x) = H(x-a) - H(x-b)$$
Equation 97

The action of the ideal Boxcar Function is illustrated in the sketch graph Figure Three:-

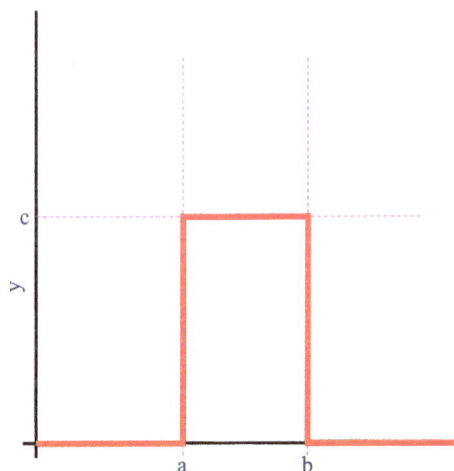

Figure Three
The Boxcar Step Function

Therefore we may declare the Composite Model Q(x) as:-

$$Q(x) = H1.E(x) + H2.F(x) + H3.G(x)$$
Equation 98

where:-

$$E(x) = a_1 + b_1.x \qquad F(x) = a_2 + b_2.x \qquad G(x) = a_3 + b_3.x$$

Equation 99a   **Equation 99b**   **Equation 99c**

and:-

$$H1(x) = c\left[H(x + x_{mid1}) - H(x - x_{mid1})\right]$$
Equation 100a

$$H2(x) = c\left[H(x - x_{mid1}) - H(x - x_{mid2})\right]$$
Equation 100b

$$H3(x) = c\left[H(x - x_{mid2}) - H(x + \{x_{max} + x_{mid2}\})\right]$$
Equation 100c

given that:-

$$H(x) = \frac{1}{2} + \frac{1}{2}erf(kx)$$
Equation 101

c is an arbitrary Scaling Constant and the crossover points x_{mid1} and x_{mid2} are respectively defined by:-

$$x_{mid1} = \frac{a_2 - a_1}{b_1 - b_2}$$ and:- $$x_{mid2} = \frac{a_3 - a_2}{b_2 - b_3}$$
Equation 102a   **Equation 102b**

The combined effect is illustrated in sketch terms by Figure Four in which the articulated trajectories E(x), F(x) and G(x) describe the course of the ideal Q(x).

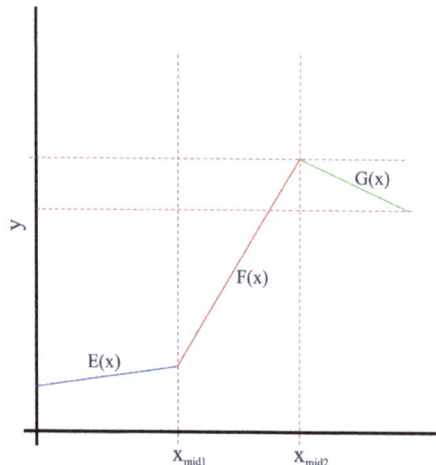

Figure Four
A Sketch of Three Disjoint Third-Degree Polynomial Lines (FDPs)

Isoperibol Calorimetry

Plot Six shows H1, H2, H3, E(x), F(x), G(x) and Q(x) in terms of their intervariation in the particular context of isoperibol bomb calorimetry though of course the method could be applied to a very large number of contexts.

Isoperibol calorimetry attempts to measure the heat yield upon combustion of a precisely-weighed sample of a given material, and to measure that heat as precisely as practicable. Accordingly, traditional procedure places the sample to be assessed in a holder sealed inside a heavy, thick-walled metal container, which is then filled with enough pressurised oxygen completely to consume the test sample when that is electrically ignited from outwith the sealed chamber. The whole assembly is placed within an insulated water jacket whose temperature is monitored on the minute every minute for several minutes before ignition. This is the Pre-Ignition Phase, which we model as E(x). The entire experimental assembly is called a calorimeter. At some minute the ignition circuit is fired, the sample explodes internally and silently within the submerged and armoured metal "bomb", and the temperature suddenly climbs within the stirred pool of its enclosure. This is the Combustion Phase, which we model as F(x), as the climbing temperature continues to be measured on the minute. Eventually, sometime after extinction, the bath temperature peaks and then gradually falls. This third, fall, episode is called the Post-Combustion Phase and is also measured on the minute for several minutes and we model it as G(x).

Classical calorimetry draws three straight line segments to characterise the experimental regimes progressed in a calorimetric exercise, though the technique depends theoretically upon Newton's Law of Cooling which is of course an exponential decay. In the literature, you may find all sorts of graphical conformations, including shallow decay curves. The relevant European Standard[21] has a sketch graph that shows the Pre-Combustion history as a straight line, followed by what appears to be a parabola that covers both the Combustion and the Post-Combustion trajectories. Many temperature correction and assessment formulae assume that the functional forms are perfect or approximate straight-line segments susceptible of linear regression. The only successful bomb calorimeter assay of my career did in fact furnish three straight-line segments on the plotted time-temperature graph, and that is what I had indeed been briefed to expect.

Equations

The heat yield upon combustion of a given material is formally defined as Heat of Combustion, C_H, L^2T^{-2}, typically quoted in KJ/kg.

To a first approximation C_H is given by:-

$$C_H = \frac{\Delta\theta_{corr} \times \sum_{j=1}^{n} m_j s_j}{m_s}$$

Equation 103

where $\Delta\theta_{corr}$ is The Corrected Calorimetric Temperature Rise (θ^1); m_j is the Mass of the jth Calorimeter Component (M^1), with its relevant Material Specific Heat, s_j, ($M^1L^2T^{-2}\theta^{-1}$) and m_s is the Mass of the Combustion Sample (M^1).

$\Delta\theta_{raw}$ is the Uncorrected (Experimental) Temperature Rise given by:-

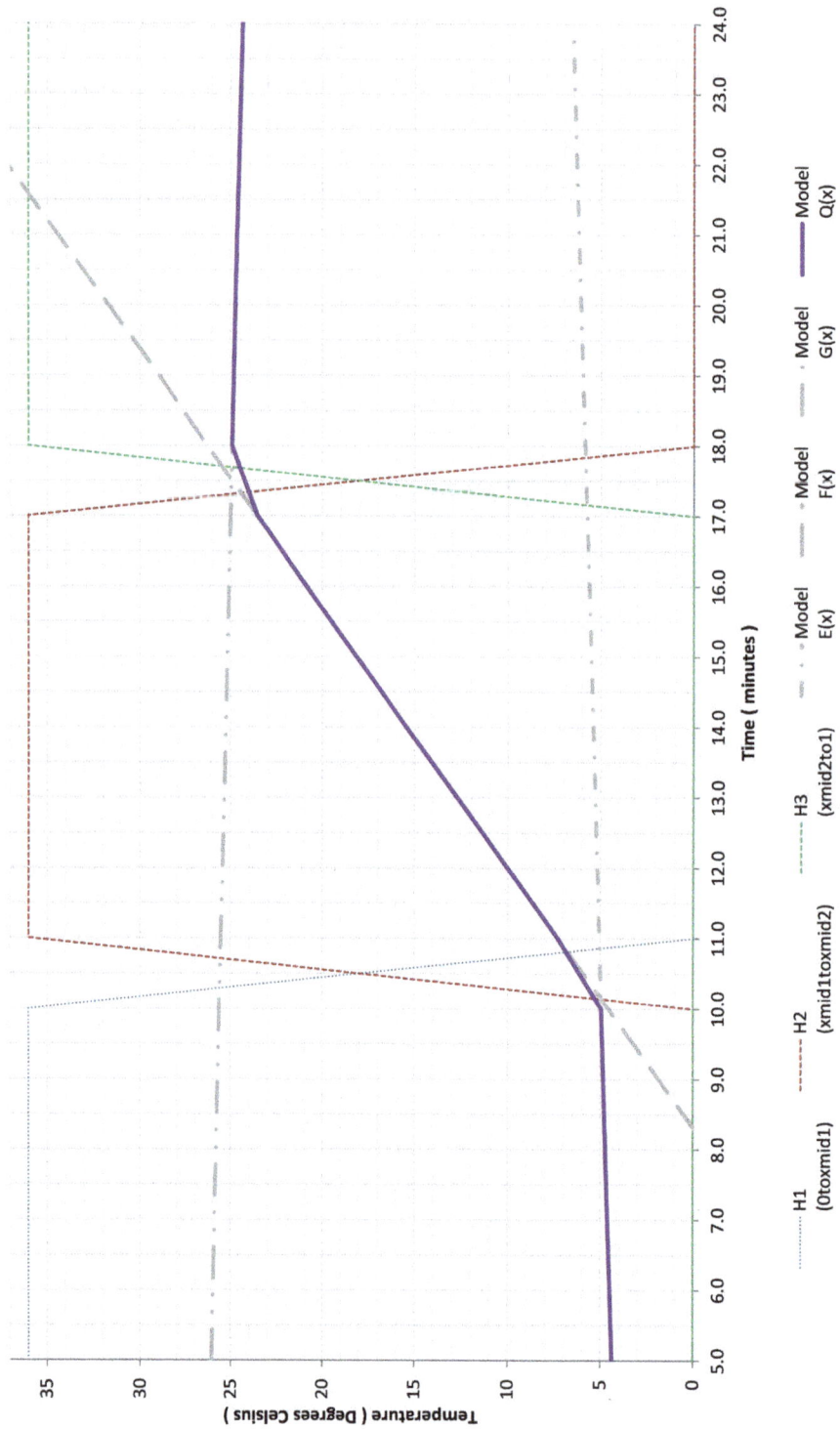

Modelled Isoperibol Calorimetry Time-Temperature Chart

Temperature (Degrees Celsius)

Time (minutes)

H1
(0toxmid1)

H2
(xmid1toxmid2)

H3
(xmid2to1)

Model
E(x)

Model
F(x)

Model
G(x)

Model
Q(x)

Plot Six
The Governing Heaviside Step Functions and their
Control of Combination Function Q(x)

$$\Delta\theta_{raw} = \theta_n - \theta_0$$
Equation 104

whilst:-

$$\Delta\theta_{corr} = \Delta\theta_{raw} - \Delta\theta_{ex}$$
Equation 105

where:-

$$\Delta\theta_{ex} = b_3\left(\tau_f - \tau_i\right) + \frac{b_1 - b_2}{\theta_{post\mu} - \theta_{pre\mu}}\left[n\theta_{post\mu} - \frac{t_i - t_f}{2} - \sum_{j=1}^{n-1}\theta_j\right]$$
Equation 106

b_1, b_2 and b_3 are the respective slopes of E(x), F(x) and G(x) as computed by linear regression or something; τ_i and τ_f are respectively the minutes when Combustion Initiates and Finishes; $\theta_{pre\mu}$ and $\theta_{post\mu}$ are respectively the Mean Temperatures in the Pre-Combustion and the Post-Combustion Phases; t_i and t_f are the respective Temperatures at Initiation (i.e. Ignition) and Exhaustion (i.e. the Finish); n is the Number of Minutes read during Combustion; and θ_j are the Internal Temperature Readings during the Combustion Phase.

The temperature rise corrector Equation One Hundred and Six is known as the Regnault-Pfaundler Heat Leak Correction and can be found on Page 16 of the European Standard FprEN 14918 "Solid Biofuels – Determination of calorific value"[21], though of course the calorimetric methodology and the physical correction are applicable to a wide range of solid and liquid combustibles: In my own bomb calorimeter experiment of 1983 the subject sample was coal.

A Hypothetical Isoperibol Calorimetry Experiment

Table Sixteen presents the EXCEL® record of a hypothetical twenty-minute bomb calorimeter experiment illustrative of several principles. Nevertheless, in good laboratory practice we would actually seek to extend the times before, during, and subsequent to the combustion. The steep Combustion curve is especially error-prone, and in the life we would try to prolong it and take more readings: But we would also like longer Pre- and Post- phases too, so that more reliable mean temperatures could be computed.

As promised, E(x), F(x) and G(x) have been treated as simple linear regressions. The x-axis is calibrated in integral minutes. Note the very large value of Steepness and its identity for H1, H2 and H3 in order to conserve as crisp and straight transitions as possible, given the short and coarse series of readings.

The Scaling Constant c is entirely arbitrary and is intended to clarify graphics and empower the Heaviside step functions. In normalised work it would of course be set to unity.

I have assigned the first five minutes (i.e. six readings) to the Pre-Combustion Phase E(x). Ignition was fired at the seventh minute (logged as Minute 12) and a further six Combustion Phase readings were taken for F(x). Finally, log Minutes 18 to 24 were assigned to

s1	Steepness	250
s2	Steepness	250

E(x)	$a_1+b_1.x$	
	a_1	3.809523809524
	b_1	0.114285714286

F(x)	$a_2+b_2.x$	
	a_2	-22.571428571429
	b_2	2.714285714286

G(x)	$a_3+b_3.x$	
	a_3	26.517857142857
	b_3	-0.089285714286

c		36.000000000000
x_{mid1}		10.146520146520
x_{mid2}		17.509554140127
t_{mid1}		4.969126111983
t_{mid2}		24.954504094632

	x	y	H1	H2	H3	Data E(x)	Data F(x)	Data G(x)	Data Q(x)	Model E(x)	Model F(x)	Model G(x)	Model Q(x)
Count	20.0	20.0	20	20	20	6	7	7	20	20	20	20	20
Mean	14.5	15.4	10.8	12.6	12.6	4.666667	15.42857	24.64286	15.425	5.466667	16.78571	25.22321	15.425
Sum	290.0	308.5	216	252	252	28	108	172.5	308.5	109.3333	335.7143	504.4643	308.5
a			43.76842	15.3473684	-23.1158	3.809524	-22.5714	26.51786	-5.218421	3.809524	-22.5714	26.51786	-5.21842
b			-2.27368	-0.18947368	2.463158	0.114286	2.714286	-0.08929	1.4236842	0.114286	2.714286	-0.08929	1.423684
R^2			0.631579	0.00404858	0.684211	0.042857	0.938882	0.066489	0.888139	1	1	1	0.901003

Q(x)= (1/c)* H1.E(x)+ H2.F(x) +H3.G(x)

Serial	x	y	H1 (0 to x_{mid1})	H2 (x_{mid1} to x_{mid2})	H3 (x_{mid2} to 1)	Data E(x)	Data F(x)	Data G(x)	Data Q(x)	Model E(x)	Model F(x)	Model G(x)	Model Q(x)
0	5.0	5.0	36	0	0	5			5	4.380952	-9	26.07143	4.380952
1	6.0	4.0	36	0	0	4			4	4.495238	-6.28571	25.98214	4.495238
2	7.0	5.0	36	0	0	5			5	4.609524	-3.57143	25.89286	4.609524
3	8.0	3.0	36	0	0	3			3	4.72381	-0.85714	25.80357	4.72381
4	9.0	6.0	36	0	0	6			6	4.838095	1.857143	25.71429	4.838095
5	10.0	5.0	36	0	0	5			5	4.952381	4.571429	25.625	4.952381
6	11.0	8.0	0	36	0		8		8	5.066667	7.285714	25.53571	7.285714
7	12.0	8.0	0	36	0		8		8	5.180952	10	25.44643	10
8	13.0	15.0	0	36	0		15		15	5.295238	12.71429	25.35714	12.71429
9	14.0	14.0	0	36	0		14		14	5.409524	15.42857	25.26786	15.42857
10	15.0	19.0	0	36	0		19		19	5.52381	18.14286	25.17857	18.14286
11	16.0	20.0	0	36	0		20		20	5.638095	20.85714	25.08929	20.85714
12	17.0	24.0	0	36	0		24		24	5.752381	23.57143	25	23.57143
13	18.0	25.0	0	0	36			25	25	5.866667	26.28571	24.91071	24.91071
14	19.0	24.0	0	0	36			24	24	5.980952	29	24.82143	24.82143
15	20.0	26.0	0	0	36			26	26	6.095238	31.71429	24.73214	24.73214
16	21.0	24.0	0	0	36			24	24	6.209524	34.42857	24.64286	24.64286
17	22.0	25.0	0	0	36			25	25	6.32381	37.14286	24.55357	24.55357
18	23.0	24.0	0	0	36			24	24	6.438095	39.85714	24.46429	24.46429
19	24.0	24.5	0	0	36			24.5	24.5	6.552381	42.57143	24.375	24.375

Table Sixteen
The Tabular Results and Regressions of a
Hypothetical Isoperibol Calorimetry Experiment

The continuous Model Values for E(x), F(x), G(x) and Q(x) are based upon the EXCEL® linear regression functions INTERCEPT(), SLOPE() and RSQ() respectively tabulated in the rows marked a, b and R^2. These values were copied manually to twelve-decimal-place precisions in the third column.

The worksheet was then used to generate Plot Seven which integrates the Step Functions with the Data, and the thermometric Functions. The blue triangles represent the E(x) Pre-Combustion temperature readings, the red squares the Main Sequence Combustion temperature readings to model F(x), and the green circles the Post-Combustion temperature readings.

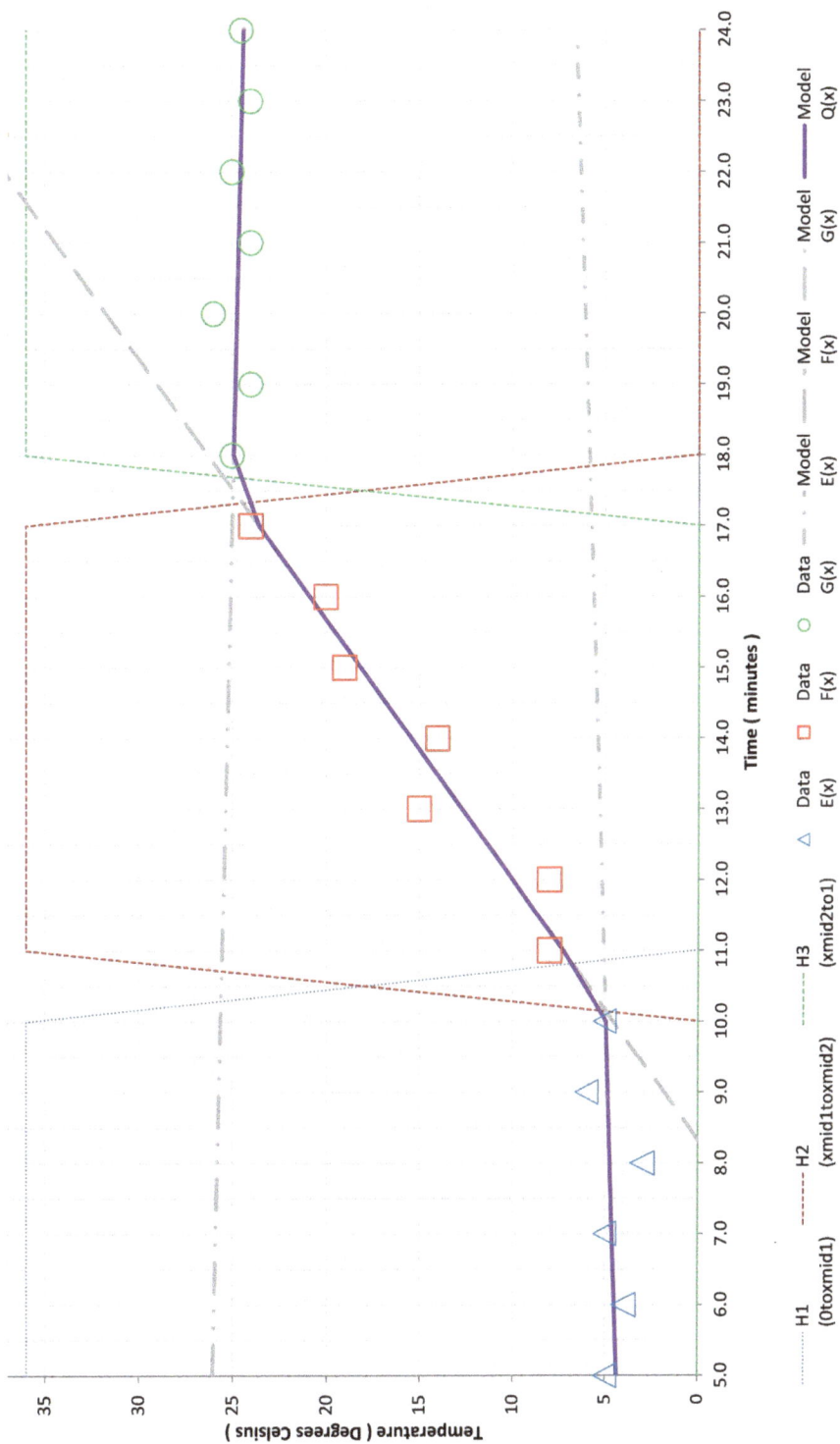

Modelled Isoperibol Calorimetry Time-Temperature Chart

Plot Seven
The Thermal History of the Hypothetical Isoperibol Calorimetry Experiment

The existence of only twenty data points leads to an unsatisfactory segmentalisation of the step functions which is reflected in the broken character of Q(x), which is depicted as the bold purple line, the interpolated trace of the experimental history.

The tripartite alternation of the step functions can be seen from the feint traces of dotted blue, dashed red and dashed green, respectively showing H1, H2 and H3. Three distinct gray pecked lines denote the component functions E(x), F(x) and G(x).

The Regnault-Pfaundler Temperature Rise Correction

Table Seventeen summarises the statistics germane to the experimental outcome, and in particular the corrected temperature rise according to the Regnault-Pfaundler formula of Equation One Hundred and Six.

The Unadjusted Rise θ_{raw} is 16°K and the Regnault-Pfaundler Correction Δt_{ex} is computed to be -0.217618764, giving a Corrected Rise θ_{corr} of +16.217618764.

I attempted to reconcile Equation One Hundred and Six with our transition methodology, and in particular with combustion initiation and termination as signalled by x_{mid1} and x_{mid2}. This issued in several absurd figures for Δt_{ex} and was abandoned. It seems to be that Equation One Hundred and Six in its current form is only good for calorimetric readings taken every minute on the minute.

Equations B9 and B10 of the eurostandard Annex B (Pages 36 and 37) specify that *for equal time intervals*:-

$$\Delta t_{ex} = G \int_{\tau_i}^{\tau_f} (t_\infty - t).d\tau = \left[\frac{1}{n} \left\{ \frac{t_0 + t_n}{2} + \sum_{k=1}^{n-1} t_k \right\} \right] . \int_{\tau_i}^{\tau_f} (t_\infty - t).d\tau$$

Equation 107

We may reasonably identify τ_i and τ_f with x_{mid1} and x_{mid2}, divide the difference x_{mid2}-x_{mid1} into integral intervals using Lagrangian Interpolation or something, and then proceed with a corrected analysis based upon irregular readings.

Variable Name (FprEN 14918:2009 (E))	Variable Name (STEP FUNCTIONS (emp))	Description	Value	Authority
τ_f	x_{min1} (proximate)	Time of Ignition	6	
τ_i	x_{min2} (proximate)	Time of Exhaustion	12	
g_f	b_3	Post-Combustion Drift Rate (K°/min)	0.114285714	
g_i	b_1	Pre-Combustion Drift Rate (K°/min)	-0.089285714	
t_{mi}		Mean Pre-Combustion Temperature	4.66666667	
t_{mf}		Mean Post-Combustion Temperature	24.64285714	
n		Internal Number of Combustion	5	
		Phase Temperature Readings		
$\sum_{k=1}^{n-1}$		Sum of Internal Combustion	76	
		Phase Temperature Readings		
t_i	t_{mid1} (proximate)	Initial Temperature	8	
t_f	t_{mid2} (proximate)	Final Temperature	24	
θ_{raw}		Uncorrected Temperature Rise	16	
Δt_{ex}		Regnault-Pfaundler Correction	-0.217618764	FprEN 14918:2009 (E)
θ_{corr}		Regnault-Pfaundler Corrected Tempera	16.21761876	

Table Seventeen
Summary Results of a Hypothetical Isoperibol Calorimetry Experiment

PART SIX
REFERENCES

1.   http://en.wikipedia.org/wiki/Heaviside.step.function

Heaviside step function. (2016, November 25). In *Wikipedia,
The Free Encyclopedia*. Retrieved 11:00, February 10, 2017, from
https://en.wikipedia.org/w/index.php?
        title=Heaviside_step_function&oldid=751453760

Error function. (2016, November 6). In *Wikipedia,
The Free Encyclopedia*.
Retrieved 11:26, February 10, 2017, from
https://en.wikipedia.org/w/index.php?
        title=Error_function&oldid=748051954

2.   False position method. (2017, February 9).
In *Wikipedia, The Free Encyclopedia*.
Retrieved 12:05, February 18, 2017, from
https://en.wikipedia.org/w/index.php?
        title=False_position_method&oldid=764561274

Converges for Model 1 after eight iterations ($x_0 = 0.5$)

3.   "Approximations for Digital Computers"
C Hastings Jr
Princeton University Press of Princeton NJ
1955
(A&S 7.1.26 p299)

4.   "Handbook of Mathematical Functions
with Formulas, Graphs and Mathematical Tables"
Edited by Milton Abramowitz and Irene A Stegun
Dover Publications of New York
after The National Bureau of Standards
December 1972
ISBN 0-486-61272-4
1046pp
£18.95 paperback
("A&S Handbook")

(A&S 7.1.5 p 297)

5   "Statistics and Data Analysis in Geology"
     John C Davis with FORTRAN programs by Robert J Sampson
     Wiley International Edition   1973
     John Wiley and Sons Inc. of New York
     ISBN 0-471-19895-1 (ISBN 0-471-19896-X International Edition)
     550pp

     p196

6   "Quantitative Methods and Statistics"
     Sonia R Wright
     Sage Publications of Beverley Hills   1979
     ISBN 0-8039-1295-1 (pbk)
     pp171

7   "Probability and Statistics for Engineers and Scientists"
     Ronald E Walpole and Raymond H Myers
     4th Edition   1989
     Macmillan Publishing Company of New York
     ISBN 0-02-946910-4   (International Paperback Edition)
     pp 765
     ("Walpole and Myers")

8   The American Phytopathological Society
     Plant Disease Epidemiology: Temporal Aspects: Mathematical Models
     http://www.apsnet.org/EDCENTER/ADVANCED/
            TOPICS/EPIDEMIOLOGYTEMPORAL/
            Pages/MathematicalModels.aspx

9   "Handbook of Mathematical Functions
     with Formulas, Graphs and Mathematical Tables"
     Edited by Milton Abramowitz and Irene A Stegun
     Dover Publications of New York
     after The National Bureau of Standards
     December 1972
     ISBN 0-486-61272-4
     1046pp
     £18.95 paperback
     ("A&S Handbook")
     (1)   (c.f. Fig 4.2 p70)

10   "Guide to Mathematical Modelling"
     Macmillan Mathematical Guides
     Macmillan Educational Ltd of Basingstoke   1989
     Dilwyn Edwards and Mike Hamson
     ISBN 0-333-45935-0

pp 277

p75   (Fig 5.4 p77 ()

11   "Solving Equations with Physical Understanding"
        John R Acton and Patrick T Squire
        Adam Hilger Ltd of Bristol   1985
        ISBN 0-85274-999-3 (pbk)
        pp219
        ("Acton and Squire")

        Exponential Rise Function p29   (ERF)

12   "Acton and Squire"
        Hot Filament Model p55

13   "Walpole and Myers"
        Correlation Coefficient (p 393)
        Sums of Squares (p338)

14   "Walpole and Myers"
        Linear Regression (p362)

15   "A Spectral and Statistical Study of River Meander Morphometry"
        James R Warren
        Unpublished PhD Thesis
        Department of Civil Engineering
        The University of Strathclyde   1980

16   "Relationships Between Young's Modulus, Compressive Strength,
        Poisson's Ratio, and Time for Early Age Concrete"
        Ryan P Carmichael
        ENGR 082 Project Final Report
        Swarthmore College
        Department of Engineering
        May 2009

        https://www.researchgate.net/...

16   Concrete Stress Equations
        "Al-Darzi"
        Cited in:-
        "Effects of Concrete Non-linear Modelling on the
         Analysis of Push-out Test by Finite Element Method"
        Suhaib Yahya Kasim Al-Darzi
        Journal of Applied Sciences 7(5):743-747,2007

ISSN 1812-5654
Asian Network for Scientific Information
http://ansinet.com

17   (a)   Hognestad (1951) Relations
             "A study of combined bending and axial load in
             reinforced concrete members"
             EA Hognestad
             Eng.Exp.Station, Bull. No. 299, , 49:22, 1951
             University of Illinois, Urbana, Illinois
             (Cited in "Al-Darzi")

18   (b)   Desayi-Krishnan (1964) Relation
             "Equation for Stress Strain Curve of Concrete"
             Int. ACI. Proc., 3:345-350, 1964
             (Cited in "Al-Darzi")

19   "Stress-Strain Relationship for Concrete in
             Compression Made of Local Materials"
             Anis Mohamad Ali, BJ Farid and AIM Al-Janabi
             JKAU: Eng.Sci., Vol.2 pp.183-194 (1410AH/1990AD)

             The Ali-Farid Al-Janabi Relations

20   "Relationships Between Young's Modulus, Compressive Strength,
             Poisson's Ratio, and Time for Early Age Concrete"
             Ryan P. Carmichael
             ENGR 082 Project Final Report
             Advisor: Prof. Frederick L. Orthlieb

             Swarthmore College
             Department of Engineering
             May 2009

             https://www.researchgate.net/

21   EUROPEAN STANDARD
             NORME EUROPÉENNE
             EUROPÄISCHE NORM
             FINAL DRAFT
             FprEN 14918
             April 2009

             English Version
             "Solid biofuels - Determination of calorific value"
             Biocombustibles solides - Détermination du pouvoir

calorifique
Feste Biobrennstoffe - Bestimmung des Heizwertes

FprEN 14918:2009 (E)

Full explanation of Regnault and Pfaundler Method and Equation on Page 16.

CHAPTER FOUR

Density, Chemistry and The Harmonic Mean

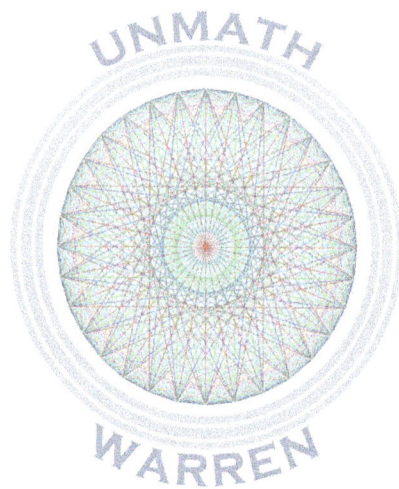

Mixture Density
and the
Harmonic Mean

by
James R Warren BSc MSc PhD PGCE

We require to compute the inertial Mass Density, ρ_{alloy}, of a simple mixture of n Components that respectively present the Mass Densities ρ_i, where i denotes a particular elementary component. These components may not of course be chemical elements or other conceptual indivisibles: They may be any arbitrary part, such as sand, cement and water in a certain mortar mix. The parts will be combined by mass in the n Fractions, f_i, for each Component i: That is unity will represent the mixture or alloy, the sum of all n Fractions.

Notwithstanding that, our present illustrations will involve chemical elements selected from Gold, Silver, Copper, Palladium, Nickel and Zinc as components of various common jewellers' alloys.

We must naturally exercise due caution in what we treat and to what extent as the simple mathematics of mixtures can be modified by bonding, whether in the strict chemical sense of compound formation, or in the sense of mechanical adhesion.

Clearly, the several and summative densities of materials are also a matter of temperature, pressure and structure. We will assume standard conditions of temperature and pressure, say 20°C and 1 atmosphere pressure.

A fundamental formula for Composite Density calculation is Dr Roy's Equation:-

$$\rho_{mix} = \frac{1}{\frac{f_1}{\rho_1} + \frac{f_2}{\rho_2} + \cdots + \frac{f_n}{\rho_n}} = \frac{1}{\sum_{i=1}^{n} \frac{f_i}{\rho_i}}$$
Equation 1

The Harmonic Mean, H, of a series of n Values, x_i, is defined as:-

$$H = \frac{n}{\frac{1}{x_1} + \frac{1}{x_2} + \cdots + \frac{1}{x_n}} = \frac{n}{\sum_{i=1}^{n} \frac{1}{x_i}}$$
Equation 2

Accordingly, let us specify:-

$$x_i = \frac{\rho_i}{f_i}$$
Equation 3

Then analogously:-

$$\rho_{mix} = \cfrac{1}{\cfrac{f_1}{\rho_1} + \cfrac{f_2}{\rho_2} + \cdots + \cfrac{f_n}{\rho_n}} = \cfrac{1}{\sum_{i=1}^{n} \cfrac{f_i}{\rho_i}} = \cfrac{1}{\cfrac{1}{x_1} + \cfrac{1}{x_2} + \cdots + \cfrac{1}{x_n}} = \cfrac{1}{\sum_{i=1}^{n} \cfrac{1}{x_i}}$$

Equation 4

Or:-

$$\rho_{mix} = \frac{H}{n}$$

Equation 5

Developments for a Ternary Alloy

Consider the 18-carat yellow gold formulation that is a simple mixture and comprises by weight 75% Au, 16% Ag and 9% Cu. The respective densities of these metals are taken to be 19.282, 10.501, and 8.96 grams per cubic centimeter. (We take it that cubic centimeters and milliliters are interchangeable).

This alloy comprises three parts.

In this case:-

$$\rho_{mix} = \cfrac{1}{\cfrac{f_{Au}}{\rho_{Au}} + \cfrac{f_{Ag}}{\rho_{Ag}} + \cfrac{f_{Cu}}{\rho_{Cu}}} = \cfrac{1}{\sum_{i=1}^{3} \cfrac{f_i}{\rho_i}} = \cfrac{1}{\cfrac{1}{x_{Au}} + \cfrac{1}{x_{Ag}} + \cfrac{1}{x_{Cu}}} = \cfrac{1}{\sum_{i=1}^{3} \cfrac{1}{x_i}}$$

Equation 6

which elaborates as:-

$$\rho_{mix} = \cfrac{1}{\cfrac{0.75}{19.282} + \cfrac{0.16}{10.501} + \cfrac{0.09}{8.96}} = \cfrac{1}{\sum_{i=1}^{3} \cfrac{f_i}{\rho_i}}$$

$$= \cfrac{1}{\cfrac{1}{25.7093'} + \cfrac{1}{65.63125} + \cfrac{1}{99.5'}} = \cfrac{1}{\sum_{i=1}^{3} \cfrac{1}{x_i}}$$

$$= 15.5817443400534$$

Equation 7

In terms of the Harmonic Mean H, Equation Six may be recast as:-

$$H = \cfrac{3}{\cfrac{1}{x_{Au}} + \cfrac{1}{x_{Ag}} + \cfrac{1}{x_{Cu}}} = \cfrac{3 \cdot x_{Au} \cdot x_{Ag} \cdot x_{Cu}}{x_{Au} \cdot x_{Ag} + x_{Au} \cdot x_{Cu} + x_{Ag} \cdot x_{Cu}}$$

Equation 8

From which it is clear that by elimination of n:-

$$\rho_{mix} = \frac{1}{\frac{1}{x_{Au}} + \frac{1}{x_{Ag}} + \frac{1}{x_{Cu}}} = \frac{x_{Au} \cdot x_{Ag} \cdot x_{Cu}}{x_{Au} \cdot x_{Ag} + x_{Au} \cdot x_{Cu} + x_{Ag} \cdot x_{Cu}}$$
Equation 9

Substitution of Equation Three yields:-

$$\rho_{mix} = \frac{1}{\frac{1}{x_{Au}} + \frac{1}{x_{Ag}} + \frac{1}{x_{Cu}}} = \frac{\frac{\rho_{Au}}{f_{Au}} \cdot \frac{\rho_{Ag}}{f_{Ag}} \cdot \frac{\rho_{Cu}}{f_{Cu}}}{\frac{\rho_{Au}}{f_{Au}} \cdot \frac{\rho_{Ag}}{f_{Ag}} + \frac{\rho_{Au}}{f_{Au}} \cdot \frac{\rho_{Cu}}{f_{Cu}} + \frac{\rho_{Ag}}{f_{Ag}} \cdot \frac{\rho_{Cu}}{f_{Cu}}}$$
Equation 10

The Fallacy of the Efficient Complement

It may naturally be anticipated that since the third fraction must wholly be determined by the other two then a substitution of the type:-

$$f_{Cu} = 1 - f_{Au} - f_{Ag}$$
Equation 11

should facilitate savings.
Therefore we could posit:-

$$\rho_{mix} = \frac{1}{\frac{1}{x_{Au}} + \frac{1}{x_{Ag}} + \frac{1}{x_{Cu}}} = \frac{\frac{\rho_{Au}}{f_{Au}} \cdot \frac{\rho_{Ag}}{f_{Ag}} \cdot \frac{\rho_{Cu}}{1 - f_{Au} - f_{Ag}}}{\frac{\rho_{Au}}{f_{Au}} \cdot \frac{\rho_{Ag}}{f_{Ag}} + \frac{\rho_{Au}}{f_{Au}} \cdot \frac{\rho_{Cu}}{1 - f_{Au} - f_{Ag}} + \frac{\rho_{Ag}}{f_{Ag}} \cdot \frac{\rho_{Cu}}{1 - f_{Au} - f_{Ag}}}$$
Equation 10

Unfortunately, the stratagem only increases computational complexity and expense. Moreover, it disastrously leads to division-by-zero when the third component is absent.
A specific two-component (secondary or binary) formulation will obviate the latter problem.

The Two-Component Mixture

Let the Geometric Mean, G, be defined by:-

$$G = \sqrt{x_1 \cdot x_2}$$
Equation11

whilst the Arithmetic Mean, A, is given by:-

$$A = \frac{x_1 + x_2}{2}$$
Equation 12

Then:-

$$\rho_{mix} = \frac{1}{\frac{1}{x_1} + \frac{1}{x_1}} = \frac{H}{2} = \frac{x_1 \cdot x_2}{x_1 + x_2} = \frac{1}{2} \cdot \frac{G^2}{A}$$
Equation 13

The Quaternary (Four-Component) Mixture

The four-component mixture's aggregate density, ρ_{mix}, is given by:-

$$\rho_{mix} = \frac{1}{\frac{1}{x_1} + \frac{1}{x_2} + \frac{1}{x_3} + \frac{1}{x_4}}$$
Equation 14

whilst the relevant Harmonic Mean, H, is given by:-

$$H = \frac{4}{\frac{1}{x_1} + \frac{1}{x_2} + \frac{1}{x_3} + \frac{1}{x_4}}$$
Equation 15

Therefore:-

$$\rho_{mix} = \frac{H}{4}$$
Equation 16

The Alloy Densities of Some Varieties of Jewellers' Gold

Table One shows the partial and summative values for four varieties of jewellers' gold alloys in some of their typical formulations.

18 Carat Gold

Metal	Component Mass (grams)	Density (grams/cc)	Eqivalent Volume (cc)
Au	0.75	19.282	0.03889638
Ag	0.16	10.501	0.015236644
Cu	0.09	8.96	0.010044643
	1		0.064177667

15.58174434 **Alloy Density (g/cc)**

22 Carat Gold

Metal	Component Mass (grams)	Density (grams/cc)	Eqivalent Volume (cc)
Au	0.916666667	19.282	0.04754002
Ag	0.083333333	10.501	0.007935752
Cu	0	8.96	0
	1		0.055475772

18.02588698 **Alloy Density (g/cc)**

24 Carat Gold

Metal	Component Mass (grams)	Density (grams/cc)	Eqivalent Volume (cc)
Au	1	19.282	0.05186184
Ag	0	10.501	0
Cu	0	8.96	0
	1		0.05186184

19.282 **Alloy Density (g/cc)**

18 Carat White Gold

Metal	Component Mass (grams)	Density (grams/cc)	Eqivalent Volume (cc)
Au	0.75	19.282	0.03889638
Pd	0.1	12.023	0.008317392
Ni	0.1	8.908	0.011225864
Zn	0.05	7.14	0.007002801
	1		0.065442437

15.28060449 **Alloy Density (g/cc)**

Table One
The Mass Densities of Typical Gold Formulations

A Degenerate Case: A Binary Alloy in Ternary Expression

Consider the case in which the ternary alloy, cuprous 18-carat gold, is re-expressed as an alloy of 75% gold with 16% silver and again 9% silver to make one hundred percent. This is of course the same as a binary alloy of 18-carat non-cuprous silver gold of 75% gold and 25% silver.

The additivity of the denominators guarantees equivalence.

In particular, for the pseudo-ternary alloy:-

$$H = \frac{3}{\dfrac{1}{x_{Au}} + \dfrac{1}{x_{Ag1}} + \dfrac{1}{x_{Ag2}}}$$

Equation 17

Or:-

$$\rho_{mix} = \frac{1}{\dfrac{1}{x_{Au}} + \dfrac{1}{x_{Ag1}} + \dfrac{1}{x_{Ag2}}}$$

Equation 18

Both Equation Seventeen and Equation Eighteen yield ρ_{mix} = 15.9480383577828 g/cc. And division by zero is theoretically impossible as long as the finitude of all the fractions is preserved.

Now for the equivalent binary alloy:-

$$H = \frac{2}{\dfrac{1}{x_{Au}} + \dfrac{1}{x_{Ag}}}$$

Equation 19

Or:-

$$\rho_{mix} = \frac{1}{\dfrac{1}{x_{Au}} + \dfrac{1}{x_{Ag}}}$$

Equation 20

And both Equation Nineteen and Equation Twenty yield ρ_{mix} = 15.9480383577828 g/cc.

Table Two lists the intermediate and summative results for both instances:-

18 Carat Pale Silver Gold   (as degenerate ternary alloy)

Metal	Component Mass (grams)	Density (grams/cc)	Eqivalent Volume (cc)
Au	0.75	19.282	0.03889638
Ag1	0.16	10.501	0.015236644
Ag2	0.09	10.501	0.008570612
	1		0.062703636

15.94803836 **Alloy Density (g/cc)**

18 Carat Pale Silver Gold   (as binary alloy)

Metal	Component Mass (grams)	Density (grams/cc)	Eqivalent Volume (cc)
Au	0.75	19.282	0.03889638
Ag	0.25	10.501	0.023807256
	1		0.062703636

15.94803836 **Alloy Density (g/cc)**

Table Two
Pseudo-Ternary and Binary Gold Alloy Formulations

Crystal Chemistry
and the
Harmonic Mean

by
James R Warren BSc MSc PhD PGCE

In the Nineteen-Eighties I did some work on the mass density of mixtures as a corollary of combinatorial mathematics.

More recently I returned to the topic to explore the relationship between the mass density of a mixture and the harmonic mean of the ratios of density to fraction for the individual components. I summarised the arithmetic in my 16 June 2017 paper "Mixture Density and the Harmonic Mean".

I wondered to what degree (if any) these elementary principles would hold good for ordered structures of matter.

It seemed to me that the very coalescence of atoms, involving the interplay of force and the condensation of matter, would of necessity engage mass densities variant to the simple laws of classical arithmetic.

In days gone by the late Mr Don Braggins and his friends strove to instil in crystal metal the most perfect purity of form and substance, whilst for centuries men of diverse culture and breeding have thought *any* crystal the epitome of perfection.

But whilst man and crystal were conceived pure and simple in the Mind of God, both Fell to Earth in disarray and fractured.

Defective Structures

The form of crystals is disrupted by internal dislocations: Tears and omissions in the structure of translational or helical geometry at the interatomic scale.

The substance of crystals is corrupted not only by simple inclusions of extraneous stuff but also by foreign atoms of substitution held in more or less tension within the very atomic order itself.

Therefore, one might be forgiven for guessing that any found object, even the most refulgent diamond, is some blend of order and chaos, of curt arithmetic and temporising stochastics.

For near-perfect corundum is lapidary but colourless, like many near-perfect things, whilst polluted rubies and sapphires are the envy of women and a delight to the eye.

A Choice of Crystalline Minerals

Further to investigate I intentionally noted minerals at random. I consulted the memories of boyhood and especially my year in the mineralogical paradise of Cornwall.

Old friends like cassiterite, cuprite and pyrite sprang readily to recollection, along with galena, cerussite and fluorite. It would have been remiss to forget olivine and hornblende, especially as the former is elastic of cation substitution and the latter so ambiguous as to challenge definition.

I realised that fair tests would include much more complex formulas and sentimentally included in my study the elaborate but exceedingly rare phyllosilicate zussmanite as well as involved hydrated ferrosilicates incorporating water of crystallisation in various proportions.

Table A1 in Appendix A presents some fundamental data for the forty-seven rock-forming elements that account for well over 99% of the Earth's Crust and contribute to the composition of our fifty-two studied minerals[1-4].

Table B1 in Appendix B presents the fifty-two crystalline minerals as mineral forms together with their nominal chemical formulae and physical properties. This source data is drawn from a variety of Internet sources listed in the References[5-10].

<u>Density and Harmonic Mean Algebra</u>

The relationships between the Mixture Mass Density, ρ_{mix}, and the Harmonic Mean, H, are defined by these equations:-

$$\rho_{mix} = \cfrac{1}{\cfrac{f_1}{\rho_1} + \cfrac{f_2}{\rho_2} + \cfrac{f_3}{\rho_3} + ... + \cfrac{f_n}{\rho_n}} = \frac{H}{n} = \frac{1}{n} \cdot \cfrac{1}{\displaystyle\sum_{i=1}^{n} \cfrac{f_i}{\rho_i}}$$

Equation 1

where:-

$$H = \cfrac{n}{\cfrac{1}{x_1} + \cfrac{1}{x_2} + \cfrac{1}{x_3} + ... + \cfrac{1}{x_n}} = \cfrac{n}{\displaystyle\sum_{i=1}^{n} \cfrac{1}{x_i}}$$

Equation 2

and:-

$$x_i = \frac{\rho_i}{f_i}$$

Equation 3

f_i and ρ_i are respectively the Mass Fraction and the Mass Density of the ith component of the mixture.

In the context of the mineral crystal studies of this disquisition, f_i is the Fraction of the Chemical Formula Weight, W_{cry}, accounted for by constituent element i; and ρ_i is the Mass Density of Element i at or near STP, or in the case of STP gases, near to absolute zero. The use of the densities of liquid, or better still, solid STP gases was chosen to be more representative of the predicament of gaseous elements within a solid-state lattice. Such is of course a further source of error.

<u>Calculation of the Formula Weight, W_{cry}</u>

Straight-forward application of Equations One, Two and Three uses the Formula Weight, W_{cry} (or m), of a mineral in which the chemical formula is considered a "molecule" without reference to the crystallographic Unit Cell or to the cell's Formula Population, Z.

Furthermore, I arbitrarily and naïvely distributed substitutional elements within formulaic elaborations. For example, there are thirteen substitutional atoms of iron, magnesium and manganese in the formula of zussmanite. In that case, my tactic was equally but arbitrarily to award 4⅓ of each of the three atoms to the computation. This of course subverts the very concept of the integral unit cell and reduces the chemical description to little more than an empirical formula. But this scheme of equality was not followed in all cases.

For each of the fifty-two minerals the Formula Weight, W_{cry}, was computed as:-

$$W_{cry} = \sum_{i=1}^{n_e} A_i f_i = \sum_{i=1}^{n_e} A_i . \frac{j_e}{j_{tot}}$$

Equation 4

where A_i is the Atomic Weight of component Element i; f_i is the weight Fraction of Element i in the chemical formula; j_e is the Number of relevant Element Atoms in the formula; and j_{tot} is the Total Number of Atoms in the formula.

ρ_{meas} in the Measured Mass Density of the mineral as selected from a trusted Web source (usually www.handbookofmineralogy.org) and ρ_H is the mineral's Harmonic Density as:-

$$\rho_H = \frac{H(x_i)}{j_{tot}}$$

Equation 5

The overall scheme of computation and an individual mineral's results is illustrated for the case of zussmanite by Table One.

The apparent volumetric Condensation of the Crystal Lattice is then defined as:-

$$\kappa = 1 - \frac{\rho_{meas}}{\rho_H}$$

Equation 6

Table C1 in Appendix C presents given and computed data for each of the fifty-two mineral species. The order is arbitrary.

SERIAL:		13
MINERAL:		ZUSSMANITE
CHEMICAL FORMULA:		K(Fe2+,Mg,Mn2+)13(AlSi)18O42(OH)14
MEASURED DENSITY:		3.146
HARMONIC DENSITY:		1.821431229
CONDENSATION:		-0.727213166
FORMULA WEIGHT:		2205.753517
WEIGHTED PSD OF ELECTRONEGATIVITY:		37.57896207

Atomic Number Z	Atomic Weight	Element Name	Element Symbol	Density ρ_i (grams/cc) (STP)	Electro negativity χ	Formula Element Count	Partial Formula Weight	Formula Weight Fraction f_i	$x_i = \rho_i / f_i$	Weighted Electro negativity
1	1.008	Hydrogen	H	0.088	2.2	14	14.112	0.006398	13.7547	30.8
5	10.811	Boron	B	2.46	2.04					
6	12.011	Carbon	C	2.2	2.55					
7	14.007	Nitrogen	N	1.03	3.04					
8	15.999	Oxygen	O	1.19	3.44	36	575.964	0.261119	4.55731	123.84
9	18.9984032	Flourine	F	1.706	3.98					
11	22.98977	Sodium	Na	0.968	0.93					
12	24.305	Magnesium	Mg	1.738	1.31	4.333333	105.3217	0.047749	36.39896	5.676666667
13	26.9815385	Aluminium	Al	2.698	1.61	18	485.6677	0.220182	12.25349	28.98
14	28.085	Silicon	Si	2.3296	1.9	18	505.53	0.229187	10.16463	34.2
15	30.973761998	Phosphorous	P	1.82	2.19					
16	32.06	Sulfur	S	2.067	2.58					
17	35.453	Chlorine	Cl	2.03	3.16					
19	39.0983	Potassium	K	0.856	0.82	1	39.0983	0.017726	48.29174	0.82
20	40.078	Calcium	Ca	1.55	1					
22	47.867	Titanium	Ti	4.506	1.54					
23	50.9415	Vanadium	V	6	1.63					
24	51.9961	Chromium	Cr	7.15	1.66					
25	54.938044	Manganese	Mn	7.3	1.55	4.333333	238.0649	0.107929	67.63703	6.716666667
26	55.845	Iron	Fe	7.874	1.83	4.333333	241.995	0.109711	71.7705	7.93
27	58.9332	Cobalt	Co	8.9	1.88					
28	58.6934	Nickel	Ni	8.912	1.91					
29	63.546	Copper	Cu	8.96	1.9					
30	65.38	Zinc	Zn	7.134	1.65					
31	69.723	Gallium	Ga	5.904	1.81					
33	74.9216	Arsenic	As	5.778	2.18					
38	87.62	Strontium	Sr	2.63	0.95					
41	92.90637	Niobium	Nb	8.61	1.6					
42	95.95	Molybdenum	Mo	10.2	2.16					
44	101.07	Ruthenium	Ru	12.1	2.2					
46	106.42	Palladium	Pd	12	2.2					
47	107.8602	Silver	Ag	10.501	1.93					
48	112.411	Cadmium	Cd	8.65	42					
49	114.818	Indium	In	7.31	1.78					
50	118.71	Tin	Sn	7.287	1.96					
51	121.76	Antimony	Sb	6.685	2.05					
56	137.327	Barium	Ba	3.51	0.89					
73	180.94788	Tantalum	Ta	16.4	1.5					
74	183.84	Tungsten	W	19.25	2.36					
76	190.23	Osmium	Os	22.59	2.2					
77	192.217	Iridium	Ir	22.562	2.2					
78	195.084	Platinum	Pt	21.5	2.28					
79	196.966569	Gold	Au	19.282	2.54					
82	207.2	Lead	Pb	11.342	1.87					

Formula Element Count	Partial Formula Weight	Formula Weight Fraction	x_i	Weighted Electronegativity	
8	8	8	8	8	Count
100	2205.754	1	264.8284	238.9633333	Sum
12.5	275.7192	0.125	33.10355	29.87041667	Arithmetic Mean
7.770165	161.9593	0.073426	22.53982	13.22768356	Geometric Mean
4.204717	66.64491	0.030214	14.57145	4.507175608	Harmonic Mean
				37.57896207	Population SD

Table One
Harmonic Density Computed for Zussmanite
Using MicroSoft® EXCEL

Computation of the Crystallographic Unit Cell Volume[11]

There are a finite number of possible spacial configurations of atoms in a crystal structure. These are called the fourteen (three-dimensional) Bravais Lattices, and they constitute seven fundamental Symmetry Systems: Cubic (sometimes known as Isometric), tetragonal, orthorhombic, trigonal, hexagonal, monoclinic and triclinic.

A Unit Cell is a nominally-fundamental element of constituent atom arrangement having a pattern of one of the lattice systems. Because atoms are very little, unit cells are inevitably so small as only to be perceptible using x-ray or electron beams. Intuitively, a unit cell could be visualised as a right or distorted cuboid or (in two cases) trigonal or hexagonal prism.

To assist scaling calculations it is helpful to use Avogadro's Number, the population of atoms or molecules in a mole-weight of substance. Avogadro's Number, N_A or L, is taken to be $6.022140857 \times 10^{+23}$ particles per (gram) mol^{-1}.

The three axial Edges of a Unit Cell a, b and c are so short that they are conveniently measured in ångströms. An ångström is 0.1 nanometers, i.e. 10^{-10} meters.

The Volume of a Unit Cell, V, depends not only upon the axial lengths a, b and c but also upon the corresponding Interaxial Angles, α, β, and γ which are recorded in degrees but are of course re-expressed in radians for computation.

Table Two represents the Edge and Interangle conformations that characterise the seven Symmetry Systems:-

Length	Crystal Symmetry System	x	y	z			
			Edge			Angle	
		a	b	c	α	β	γ
	Cubic	x	x	x	$\pi/2$	$\pi/2$	$\pi/2$
	Tetragonal	x	x	z	$\pi/2$	$\pi/2$	$\pi/2$
	Orthorhombic	x	y	z	$\pi/2$	$\pi/2$	$\pi/2$
	Trigonal	x	x	z	$\pi/2$	$\pi/2$	$2\pi/3$
	Hexagonal	x	x	z	$\pi/2$	$\pi/2$	$2\pi/3$
	Monoclinic	x	y	z	$\pi/2$	β	$\pi/2$
	Triclinic	x	y	z	α	β	γ

x,y,z and α,β,χ all differ

Table Two
Unit Cell Geometries for the Symmetry Systems
(Angles for General Cosine Solutions)

To quantify V we use an assemblage of trigonometrical equations described below. The units of volume are in practice cubic ångströms.

The Triclinic Case

In a triclinic crystal a, b, c, α, β, γ are, or are capable of being, numerically different. The applicable formula is[12]:-

$$V = abc\sqrt{1 + 2\,Cos\alpha\,Cos\beta\,Cos\gamma - Cos^2\alpha - Cos^2\beta - Cos^2\gamma}$$

Equation 7

The Cubic, Tetragonal and Orthorhombic Cases

The commonality of these lattice systems is that $\alpha = \beta = \gamma = \pi/2$; that is all angles are orthogonal.

In the Cubic system a = b = c and accordingly:-

$$V = a^3$$
Equation 8

The Tetragonal system forms a square cuboid where a = b ≠ c. Therefore:-

$$V = a^2 c$$
Equation 9

The Orthorhombic system is a general cuboid cell like a house brick in which a ≠ b ≠ c. Therefore:-

$$V = abc$$
Equation 10

It is notable that Equation Seven is sufficiently general to apply to the cubic, Tetragonal and Orthorhombic systems because:-

$$Cos(\alpha) = Cos(\beta) = Cos(\gamma) = 0$$
Equation 11

The Hexagonal and Trigonal Cases

The cells are respectively hexagonal and triangular prisms. Accordingly:-

$$V = a^2 c.Sin\left(\frac{\pi}{3}\right)$$
Equation 12

The Monoclinic Case

In this final case a ≠ b ≠ c but $\alpha = \gamma = \pi/2$. The internal angle β may be other than orthogonal.

The required formula for Cell Volume is:-

$$V = abc.Sin(\beta)$$
Equation 13

The Theoretical Cell Density of a Crystal[11]

Having now the means of knowing Cell Volume and also constituent Atomic and Formula Weights, we may also incorporate the constant Avogadro's Number to establish the Theoretical Cell Mass Density, ρ_{theo}, as:-

$$\rho_{theo} = \frac{ZW_{cry}}{VL} \times k$$

Equation 14

where W_{cry} is the crystal's chemical Formula Weight; V is Cell Volume and L (or N_A) is Avogadro's Number. k is an arbitrary Scaling Constant used to bring ρ_{theo} to co-ordinate units. In my calculations k was ten.

Measured Density and Harmonic Density

Measured Mass Density, ρ_{meas}, is actually assessed using inertial (in practice gravimetric) laboratory methods. Traditionally, that was done by comparing the weight of mineral to that of water in a glass density jar (or more properly specific gravity jar).

Theoretical Mass Density, ρ_{theo}, is indirectly established from radiological determinations.

Our third approach, Harmonic Mass Density, ρ_H, is defined by:-

$$\rho_H = \frac{H(x_e)}{n_e} = \frac{H(\rho_e / f_e)}{n_e}$$

Equation 15

where n_e is the Number of Different Elements in the Mineral Formula; ρ_e is the Mass Density of Element e; f_e is the Fraction of the Formula Weight accounted for by Element e; x_e is the Density/Fraction Ratio for Element e; and $H(x_e)$ is the Harmonic Mean of the Density/Fraction Ratios of all the constituent elements.

We cannot of course expect ρ_H closely to resemble inertial density because of the action of the interatomic forces that condense solid matter. Notwithstanding, Harmonic Density has its uses as we shall see.

Lattice Strain

In our context, the phrase "Lattice Strain" denotes the apparent isometric volume contraction engendered by interatomic binding forces.

The Z of Equation Fourteen is the Number of Formula Units in the Crystallographic Unit Cell. It is almost always an even integer, often 2 or 4. Deviant odd Z are known, but very rare.

Alongside this parameter we may define a more flexible analog Ξ (big Xi) which is the ratio of Measured Density to Harmonic Density. That is:-

$$\Xi = \frac{\rho_{meas}}{\rho_H}$$
Equation 16

Ξ is apparent Cell Lattice Strain (not necessarily the number of formula weights in the repeating lattice element, which is of course Z).

Appendix D presents tabulations and column plots of Ξ for all fifty-two minerals. In the tables Ξ is called "Precise Harmonic ρ Multiplier": In the plots it is named "Precise Density Ratio".

Table D1 lists; Serial Number, Mineral, Chemical Formula, Ξ, Z, and Lattice System in *arbitrary order*. It is clear that Ξ is near to, but not at, low integral values for many simple chemical formulae, especially heavy metal sulphides.

Figure D1 represents Mineral Formula versus Ξ as a column chart *in the arbitrary order* of Table D1. This clarifies that Ξ tends to cluster around one, near to two, and also to some extent near 1.7

Table D2 and Figure D2 are a tabulation of the same data and its column chart but *in order of ascending Ξ*. The approximate quantisation of Ξ is now striking.

Inspection infers that:-

$$\Xi \approx \sqrt{n}$$
Equation 17

where n is a low positive integer. n = 1 for simple or complex heavy metal sulphides; n = 2 for some hydrated phyllosilicates; n = 3 for some simple silicates and also zussmanite; n = 4 for several light metal silicates and alkaline carbonates, nitrates and sulphates.

There is a level of $\Xi = 5^{1/2} \approx 2.236$ very near to the Ξ of the hydrated ferro-aluminium phosphate childrenite. I speculate that this may relate to the relation of mass density to the Phidian Ratio $[(1+5^{1/2})/2]$ proposed by some researchers[13,14].

It is of course possible statistically to assess the agreement of mineral Ξ values with our $\Xi = n^{1/2}$ theory. We can do this using simple or sophisticated methods. I prefer simple things. I can understand them.

So I adopted a simple criterion:-

$$\Xi_{test} - t < \Xi_{mineral} < \Xi_{test} + t$$
Criterion 1

where t is the Tolerance and Ξ_{test} is some exact Ξ for $\Xi = n^{1/2}$ where n = 1,2,3,4,5,…

Table Three shows how the fifty-two minerals group into five Ξ classes when the tolerance is set to 0.07 (i.e. ±7%).

Some 44% of minerals have Ξ = 1 or 2, whilst just shy of 10% have $\Xi = 3^{1/2}$. Strikingly, three minerals (5.7%) have $\Xi = 5^{1/2}$; not only childrenite, but also the very different though simpler spodumene and goethite.

More than 61.5% of minerals fall into some $\Xi = n^{1/2}$ category.

Basis Integer	1	2	3	4	5
Test Ξ	1	1.25992105	1.44224957	1.587401052	1.709975947
Tolerance	0.07	0.07	0.07	0.07	0.07
LB	0.93	1.18992105	1.37224957	1.517401052	1.639975947
UB	1.07	1.32992105	1.51224957	1.657401052	1.779975947
Count	11	2	1	3	5
Mean	1.02052213	1.228915457	1.433495267	1.562760409	1.712334719
Pop.SD	0.035195671	0.023600686	0	0.029530443	0.014183739
Fractional SpDef(Ξ,Mean)	-0.02052213	0.024609155	0.006069895	0.015522632	-0.001379419
Prob(Ξ)	0.211538462	0.038461538	0.019230769	0.057692308	0.096153846
Prob(Ξ=1,2,3,4)	0.423076923				
F_R	0.487765693				

Table Three
Mineral Ξ Groupings According to
The Model $\Xi = n^{\frac{1}{2}}$

Quandaries

The obvious corollary is that 38.5% of minerals do not conform to our $\Xi = n^{\frac{1}{2}}$ model.

This is not because the 38.5% are deviant: In fact they are almost typical. It is because they are defective, as are easily 38.5% of men.

Or else Ξ is merely a chimera, a trick of cognition, a fallacy of judgement.

Figure One is a histogram representing the frequencies of minerals in twelve subranges of the total range of Equation Sixteen Density Ratio Ξ. The quantisation of Ξ is dramatically pronounced but so too is a groundmass of non-conformable crystal constitutions: And it is difficult to say whether the peaks are culminations of statistical distributions, in particular Gaussian Distributions, or whether they are not.

Distribution of Ξ Values

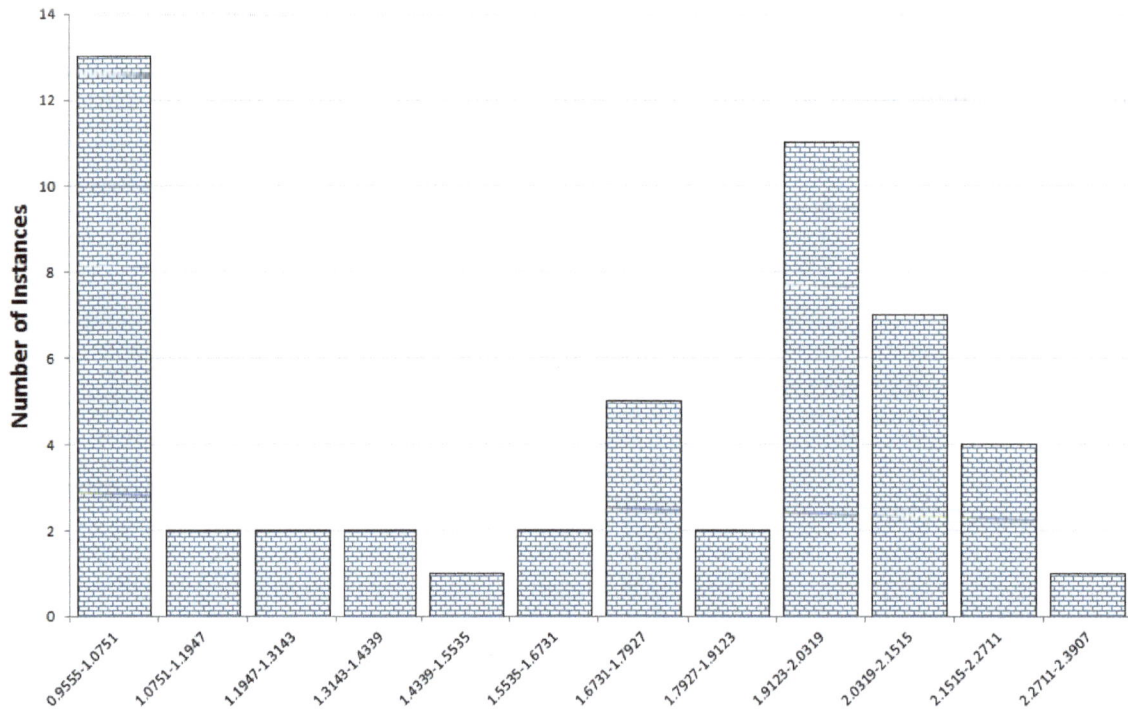

Figure One
Density Ratio (Ξ) Subranges for Fifty-Two Crystalline Minerals

Allow that R is the Total Range of Ξ Density Ratios, and that the n values of subranges r_j represent the clusters of minerals conformable to $\Xi = n^{1/2}$.

Then, neglecting any end-of-range overlaps, we may write:-

$$F_R = \frac{2tn}{R_+ - R_-}$$

Equation 18

where F_R is the Fraction of the Total Range defined by $R_+ - R_-$, given that R_+ is the Upper Limit of R and R_- the Lower Limit. t is the Tolerance.

The data of Table Three show that for our assemblage of fifty-two minerals F_R = 0.487765693 or almost half the total range R is conformable to $\Xi = n^{1/2}$. (Where t varies from point to point, mean t is employed in Equation Eighteen).

Summary Equations for the Lattice Strain Density Ratio

Should we be willing to identify the experimentally-measured mass density with that determined by radiographic studies of the crystal lattice, i.e.:-

$$\rho_{meas} \equiv \rho_{theo}$$

Equation 19

then:-

$$\rho_{theo} = \frac{kZW_{cry}}{VL}$$

Equation 14

Also:-

$$\Xi = \frac{\rho_{meas}}{\rho_H}$$

Equation 16

So by incorporation of Equation Fifteen:-

$$\Xi = \frac{kZW_{cry}}{VL} \times \frac{n_e}{H(\rho_e / f_e)}$$

Equation 20

Or:-

$$\Xi = \frac{kZ\sum_{j=1}^{n_e} A_j}{VL} \cdot \sum_{j=1}^{n_e} \frac{1}{x_j} = \frac{kZ\sum_{j=1}^{n_e} A_j}{VL} \cdot \sum_{j=1}^{n_e} \frac{f_j}{\rho_j}$$

Equation 21

which allows us to write:-

$$\Xi = \frac{kZ}{VL} \sum_{j=1}^{n_e} A_j \sum_{j=1}^{n_e} \left[\frac{1}{\rho_j} \cdot \frac{A_j}{\sum_{j=1}^{n_e} A_j} \right]$$

Equation 22

Equation Twenty-Two may be re-arranged as:-

$$\Xi = \frac{kZ}{VL} \cdot \sum_{j=1}^{n_e} \left(\sum_{j=1}^{n_e} A_j \right) \cdot \left[\sum_{j=1}^{n_j} \left(\frac{1}{\rho_j} \cdot \frac{A_j}{\sum_{j=1}^{n_v} A_j} \right) \right]$$

Equation 23

from which simplification yields:-

$$\Xi = \frac{kZ}{VL} \cdot \sum_{j=1}^{n_e} \frac{A_j}{\rho_j}$$

Equation 24

As we aforenoted Ξ is sometimes equal to $n^{\frac{1}{2}}$ where n = 1,2,3,4,5,…?

$$\Diamond \left(\Xi = \frac{kZ}{VL} \cdot \sum_{j=1}^{n_e} \frac{A_j}{\rho_j} \right)$$

Equation 25

where $\Diamond$ is the Logical Possibility Operator.

Formula Weight

We know from Equation Sixteen that:-

$$\Xi = \frac{\rho_{meas}}{\rho_H}$$

Equation 16

and from Equation Fourteen that:-

$$\rho_{theo} = \frac{kZW_{cry}}{VL}$$

Equation 14

If we identify ρ_{meas} and ρ_{theo} as a Synthetic Mass Density, ρ_{syn}, it is possible to write:-

$$\rho_{syn} = \frac{1}{\sum_{j=1}^{n_e} \frac{1}{x_j}} . K . \sum_{j=1}^{n_e} \left(\frac{A_j}{\rho_j} \right)$$

Equation 26

Also:-

$$\rho = K.W_{cry}$$

Equation 27

where W_{cry} is the chemical Formula Weight of the mineral.

By elimination of the crystallographic construct $K = kZ/VL$ we may expand Formula Weight as:-

$$W_{cry} = \frac{\sum_{j=1}^{n_e} \left(\frac{A_j}{\rho_j} \right)}{\sum_{j=1}^{n_e} \frac{1}{x_j}} = \frac{\sum_{j=1}^{n_e} \left(\frac{A_j}{\rho_j} \right)}{\sum_{j=1}^{n_e} \frac{f_j}{\rho_j}}$$

Equation 28

which simplifies to:-

$$W_{cry} = \frac{1}{n_e} \sum_{j=1}^{n_e} \left(\frac{A_j}{f_j} \right)$$

Equation 29

References

Element Properties Data

1   https://en.wikipedia.org/wiki/list_of_chemical_elements
2   http://periodictable.com
3   http://www.knowledgedoor.com/2/elements_handbook
4   http://www.espimetals.com

Minerals Properties Data

5   http://www.handbookofmineralogy.org
6   http://webmineral.com
7   http://som.web.cmu.edu/structures/
8   http://www.spec2000.net/05-mineralprops.htm
9   http://database.iem.ac.ru/mincryst/

10 http://www.minweb.co.uk/

Theoretical Cell Volume

11 http://webmineral.com/help/
    CellDimensions.shtml#WZg2nE1K0r9

General Equation of Cell Volume

12 Lattice constant. (2017, July 31).
   In *Wikipedia, The Free Encyclopedia*.
   Retrieved 08:19, August 29, 2017
   from https://en.wikipedia.org/w/index.php?
     title=Lattice_constant&oldid=793291329

The Crystal Lattice and the Golden Mean (The Ratio of Phidias)

13 "X-Ray Multiple-Wave Diffraction: Theory and Application"
   Shih-Lin Chang
   Springer-Verlag of Berlin   2004 Edition
   ISBN 978-3540211969
   pp 452

14 "The Golden Ratio, ionic and atomic radii and bond lengths"
   Raji Heyrovska
   Molecular Physics; V103; No.6-8;
     20 March – 20 April 2005; 877-882
   ISSN 0026 8976
   Taylor and Francis Group Ltd

APPENDIX A
DATA FOR THE
PRINCIPAL ROCK-FORMING ELEMENTS

Atomic Number 7	Atomic Weight	Element Name	Element Symbol	Density ρ_l (grams/cc) (STP)	Density Reference Temperature (°K)	Coefficient of Linear Expansion α (10^{-6}/°K)	Bulk Modulus K (Gpa)	Electro negativity χ	Atomic Radius R (pm)
1	1.008	Hydrogen	H	0.088	4		0.2	2.2	53
3	6.997	Lithium	Li	0.534	298.15	46	11.6	0.98	152
5	10.811	Boron	B	2.46	293.15	6	320	2.04	87
6	12.011	Carbon	C	2.2	293.15	25	33	2.55	67
7	14.007	Nitrogen	N	1.03	20		1.2	3.04	56
8	15.999	Oxygen	O	1.19	80			3.44	48
9	18.9984032	Flourine	F	1.706	53.48			3.98	42
11	22.98977	Sodium	Na	0.968	293.15	71	6.3	0.93	190
12	24.305	Magnesium	Mg	1.738	293.15	8.2	45	1.31	145
13	26.9815385	Aluminium	Al	2.698	293.15	23.1	76	1.61	118
14	28.085	Silicon	Si	2.3296	293.15	2.6	100	1.9	111
15	30.973761998	Phosphorous	P	1.82	293.15	124.5	11	2.19	98
16	32.06	Sulfur	S	2.067	293.15	64	7.7	2.58	88
17	35.453	Chlorine	Cl	2.03	93		1.1	3.16	79
19	39.0983	Potassium	K	0.856	293.15	83.3	3.1	0.82	243
20	40.078	Calcium	Ca	1.55	293.15	22.3	17	1	194
22	47.867	Titanium	Ti	4.506	293.15	8.6	110	1.54	176
23	50.9415	Vanadium	V	6	293.15	8.4	160	1.63	171
24	51.9961	Chromium	Cr	7.15	293.15	4.9	160	1.66	166
25	54.938044	Manganese	Mn	7.3	293.15	21.7	120	1.55	161
26	55.845	Iron	Fe	7.874	293.15	11.8	170	1.83	156
27	58.9332	Cobalt	Co	8.9	298.15	13	180	1.88	152
28	58.6934	Nickel	Ni	8.912	293.15	13.4	180	1.91	149
29	63.546	Copper	Cu	8.96	293.15	16.5	140	1.9	145
30	65.38	Zinc	Zn	7.134	293.15	30.2	70	1.65	142
31	69.723	Gallium	Ga	5.904	298.15	120	22	1.81	136
33	74.9216	Arsenic	As	5.778	299.15	5.6	22	2.18	114
38	87.62	Strontium	Sr	2.63	298.15	22.5	11.6	0.95	219
41	92.90637	Niobium	Nb	8.61	298.15	7.3	170.2	1.6	198
40	91.224	Zirconium	Zr	6.52	298.15	5.7	91.1	1.33	160
42	95.95	Molybdenum	Mo	10.2	293.15	4.8	230	2.16	190
44	101.07	Ruthenium	Ru	12.1	298.15	6.4	320.8	2.2	178
46	106.42	Palladium	Pd	12	298.15	11.8	180.8	2.2	169
47	107.8602	Silver	Ag	10.501	293.15	18.9	100	1.93	165
48	112.411	Cadmium	Cd	8.65	298.15	30.8	46.7	42	161
49	114.818	Indium	In	7.31	298.15	32.1	41.1	1.78	156
50	118.71	Tin	Sn	7.287	293.15	22	58	1.96	145
51	121.76	Antimony	Sb	6.685	293.15	11	42	2.05	133
56	137.327	Barium	Ba	3.51	298.15	20.6	10.3	0.89	253
73	180.94788	Tantalum	Ta	16.4	298.15	6.3	200	1.5	200
74	183.84	Tungsten	W	19.25	293.15	4.5	310	2.36	193
76	190.23	Osmium	Os	22.59	298.15	5.1	418	2.2	185
77	192.217	Iridium	Ir	22.562	298.15	6.4	355	2.2	180
78	195.084	Platinum	Pt	21.5	298.15	8.8	278.3	2.28	177
79	196.966569	Gold	Au	19.282	293.15	14.2	220	2.54	174
80	200.592	Mercury	Hg	13.534	298.15	60.4	38.2	2	151
82	207.2	Lead	Pb	11.342	293.15	28.9	46	1.87	154

Table A1
Selected Element Data

| | | | Density (g/cc) | | | | | | | | | | | | | | | |
| | | | Handbook of mineralogy | | | | Webmineral Com | | | Lattice System | Space Group* | a | b | c | α | β | γ | Z |
Serial	Mineral	Chemical Formula	LB	UB	mean	calc	Bulk Electron Density	Measured Density	Calculated Formula Weight									
1	ARGENTOPYRITE	$AgFe_2S_3$	4.25	4.25	4.25	4.27	3.98	4.25	315.7302	orthorhombic	Pmmn	6.639	11.463	6.452	1.570796	1.570796	1.570796	4
2	ARGENTOTENNANTITE	$(Ag,Cu)_{10}(Zn,Fe)_2(As,Sb)_4S_{13}$	4.71	4.71	4.71	5.05	4.71	4.71	1788.399	cubic	I43m	10.584	10.584	10.584	1.570796	1.570796	1.570796	2
3	BARITE	$BaSO_4$	4.5	4.5	4.5	4.47	3.99	4.5	233.383	orthorhombic	Pnma	8.884	5.457	7.157	1.570796	1.570796	1.570796	4
4	CASSITERITE	SnO_2	6.98	7.01	6.995	6.993	6.26	6.995	150.708	tetragonal	P4$_2$/mnm	4.7382	4.7382	3.1871	1.570796	1.570796	1.570796	2
5	CERUSSITE	$PbCO_3$	6.552	6.552	6.552	6.577	5.52	6.552	267.208	orthorhombic	Pmcn	5.179	8.492	6.141	1.570796	1.570796	1.570796	4
6	CHALCOPYRITE	$CuFeS_2$	4.1	4.3	4.2	4.283	3.98	4.2	183.511	tetragonal	I42d	5.281	5.281	10.401	1.570796	1.570796	1.570796	4
7	CHILDRENITE	$Fe^{2+}Al(PO_4)(OH)_2 \cdot H_2O$	3.11	3.19	3.15	3.135	3.14	3.15	228.8173	orthorhombic	Bba2	10.395	13.394	6.918	1.570796	1.570796	1.570796	8
8	CUPRITE	$Cu_2^{3+}O$	6.14	6.14	6.14	6.15	5.64	6.14	143.091	cubic	Pn3m	4.2685	4.2685	4.2685	1.570796	1.570796	1.570796	2
9	EDGARITE	$FeNb_3S_6$	4.98	4.98	4.98	4.99	3.85	4.98	526.9241	hexagonal	P6$_3$22	5.771	5.771	12.19	1.570796	1.570796	2.094395	2
10	GALENA	PbS	7.58	7.58	7.58	7.57	6.23	7.58	239.26	cubic	Fm3m	5.936	5.936	5.936	1.570796	1.570796	1.570796	4
11	SMITHSONITE	$ZnCO_2$	4.431	4.431	4.431	4.43	4.24	4.431	125.388	hexagonal	R3c	4.6526	4.6526	15.9257	1.570796	1.570796	2.094395	6
12	SPHALERITE	$(Zn,Fe)S$	3.9	4.1	4	4.096	3.85	4	92.6725	cubic	F43m	5.406	5.406	5.406	1.570796	1.570796	1.570796	4
13	ZUSSMANITE	$K(Fe^{2+},Mg,Mn^{2+})_{13}(AlSi)_{18}O_{42}(OH)_{14}$	3.146	3.146	3.146	3.14	3.09	3.146	2205.754	hexagonal	R3	11.66	11.66	28.69	1.570796	1.570796	2.094395	3
14	TENNANTITE	$(Cu,Ag,Sn,Fe)_{12}As_4S_{13}$	4.72	4.72	4.72	4.62	4.3	4.72	1722.29	cubic	I43m	10.19	10.19	10.19	1.570796	1.570796	1.570796	2
15	WOLLASTONITE 2M	$CaSiO_3$	2.86	3.09	2.975	2.9	2.91	2.84	116.16	monoclinic	2/m	15.409	7.322	7.063	1.570796	1.663299	1.570796	12
16	WOLLASTONITE 1A	$CaSiO_3$	2.86	3.09	2.975	2.9	2.91	2.84	116.16	triclinic	P1	7.94	7.32	7.063	1.567306	1.47969	1.339366	6
17	PYRITE	FeS_2	5.018	5.018	5.018	5.013	4.84	5.018	119.98	cubic	Pa3	5.4179	5.4179	5.4179	1.570796	1.570796	1.570796	4
18	CALCITE	$CaCO_3$	2.7102	2.7102	2.7102	2.711	2.71	2.7102	100.086	hexagonal	R3c	4.9896	4.9896	17.061	1.570796	1.570796	2.094395	6
19	STRONTIANITE	$SrCO_3$	3.76	3.76	3.76	3.78	3.48	3.76	147.628	orthorhombic	Pmcn	5.1059	8.4207	6.0319	1.570796	1.570796	1.570796	4
20	QUARTZ-α	SiO_2	2.65	2.65	2.65	2.66	2.65	2.65	60.083	trigonal	P3$_1$21	4.9135	4.9135	5.405	1.570796	1.570796	2.094395	3
21	QUARTZ-β	SiO_2	2.65	2.65	2.65	2.66	2.53	2.65	60.083	hexagonal	P3$_1$21	4.9965	4.9965	5.4546	1.570796	1.570796	2.094395	6
22	CORUNDUM	Al_2O_3	3.98	4.1	4.04	3.997	3.93	4.04	101.9601	hexagonal	R3c	4.754	4.754	12.982	1.570796	1.570796	2.094395	6
23	ACANTHITE	Ag_2S	7.2	7.22	7.21	7.24	6.43	7.21	247.7804	monoclinic	P2$_1$/n	4.229	6.931	7.862	1.570796	1.738522	1.570796	4
24	ARGENTITE	Ag_2S	7.2	7.4	7.3	7.04	6.25	7.3	247.7804	cubic	Im3m	4.88	4.88	4.88	1.570796	1.570796	1.570796	2
25	PYRARGYRITE	Ag_3SbS_3	5.82	5.82	5.82	5.855	5.19	5.82	541.5206	hexagonal	R3c	11.047	11.047	8.719	1.570796	1.570796	2.094395	6
26	HALITE	$NaCl$	2.168	2.168	2.168	2.165	2.07	2.16	58.44	cubic	Fm3m	5.6404	5.6404	5.6404	1.570796	1.570796	1.570796	4
27	SYLVITE	KCl	1.993	1.993	1.993	1.987	1.92	1.99	74.55	cubic	Fm3m	6.2931	6.2931	6.2931	1.570796	1.570796	1.570796	4
28	NATRON	$Na_2CO_3 \cdot 10H_2O$	1.478	1.478	1.478	1.448	1.55	1.46	286.14	monoclinic	Cc	12.83	9.026	13.44	1.570796	2.146755	1.570796	4
29	NITRATITE	$NaNO_3$	2.24	2.29	2.265	2.25	2.23	2.26	84.99	hexagonal	R3c	5.07	5.07	16.829	1.570796	1.570796	2.094395	6
30	SPODUMENE	$LiAlSi2O6$	3.03	3.23	3.13	3.184	3.11	3.15	186.09	monoclinic	C2/c	9.45	8.39	5.215	1.570796	1.919862	1.570796	4
31	GYPSUM	$CaSO_4 \cdot 2H_2O$	2.317	2.317	2.317	2.31	2.36	2.31	172.17	monoclinic	I2/a	5.679	15.202	6.522	1.570796	2.066993	1.570796	4
32	ANHYDRITE	$CaSO_4$	2.98	2.98	2.98	2.95	2.97	2.96	136.14	orthorhombic	Amma	6.993	6.995	6.245	1.570796	1.570796	1.570796	4
33	BIOTITE	$K(Mg,Fe^{2+})_3(Al,Fe^{3+})Si_3O_{10}(OH,F)_2$	2.7	3.3	3	3.25	3.07	2.89	433.53	monoclinic	C2/m	5.3	9.2	10.2	1.570796	1.745329	1.570796	2
34	PENNANTITE	$Mn_5^{2+}Al(Si_3Al)O_{10}(OH)_8$	2.89	3.07	2.98	3.18	2.98	3.2	708.06	monoclinic	C1	5.45	9.5	14.4	1.570796	1.698205	1.570796	4
35	MUSCOVITE	$KAl_2(AlSi_3O_{10})(F,OH)_2$	2.77	2.88	2.825	2.83	2.81	2.83	398.71	monoclinic	C2/c	5.199	9.027	20.106	1.570796	1.666789	1.570796	4
36	CINNABAR	HgS	8.176	8.176	8.176	8.2	6.76	8.19	232.66	hexagonal	P3$_1$21	4.145	4.145	9.496	1.570796	1.570796	2.094395	3
37	GOETHITE	$\alpha\text{-}Fe^{3+}O(OH)$	4.28	4.28	4.28	4.18	4.13	4.27	88.85	orthorhombic	Pbnm	4.608	9.956	3.0215	1.570796	1.570796	1.570796	4
38	CHLORARGYRITE	$AgCl$	5.556	5.556	5.556	5.57	4.96	5.55	143.32	cubic	Fm3m	5.554	5.554	5.554	1.570796	1.570796	1.570796	4
39	MAGNETITE	$Fe^{2+}Fe^{3+}_2O_4$	5.175	5.175	5.175	5.2	4.89	5.21	231.54	cubic	Fd3m	8.397	8.397	8.397	1.570796	1.570796	1.570796	8
40	HEAMATITE	$\alpha\text{-}Fe_2O_3$	5.26	5.26	5.26	5.255	5.04	5.28	159.69	hexagonal	R3c	5.038	5.038	13.772	1.570796	1.570796	2.094395	6
41	OLIVINE	$(Mg,Fe)_2SiO_4$	3.27	3.37	3.32	3.3	3.25	3.3	153.31	orthorhombic	Pbnm	4.78	10.25	6.3	1.570796	1.570796	1.570796	4
42	MAGNESIOHORNBLENDE	$Ca_2[Mg_4(Al,Fe^{4+})]Si_7AlO_{22}(OH)_2$	3	3.47	3.235	2.96	3.22	2.96	821.16	monoclinic	C2/m	9.887	18.174	5.308	1.570796	1.832596	1.570796	2
43	VERMICULITE	$(MgFe,Al)_3(Al,Si)_4O_{10}(OH)_2 \cdot 4H_2O$	2.2	2.6	2.4	2.26	2.51	2.32	504.19	monoclinic	C2/c	5.349	9.255	28.89	1.570796	1.695006	1.570796	4
44	MAGNESITE	$MgCO_3$	3	3	3	3.01	2.97	2.98	84.31	hexagonal	R3c	4.6632	4.6632	15.015	1.570796	1.570796	2.094395	6
45	CHALCOCITE	Cu_2S	5.5	5.8	5.65	5.8	6.01	6.46	159.16	monoclinic	P2$_1$/c	15.246	11.884	13.494	1.570796	2.030691	1.570796	48
46	COVELLITE	CuS	4.6	4.76	4.68	4.602	4.41	4.68	95.61	hexagonal	P63/mmc	3.7938	3.7938	16.341	1.570796	1.570796	2.094395	6
47	NATIVE PALLADIUM	Pd	11.9	11.9	11.9	12.04	10.4	12.19	107.44	cubic	Fm3m	3.8898	3.8898	3.8898	1.570796	1.570796	1.570796	4
48	SILLIMANITE	Al_2SiO_5	3.23	3.24	3.235	3.24	3.2	3.25	162.05	orthorhombic	Pbnm	7.4883	7.6808	5.7774	1.570796	1.570796	1.570796	4
49	TITANITE	$CaTiSiO_5$	3.48	3.6	3.54	3.53	3.47	3.55	197.76	monoclinic	P2$_1$/a	7.057	8.707	6.555	1.570796	1.986359	1.570796	4
50	ZIRCON	$ZrSiO_4$	4.6	4.7	4.65	4.714	4.52	4.85	190.31	tetragonal	I4$_1$/amd	6.607	6.607	5.982	1.570796	1.570796	1.570796	4
51	ALLARGENTUM	Ag_6Sb	10	10	10	10.12	8.72	10.01	108.01	hexagonal	nd	2.952	2.952	4.773	1.570796	1.570796	2.094395	2
52	FLUORITE	CaF_2	3.175	3.184	3.1795	3.18	3.1	3.18	78.07	cubic	Fm3m	5.4626	5.4626	5.4626	1.570796	1.570796	1.570796	4

* Rotoinversion Axes Neglected

Table B1
Selected Mineral Data

APPENDIX C
GIVEN AND COMPUTED DATA
FOR THE
ROCK-FORMING MINERALS

Serial	Mineral	Chemical Formula	Formula Weight	Space Group	a	b	c	α	β	γ	Z	V	V by Equation Set	Lattice System	Theoretical ρ	Measured Density	Harmonic Density	Fractional Defect for Measured Density	Fractional Defect for Harmonic Density	Notional Harmonic ρ Multiplier	Precise Harmonic ρ Multiplier
1	ARGENTOPYRITE	$AgFe_2S_3$	315.7302	Pmmn	6.639	11.463	6.452	1.570796	1.570796	1.570796	4	491.0156	491.0156	orthorhombic	4.2710031	4.25	4.447701	0.0049176	-0.04137[illegible]464	1	0.9555499
2	ARGENTOTENNANTITE	$(Ag,Cu)_{10}(Zn,Fe)_2(As,Sb)_4S_{13}$	1788.3992	I43m	10.584	10.584	10.584	1.570796	1.570796	1.570796	2	1185.631	1185.631	cubic	5.00949629	4.71	4.872075	0.05978571	0.02743[illegible]224	1	0.96673395
3	BARITE	$BaSO_4$	233.383	Pnma	8.884	5.457	7.157	1.570796	1.570796	1.570796	4	346.9713	346.9713	orthorhombic	4.46770798	4.5	2.15272	-0.00722787	0.5181[illegible]999	2	2.09037824
4	CASSITERITE	SnO_2	150.708	$P4_2/mnm$	4.7382	4.7382	3.1871	1.570796	1.570796	1.570796	2	71.55211	71.55211	tetragonal	6.99508386	6.995	3.490249	1.1989E-05	0.50104[illegible]614	2	2.00415514
5	CERUSSITE	$PbCO_3$	267.208	Pmcn	5.179	8.492	6.141	1.570796	1.570796	1.570796	4	270.0816	270.0816	orthorhombic	6.57148536	6.552	4.171114	0.00296514	0.36527[illegible]732	3	1.57080335
6	CHALCOPYRITE	$CuFeS_2$	183.511	I42d	5.281	5.281	10.401	1.570796	1.570796	1.570796	4	290.0731	290.0731	tetragonal	4.20207454	4.2	4.0595	0.0004937	0.03392[illegible]538	1	1.03461015
7	CHILDRENITE	$Fe^{2+}Al(PO_4)(OH)_2.H_2O$	228.8173	Bba2	10.395	13.394	6.918	1.570796	1.570796	1.570796	8	963.1975	963.1975	orthorhombic	3.15582265	3.15	1.407354	0.00184505	0.554045[illegible]94	2	2.23824239
8	CUPRITE	$Cu_2^{1+}O$	143.091	Pn3m	4.2685	4.2685	4.2685	1.570796	1.570796	1.570796	2	77.77246	77.77246	cubic	6.11034241	6.14	5.179031	-0.00485367	0.152415[illegible]61	6	1.18554993
9	EDGARITE	$FeNb_3S_6$	526.92411	$P6_322$	5.771	5.771	12.19	1.570796	1.570796	2.094395	2	351.59	351.59	hexagonal	4.97726393	4.98	3.975997	-0.00054971	0.20116[illegible]21	5	1.25251614
10	GALENA	PbS	239.26	Fm3m	5.936	5.936	5.936	1.570796	1.570796	1.570796	4	209.1615	209.1615	cubic	7.59796876	7.58	7.083144	0.00236494	0.06775[illegible]19	1	1.07014623
11	SMITHSONITE	$ZnCO_3$	125.388	R3c	4.6526	4.6526	15.9257	1.570796	1.570796	2.094395	6	298.5524	298.5524	hexagonal	4.18442436	4.431	2.281536	-0.05892702	0.454755[illegible]11	2	1.94211304
12	SPHALERITE	$(Zn,Fe)S$	92.6725	F43m	5.406	5.406	5.406	1.570796	1.570796	1.570796	4	157.9895	157.9895	cubic	3.89611562	4	3.920348	-0.02666358	-0.006219[illegible]03	1	1.02031771
13	ZUSSMANITE	$K(Fe^{2+},Mg,Mn^{2+})_{13}(AlSi)_{18}O_{42}(OH)_{14}$	2205.7535	R3	11.66	11.66	28.69	1.570796	1.570796	2.094395	3	3377.989	3377.989	hexagonal	3.25288751	3.146	1.821431	0.03285927	0.440057[illegible]11	2	1.72721317
14	TENNANTITE	$(Cu,Ag,Sn,Fe)_{12}As_4S_{12}$	1722.29	I43m	10.19	10.19	10.19	1.570796	1.570796	1.570796	2	1058.09	1058.09	cubic	5.40583535	4.72	4.781066	0.12686945	0.115573[illegible]95	9	0.9872276
15	WOLLASTONITE 2M	$CaSiO_3$	116.16	$P2_1/a$	15.409	7.322	7.063	1.570796	1.663299	1.570796	12	793.4739	793.4739	monoclinic	2.91711986	2.975	1.484547	-0.01984154	0.491091[illegible]24	2	2.00397868
17	WOLLASTONITE 1A	$CaSiO_3$	116.16	$P2_1/a$	7.74	7.32	7.063	1.567306	1.47969	1.339366	6	375.8138	375.7992	triclinic	3.0795288	2.975	1.484547	0.03394312	0.517930[illegible]58	2	2.00397868
18	PYRITE	FeS_2	119.965	Pa3	5.4179	5.4179	5.4179	1.570796	1.570796	1.570796	4	159.0351	159.0351	cubic	5.01038024	5.018	3.147603	-0.0015208	0.371783[illegible]57	3	1.59422913
19	CALCITE	$CaCO_3$	100.086	R3c	4.9896	4.9896	17.061	1.570796	1.570796	2.094395	6	367.8465	367.8465	hexagonal	2.71086005	2.7102	1.396875	0.00024348	0.484711[illegible]4	2	1.94018793
20	STRONTIANITE	$SrCO_3$	147.628	Pmcn	5.1059	8.4207	6.0319	1.570796	1.570796	1.570796	4	259.3431	259.3431	orthorhombic	3.7809696	3.76	1.86614	0.00554609	0.506438[illegible]59	2	2.01485418
21	QUARTZ-a	SiO_2	60.083	$P3_121$	4.9135	4.9135	5.405	1.570796	1.570796	2.094395	3	113.0078	113.0078	trigonal	2.64858371	2.65	1.542774	-0.00053473	0.417509[illegible]53	2	1.71768557
22	QUARTZ-b	SiO_2	60.083	$P3_121$	4.9965	4.9965	5.4546	1.570796	1.570796	2.094395	6	117.9303	117.9303	hexagonal	5.0760586	2.65	1.542774	0.47794141	0.696068[illegible]4	2	1.71768557
23	CORUNDUM	Al_2O_3	101.96008	R3c	4.754	4.754	12.982	1.570796	1.570796	2.094395	6	254.0918	254.0918	hexagonal	3.99797344	4.04	1.689906	-0.01051197	0.577309[illegible]8	2	2.39066518
24	ACANTHITE	Ag_2S	247.7804	$P2_1/n$	4.229	6.931	7.862	1.570796	1.738522	1.570796	4	227.2108	227.2108	monoclinic	7.24347667	7.21	6.872624	0.00462163	0.051198[illegible]5	1	1.04908983
25	ARGENTITE	Ag_2S	247.7804	Im3m	4.88	4.88	4.88	1.570796	1.570796	1.570796	2	116.2143	116.2143	cubic	7.08086917	7.3	6.872624	-0.03094688	0.0294095[illegible]9	1	1.06218526
26	PYRARGYRITE	Ag_3SbS_3	541.5206	R3c	11.047	11.047	8.719	1.570796	1.570796	2.094395	6	921.4802	921.4802	hexagonal	5.85503244	5.82	5.666849	0.0059833	0.0321404[illegible]7	1	1.02702575
27	HALITE	$NaCl$	58.44277	Fm3m	5.6404	5.6404	5.6404	1.570796	1.570796	1.570796	4	179.4443	179.4443	cubic	2.16326718	2.16	1.418022	0.0015103	0.3444999[illegible]2	3	1.52324874
28	SYLVITE	KCl	74.5513	Fm3m	6.2931	6.2931	6.2931	1.570796	1.570796	1.570796	4	249.2263	249.2263	cubic	1.98687435	1.99	1.180728	-0.00157315	0.4057359[illegible]5	3	1.68540094
29	NATRON	$Na_2CO_3·10H_20$	276.05754	Cc	12.83	9.026	13.44	1.570796	2.146755	1.570796	4	1305.307	1305.307	monoclinic	1.40474029	1.46	0.806517	-0.03933803	0.4258602[illegible]5	3	1.8102528
30	NITRATITE	$NaNO_3$	84.99377	R3c	5.07	5.07	16.829	1.570796	1.570796	2.094395	6	374.632	374.632	hexagonal	2.26038582	2.26	1.094119	0.00017069	0.5159593[illegible]6	2	2.06558946
31	SPODUMENE	LiAlSi2O6	186.14254	C2/c	9.45	8.39	5.215	1.570796	1.919862	1.570796	4	388.5384	388.5384	monoclinic	3.18215123	3.15	1.455578	0.01010362	0.5425805[illegible]	2	2.164089
32	GYPSUM	$CaSO_4·2H_2O$	172.164	I2/a	5.679	15.202	6.522	1.570796	2.066993	1.570796	4	495.1532	495.1532	monoclinic	2.30946754	2.31	1.025686	-0.00023056	0.5558778[illegible]	2	2.25215184
33	ANHYDRITE	$CaSO_4$	136.134	Amma	6.993	6.995	6.245	1.570796	1.570796	1.570796	4	305.4806	305.4806	orthorhombic	2.96000195	2.96	1.430801	6.5994E-07	0.5166217[illegible]	2	2.06877171
34	BIOTITE	$K(Mg,Fe^{2+})_3(Al,Fe^{3+})Si_3O_{10}(OH,F)_2$	566.61734	C2/m	5.3	9.2	10.2	1.570796	1.745329	1.570796	2	489.7961	489.7961	monoclinic	3.84196694	2.89	2.016051	0.24778114	0.4752554[illegible]	3	1.43349527
35	PENNANTITE	$Mn_5^{2+}Al(Si_3Al)O_{10}(OH)_8$	708.9543	C1	5.45	9.5	14.4	1.570796	1.698205	1.570796	2	739.5168	739.5168	monoclinic	3.18382559	3.2	1.658624	-0.00508018	0.4790467[illegible]	2	1.92930974
36	MUSCOVITE	$KAl_2(AlSi_3O_{10})(F,OH)_2$	436.29872	C2/c	5.199	9.027	20.106	1.570796	1.666789	1.570796	4	939.258	939.258	monoclinic	3.08537607	2.83	1.370455	0.08276984	0.5558223[illegible]	2	2.06500734
37	CINNABAR	HgS	232.652	$P3_121$	4.145	4.145	9.496	1.570796	1.570796	2.094395	3	141.2929	141.2929	hexagonal	8.20269811	8.19	7.670249	0.00154804	0.0649114[illegible]	1	1.06776195
38	GOETHITE	$\alpha\text{-}Fe^{3+}O(OH)$	88.851	Pbnm	4.608	9.956	3.0215	1.570796	1.570796	1.570796	4	138.6181	138.6181	orthorhombic	4.25746852	4.27	1.955522	-0.00294341	0.54068433[illegible]	2	2.1835602
39	CHLORARGYRITE	$AgCl$	143.3132	Fm3m	5.554	5.554	5.554	1.570796	1.570796	1.570796	4	171.3238	171.3238	cubic	5.55619717	5.55	5.167055	0.00111536	0.07003749[illegible]	1	1.07411281
40	MAGNETITE	$Fe^{2+}Fe^{3+}O_4$	175.686	Fd3m	8.397	8.397	8.397	1.570796	1.570796	1.570796	8	592.0692	592.0692	cubic	3.94188339	5.21	2.585031	-0.32170323	0.34421412[illegible]	2	2.01544937
41	HEAMATITE	$\alpha\text{-}Fe_2O_3$	159.687	R3c	5.038	5.038	13.772	1.570796	1.570796	2.094395	6	302.722	302.722	hexagonal	5.25564389	5.28	2.929055	-0.00463428	0.4426839[illegible]	2	1.80262936
42	OLIVINE	$(Mg,Fe)_2SiO_4$	144.146	Pbnm	4.78	10.25	6.3	1.570796	1.570796	1.570796	4	308.6685	308.6685	orthorhombic	3.10183981	3.3	1.925671	-0.06388473	0.37918424[illegible]	2	1.71368834
43	MAGNESIOHORNBLENDE	$Ca_2[Mg_4(Al,Fe^{++})]Si_7AlO_{22}(OH)_2$	869.77108	C2/m	9.887	18.174	5.308	1.570796	1.832596	1.570796	2	921.276	921.276	monoclinic	3.1354097	2.96	1.5402	0.05594475	0.50877217[illegible]	2	1.92182774
44	VERMICULITE	$(MgFe,Al)_3(Al,Si)_4O_{10}(OH)_2·4H_2O$	803.69277	C2/c	5.349	9.255	28.89	1.570796	1.695006	1.570796	4	1419.181	1419.181	monoclinic	3.76150259	2.32	1.727268	0.3832252	0.54080375[illegible]	4	1.34316167
45	MAGNESITE	$MgCO_3$	84.313	R3c	4.6632	4.6632	15.015	1.570796	1.570796	2.094395	6	282.764	282.764	hexagonal	2.97078232	2.98	1.410444	-0.00310278	0.52522803[illegible]	2	2.11280964
46	CHALCOCITE	Cu_2S	159.152	$P2_1/c$	15.246	11.884	13.494	1.570796	2.030691	1.570796	48	2190.864	2190.864	monoclinic	5.79011194	6.46	5.359596	-0.11569518	0.07435368[illegible]	1	1.20531477
47	COVELLITE	CuS	95.606	P63/mmc	3.7938	3.7938	16.341	1.570796	1.570796	2.094395	6	203.6846	203.6846	hexagonal	4.67656918	4.68	4.229869	-0.00073362	0.095518692	1	1.10641714
48	NATIVE PALLADIUM	Pd	106.42	Fm3m	3.8898	3.8898	3.8898	1.570796	1.570796	1.570796	4	58.85479	58.85479	cubic	12.0102078	12.19	12	-0.01496995	0.000849925	1	1.01583333
49	SILLIMANITE	Al_2SiO_5	162.04308	Pbnm	7.4883	7.6808	5.7774	1.570796	1.570796	1.570796	4	332.2937	332.2937	orthorhombic	3.23904838	3.25	1.63219	-0.00338112	0.496089673	2	1.99118984
50	TITANITE	$CaTiSiO_5$	196.025	$P2_1/a$	7.057	8.707	6.555	1.570796	1.986359	1.570796	4	368.4935	368.4935	monoclinic	3.53338267	3.55	1.693402	-0.00470295	0.520742088	2	2.09637217
51	ZIRCON	$ZrSiO_4$	183.305	$I4_1/amd$	6.607	6.607	5.982	1.570796	1.570796	1.570796	4	261.1289	261.1289	tetragonal	4.66260227	4.85	2.296328	-0.04019166	0.507500845	2	2.11206791
52	ALLARGENTUM	Ag_6Sb	768.9212	nd	2.952	2.952	4.773	1.570796	1.570796	2.094395	2	36.02092	36.02092	hexagonal	70.8934561	10.01	9.630482	0.8588022	0.864155558	1	1.039408
53	FLUORITE	CaF_2	78.074806	Fm3m	5.4626	5.4626	5.4626	1.570796	1.570796	1.570796	4	163.004	163.004	cubic	3.18142579	3.18	1.622191	0.00044816	0.490105623	2	1.96031156

Table C1
Given and Computed Data
for the
Rock-Forming Mineral Species

APPENDIX D
TABULATIONS AND PLOTS
OF THE
APPARENT LATTICE STRAIN

Serial	Mineral	Chemical Formula	Precise Harmonic ρ Multiplier	Z	Lattice System
1	ARGENTOPYRITE	$AgFe_2S_3$	0.955549898	4	orthorhombic
2	ARGENTOTENNANTITE	$(Ag,Cu)_{10}(Zn,Fe)_2(As,Sb)_4S_{13}$	0.966733952	2	cubic
3	BARITE	$BaSO_4$	2.090378237	4	orthorhombic
4	CASSITERITE	SnO_2	2.00415514	2	tetragonal
5	CERUSSITE	$PbCO_3$	1.570803354	4	orthorhombic
6	CHALCOPYRITE	$CuFeS_2$	1.034610149	4	tetragonal
7	CHILDRENITE	$Fe_2+Al(PO_4)(OH)_2.H_2O$	2.238242387	8	orthorhombic
8	CUPRITE	Cu_21+O	1.185549928	2	cubic
9	EDGARITE	$FeNb_3S_6$	1.252516143	2	hexagonal
10	GALENA	PbS	1.07014623	4	cubic
11	SMITHSONITE	$ZnCO_3$	1.942113045	6	hexagonal
12	SPHALERITE	$(Zn,Fe)S$	1.020317707	4	cubic
13	ZUSSMANITE	$K(Fe2+,Mg,Mn2+)_{13}(AlSi)_{18}O_{42}(OH)_{14}$	1.727213166	3	hexagonal
14	TENNANTITE	$(Cu,Ag,Sn,Fe)_{12}As_4S_{12}$	0.987227599	2	cubic
15	WOLLASTONITE 2M	$CaSiO_3$	2.00397868	12	monoclinic
17	WOLLASTONITE 1A	$CaSiO_3$	2.00397868	6	triclinic
18	PYRITE	FeS_2	1.594229132	4	cubic
19	CALCITE	$CaCO_3$	1.940187925	6	hexagonal
20	STRONTIANITE	$SrCO_3$	2.014854184	4	orthorhombic
21	QUARTZ-a	SiO_2	1.717685575	3	trigonal
22	QUARTZ-b	SiO_2	1.717685575	6	hexagonal
23	CORUNDUM	Al_2O_3	2.39066518	6	hexagonal
24	ACANTHITE	Ag_2S	1.049089827	4	monoclinic
25	ARGENTITE	Ag_2S	1.062185262	2	cubic
26	PYRARGYRITE	Ag_3SbS_3	1.027025752	6	hexagonal
27	HALITE	$NaCl$	1.523248743	4	cubic
28	SYLVITE	KCl	1.68540094	4	cubic
29	NATRON	$Na_2CO_3 \cdot 10H_2O$	1.810252797	4	monoclinic
30	NITRATITE	$NaNO_3$	2.065589457	6	hexagonal
31	SPODUMENE	$LiAlSi_2O_6$	2.164089004	4	monoclinic
32	GYPSUM	$CaSO_4 \cdot 2H_2O$	2.252151844	4	monoclinic
33	ANHYDRITE	$CaSO_4$	2.06877171	4	orthorhombic
34	BIOTITE	$K(Mg,Fe2+)_3(Al,Fe3+)Si_3O_{10}(OH,F)_2$	1.433495267	2	monoclinic
35	PENNANTITE	$Mn_52+Al(Si_3Al)O_{10}(OH)_8$	1.929309743	2	monoclinic
36	MUSCOVITE	$KAl_2(AlSi_3O_{10})(F,OH)_2$	2.065007342	4	monoclinic
37	CINNABAR	HgS	1.067761953	3	hexagonal
38	GOETHITE	$a-Fe3+O(OH)$	2.183560204	4	orthorhombic
39	CHLORARGYRITE	$AgCl$	1.074112807	4	cubic
40	MAGNETITE	$Fe2+Fe3+O_4$	2.015449371	8	cubic
41	HEAMATITE	$a-Fe_2O_3$	1.802629364	6	hexagonal
42	OLIVINE	$(Mg,Fe)_2SiO_4$	1.713688341	4	orthorhombic
43	MAGNESIOHORNBLENDE	$Ca_2[Mg_4(Al,Fe++)]Si_7AlO_{22}(OH)_2$	1.921827741	2	monoclinic
44	VERMICULITE	$(MgFe,Al)_3(Al,Si)_4O_{10}(OH)_2 \cdot 4H_2O$	1.343161672	4	monoclinic
45	MAGNESITE	$MgCO_3$	2.112809637	6	hexagonal
46	CHALCOCITE	Cu_2S	1.205314771	48	monoclinic
47	COVELLITE	CuS	1.106417136	6	hexagonal
48	NATIVE PALLADIUM	Pd	1.015833333	4	cubic
49	SILLIMANITE	Al_2SiO_5	1.99118984	4	orthorhombic
50	TITANITE	$CaTiSiO_5$	2.096372172	4	monoclinic
51	ZIRCON	$ZrSiO_4$	2.112067909	4	tetragonal
52	ALLARGENTUM	Ag_6Sb	1.039408002	2	hexagonal
53	FLUORITE	CaF_2	1.960311559	4	cubic

Table D1
Lattice Strain Data for Each Mineral
in
Arbitrary Order of Record

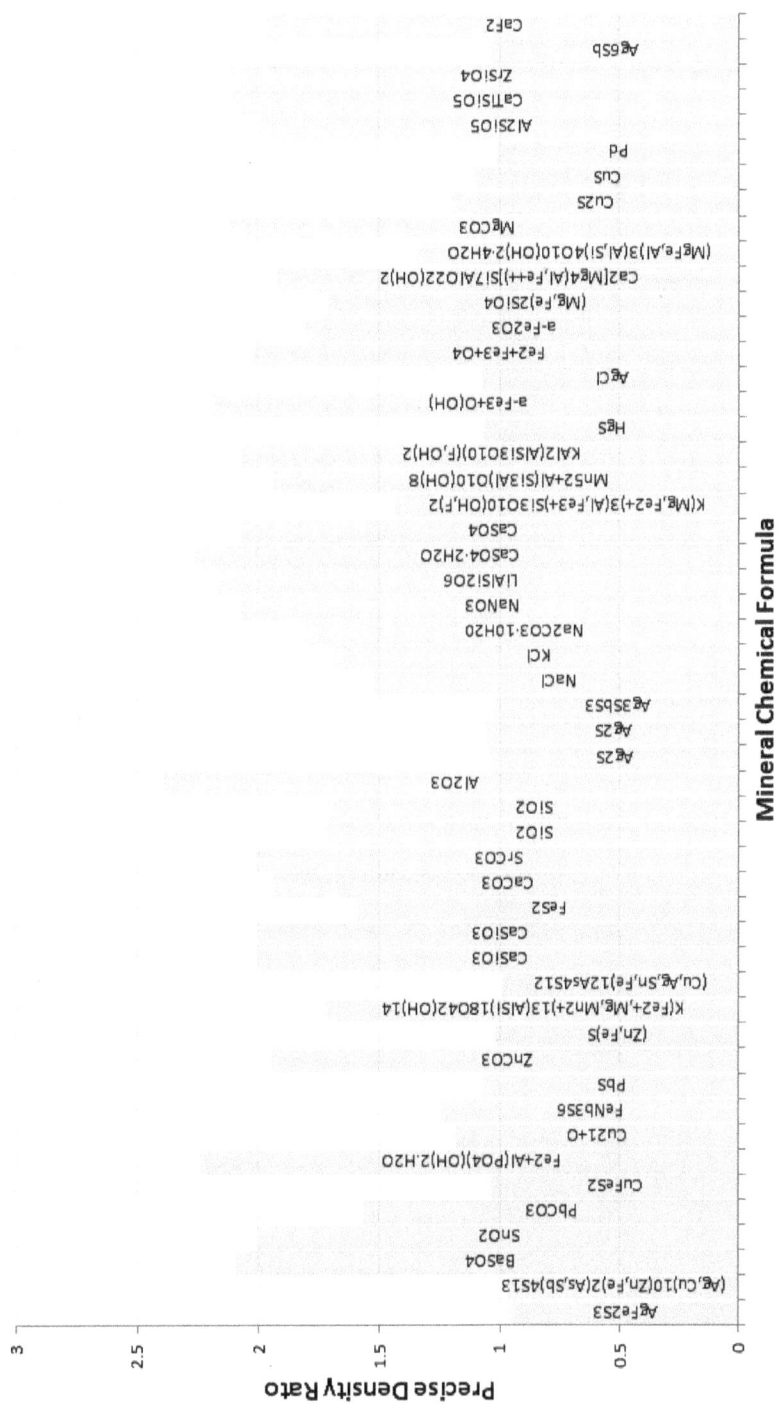

Figure D1
Lattice Strain Data for Each Mineral
in
Arbitrary Order of Record

Serial	Mineral	Chemical Formula	Precise Harmonic ρ Multiplier	Z	Lattice System
1	ARGENTOPYRITE	$AgFe_2S_3$	0.955549898	4	orthorhombic
2	ARGENTOTENNANTITE	$(Ag,Cu)_{10}(Zn,Fe)_2(As,Sb)_4S_{13}$	0.966733952	2	cubic
14	TENNANTITE	$(Cu,Ag,Sn,Fe)_{12}As_4S_{12}$	0.987227599	2	cubic
48	NATIVE PALLADIUM	Pd	1.015833333	4	cubic
12	SPHALERITE	$(Zn,Fe)S$	1.020317707	4	cubic
26	PYRARGYRITE	Ag_3SbS_3	1.027025752	6	hexagonal
6	CHALCOPYRITE	$CuFeS_2$	1.034610149	4	tetragonal
52	ALLARGENTUM	Ag_6Sb	1.039408002	2	hexagonal
24	ACANTHITE	Ag_2S	1.049089827	4	monoclinic
25	ARGENTITE	Ag_2S	1.062185262	2	cubic
37	CINNABAR	HgS	1.067761953	3	hexagonal
10	GALENA	PbS	1.07014623	4	cubic
39	CHLORARGYRITE	$AgCl$	1.074112807	4	cubic
47	COVELLITE	CuS	1.106417136	6	hexagonal
8	CUPRITE	$Cu_{21}+O$	1.185549928	2	cubic
46	CHALCOCITE	Cu_2S	1.205314771	48	monoclinic
9	EDGARITE	$FeNb_3S_6$	1.252516143	2	hexagonal
44	VERMICULITE	$(MgFe,Al)_3(Al,Si)_4O_{10}(OH)_2 \cdot 4H_2O$	1.343161672	4	monoclinic
34	BIOTITE	$K(Mg,Fe2+)_3(Al,Fe3+)Si_3O_{10}(OH,F)_2$	1.433495267	2	monoclinic
27	HALITE	$NaCl$	1.523248743	4	cubic
5	CERUSSITE	$PbCO_3$	1.570803354	4	orthorhombic
18	PYRITE	FeS_2	1.594229132	4	cubic
28	SYLVITE	KCl	1.68540094	4	cubic
42	OLIVINE	$(Mg,Fe)_2SiO_4$	1.713688341	4	orthorhombic
21	QUARTZ-a	SiO_2	1.717685575	3	trigonal
22	QUARTZ-b	SiO_2	1.717685575	6	hexagonal
13	ZUSSMANITE	$K(Fe2+,Mg,Mn2+)_{13}(AlSi)_{18}O_{42}(OH)_{14}$	1.727213166	3	hexagonal
41	HEAMATITE	$a\text{-}Fe_2O_3$	1.802629364	6	hexagonal
29	NATRON	$Na_2CO_3 \cdot 10H_2O$	1.810252797	4	monoclinic
43	MAGNESIOHORNBLENDE	$Ca_2[Mg_4(Al,Fe++)]Si_7AlO_{22}(OH)_2$	1.921827741	2	monoclinic
35	PENNANTITE	$Mn_52+Al(Si_3Al)O_{10}(OH)_8$	1.929309743	2	monoclinic
19	CALCITE	$CaCO_3$	1.940187925	6	hexagonal
11	SMITHSONITE	$ZnCO_3$	1.942113045	6	hexagonal
53	FLUORITE	CaF_2	1.960311559	4	cubic
49	SILLIMANITE	Al_2SiO_5	1.99118984	4	orthorhombic
15	WOLLASTONITE 2M	$CaSiO_3$	2.00397868	12	monoclinic
17	WOLLASTONITE 1A	$CaSiO_3$	2.00397868	6	triclinic
4	CASSITERITE	SnO_2	2.00415514	2	tetragonal
20	STRONTIANITE	$SrCO_3$	2.014854184	4	orthorhombic
40	MAGNETITE	$Fe2+Fe3+O_4$	2.015449371	8	cubic
36	MUSCOVITE	$KAl_2(AlSi_3O_{10})(F,OH)_2$	2.065007342	4	monoclinic
30	NITRATITE	$NaNO_3$	2.065589457	6	hexagonal
33	ANHYDRITE	$CaSO_4$	2.06877171	4	orthorhombic
3	BARITE	$BaSO_4$	2.090378237	4	orthorhombic
50	TITANITE	$CaTiSiO_5$	2.096372172	4	monoclinic
51	ZIRCON	$ZrSiO_4$	2.112067909	4	tetragonal
45	MAGNESITE	$MgCO_3$	2.112809637	6	hexagonal
31	SPODUMENE	$LiAlSi_2O_6$	2.164089004	4	monoclinic
38	GOETHITE	$a\text{-}Fe3+O(OH)$	2.183560204	4	orthorhombic
7	CHILDRENITE	$Fe2+Al(PO_4)(OH)_2.H_2O$	2.238242387	8	orthorhombic
32	GYPSUM	$CaSO_4 \cdot 2H_2O$	2.252151844	4	monoclinic
23	CORUNDUM	Al_2O_3	2.39066518	6	hexagonal

Table D2
Lattice Strain Data for Each Mineral
in
Ascending Order of Lattice Strain

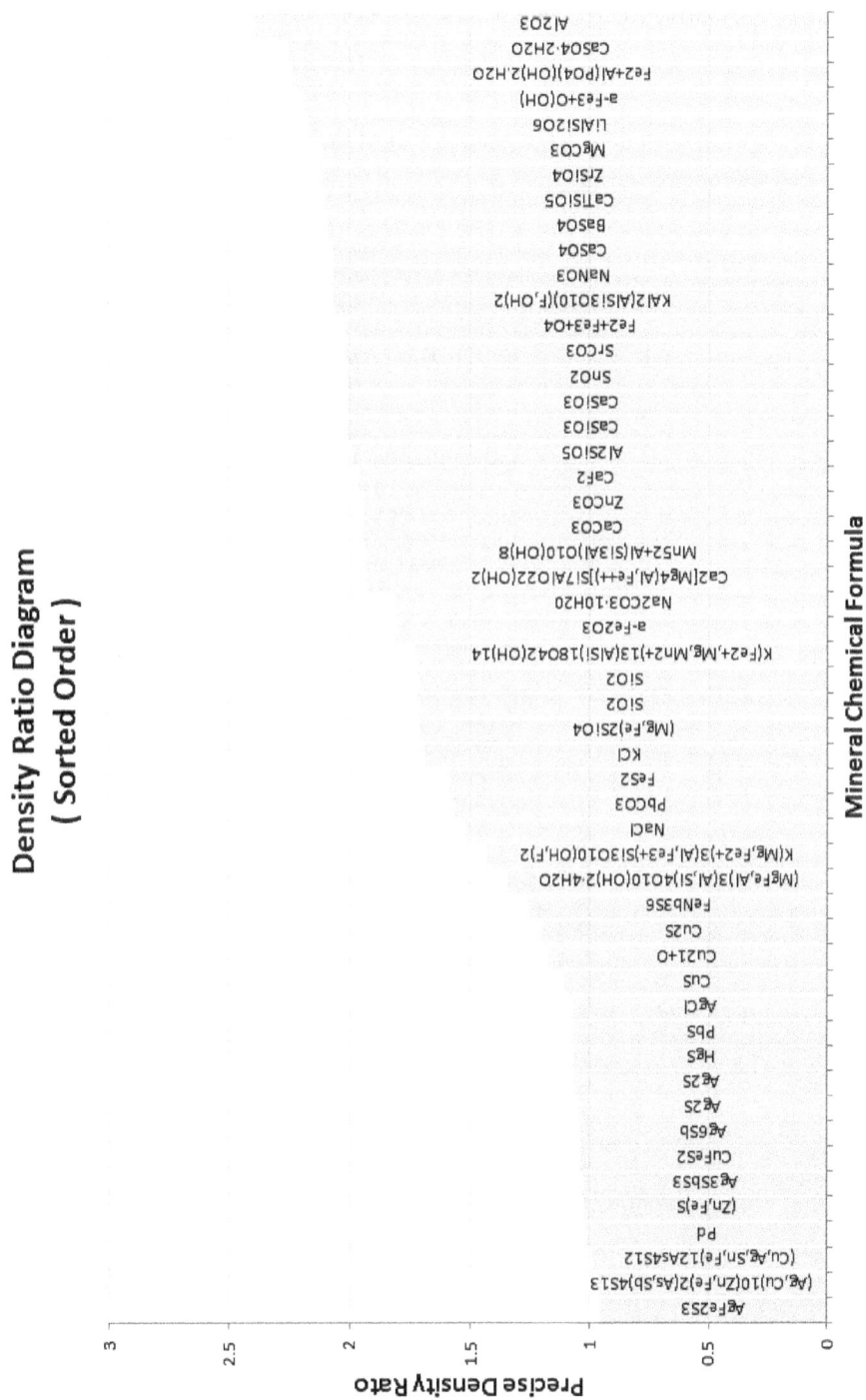

Figure D2
Lattice Strain Data for Each Mineral
in
Ascending Order of Lattice Strain

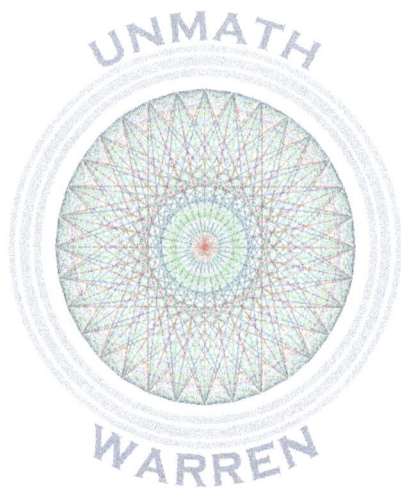

Crystal Chemistry
and the
Harmonic Mean

AN EXTENDED TREATMENT

by
James R Warren BSc MSc PhD PGCE

PART ONE
PRINCIPLES

In the Nineteen-Eighties I did some work on the mass density of mixtures as a corollary of combinatorial mathematics.

More recently I returned to the topic to explore the relationship between the mass density of a mixture and the harmonic mean of the ratios of density to fraction for the individual components. I summarised the arithmetic in my 16 June 2017 paper "Mixture Density and the Harmonic Mean".

I wondered to what degree (if any) these elementary principles would hold good for ordered structures of matter.

It seemed to me that the very coalescence of atoms, involving the interplay of force and the condensation of matter, would of necessity engage mass densities variant to the simple laws of classical arithmetic.

In days gone by the late Mr Don Braggins and his friends strove to instil in crystal metal the most perfect purity of form and substance, whilst for centuries men of diverse culture and breeding have thought *any* crystal the epitome of perfection.

But whilst man and crystal were conceived pure and simple in the Mind of God, both Fell and fractured, and There resigned their recondite perfections.

Defective Structures

The form of crystals is disrupted by internal dislocations: Tears and omissions in the structure of translational or helical geometry at the interatomic scale.

The substance of crystals is corrupted not only by simple inclusions of extraneous stuff but also by foreign atoms of substitution held in more or less tension within the very atomic order itself.

Therefore, one might be forgiven for guessing that any found object, even the most refulgent diamond, is some blend of order and chaos, of curt arithmetic and temporising stochastics.

For near-perfect corundum is lapidary but colourless, like many near-perfect things, whilst polluted rubies and sapphires are the envy of women and a delight to the eye.

A Choice of Crystalline Minerals

Further to investigate I intentionally noted minerals at random. I consulted the memories of boyhood and especially my year in the mineralogical paradise of Cornwall.

Old friends like cassiterite, cuprite and pyrite sprang readily to recollection, along with galena, cerussite and fluorite. It would have been remiss to forget olivine and hornblende, especially as the former is elastic of cation substitution and the latter so ambiguous as to challenge definition.

I realised that fair tests would include much more complex formulas and sentimentally included in my study the elaborate but exceedingly rare phyllosilicate zussmanite as well as involved hydrated ferrosilicates incorporating water of crystallisation in various proportions.

Table A1 in Appendix A presents some fundamental data for the sixty-one rock-forming elements that account for well over 99% of the Earth's Crust and contribute to the composition of our one hundred and twenty-five studied minerals[1-4].

Table B1 in Appendix B presents the one hundred and twenty-five crystalline minerals as mineral forms together with their nominal chemical formulae and physical properties. This source data is drawn from a variety of Internet sources listed in the References[5-10].

Density and Harmonic Mean Algebra

The relationships between the Mixture Mass Density, ρ_{mix}, and the Harmonic Mean, H, are defined by these equations:-

$$\rho_{mix} = \frac{1}{\dfrac{f_1}{\rho_1} + \dfrac{f_2}{\rho_2} + \dfrac{f_3}{\rho_3} + ... + \dfrac{f_n}{\rho_n}} = \frac{H}{n} = \frac{1}{n} \cdot \frac{1}{\displaystyle\sum_{i=1}^{n} \dfrac{f_i}{\rho_i}}$$

Equation 1

where:-

$$H = \frac{n}{\dfrac{1}{x_1} + \dfrac{1}{x_2} + \dfrac{1}{x_3} + ... + \dfrac{1}{x_n}} = \frac{n}{\displaystyle\sum_{i=1}^{n} \dfrac{1}{x_i}}$$

Equation 2

and:-

$$x_i = \frac{\rho_i}{f_i}$$

Equation 3

f_i and ρ_i are respectively the Mass Fraction and the Mass Density of the ith component of the mixture.

In the context of the mineral crystal studies of this disquisition, f_i is the Fraction of the Chemical Formula Weight, W_{cry}, accounted for by constituent element i; and ρ_i is the Mass Density of Element i at or near STP, or in the case of STP gases, near to absolute zero. The use of the densities of liquid, or better still, solid STP gases was chosen to be more representative of the predicament of gaseous elements within a solid-state lattice. Such is of course a further source of error.

Calculation of the Formula Weight, W_{cry}

Straight-forward application of Equations One, Two and Three uses the Formula Weight, W_{cry} (or m), of a mineral in which the chemical formula is considered a "molecule" without reference to the crystallographic Unit Cell or to the cell's Formula Population, Z.

Furthermore, I arbitrarily and naïvely distributed substitutional elements within formulaic elaborations. For example, there are thirteen substitutional atoms of iron, magnesium and manganese in the formula of zussmanite. In that case, my tactic was equally but arbitrarily to award 4⅓ of each of the three atoms to the computation. This of course subverts the very concept of the integral unit cell and reduces the chemical description to little more than an empirical formula. But this scheme of equality was not followed in all cases.

For each of the one hundred and twenty-five minerals the Formula Weight, W_{cry}, was computed as:-

$$W_{cry} = \sum_{i=1}^{n_e} A_i f_i = \sum_{i=1}^{n_e} A_i \cdot \frac{j_e}{j_{tot}}$$

Equation 4

where A_i is the Atomic Weight of component Element i; f_i is the weight Fraction of Element i in the chemical formula; j_e is the Number of relevant Element Atoms in the formula; and j_{tot} is the Total Number of Atoms in the formula.

ρ_{meas} in the Measured Mass Density of the mineral as selected from a trusted Web source (usually www.handbookofmineralogy.org) and ρ_H is the mineral's Harmonic Density as:-

$$\rho_H = \frac{H(x_i)}{j_{tot}}$$

Equation 5

The overall scheme of computation and an individual mineral's results is illustrated for the case of zussmanite by Table One.

The apparent volumetric Condensation of the Crystal Lattice is then defined as:-

$$\kappa = 1 - \frac{\rho_{meas}}{\rho_H}$$

Equation 6

Table C1 in Appendix C presents given and computed data for each of the one hundred and twenty-five mineral species. The order is arbitrary.

SERIAL: 13
MINERAL: ZUSSMANITE
CHEMICAL FORMULA: K(Fe2+,Mg,Mn2+)13(AlSi)18O42(OH)14
MEASURED DENSITY: 3.146
HARMONIC DENSITY: 1.821431229
CONDENSATION: -0.727213166
FORMULA WEIGHT: 2205.753517
WEIGHTED PSD OF ELECTRONEGATIVITY: 37.57896207

Atomic Number Z	Atomic Weight	Element Name	Element Symbol	Density ρ_i (grams/cc) (STP)	Electro negativity χ	Formula Element Count	Partial Formula Weight	Formula Weight Fraction f_i	$x_i=\rho_i/f_i$	Weighted Electro negativity
1	1.008	Hydrogen	H	0.088	2.2	14	14.112	0.006398	13.7547	30.8
3	6.997	Lithium	Li	0.534	0.98					
4	9.0121831	Beryllium	Be	1.85	1.57					
5	10.811	Boron	B	2.46	2.04					
6	12.011	Carbon	C	2.2	2.55					
7	14.007	Nitrogen	N	1.03	3.04					
8	15.999	Oxygen	O	1.19	3.44	36	575.964	0.261119	4.55731	123.84
9	18.9984032	Flourine	F	1.706	3.98					
11	22.98977	Sodium	Na	0.968	0.93					
12	24.305	Magnesium	Mg	1.738	1.31	4.333333	105.3217	0.047749	36.39896	5.676666667
13	26.9815385	Aluminium	Al	2.698	1.61	18	485.6677	0.220182	12.25349	28.98
14	28.085	Silicon	Si	2.3296	1.9	18	505.53	0.229187	10.16463	34.2
15	30.973761998	Phosphorous	P	1.82	2.19					
16	32.06	Sulfur	S	2.067	2.58					
17	35.453	Chlorine	Cl	2.03	3.16					
19	39.0983	Potassium	K	0.856	0.82	1	39.0983	0.017726	48.29174	0.82
20	40.078	Calcium	Ca	1.55	1					
22	47.867	Titanium	Ti	4.506	1.54					
23	50.9415	Vanadium	V	6	1.63					
24	51.9961	Chromium	Cr	7.15	1.66					
25	54.938044	Manganese	Mn	7.3	1.55	4.333333	238.0649	0.107929	67.63703	6.716666667
26	55.845	Iron	Fe	7.874	1.83	4.333333	241.995	0.109711	71.7705	7.93
27	58.9332	Cobalt	Co	8.9	1.88					
28	58.6934	Nickel	Ni	8.912	1.91					
29	63.546	Copper	Cu	8.96	1.9					
30	65.38	Zinc	Zn	7.134	1.65					
31	69.723	Gallium	Ga	5.904	1.81					
33	74.9216	Arsenic	As	5.778	2.18					
34	78.971	Selenium	Se	4.809	2.55					
35	79.904	Bromine	Br	3.122	2.96					
38	87.62	Strontium	Sr	2.63	0.95					
41	92.90637	Niobium	Nb	8.61	1.6					
40	91.224	Zirconium	Zr	6.52	1.33					
42	95.95	Molybdenum	Mo	10.2	2.16					
44	101.07	Ruthenium	Ru	12.1	2.2					
45	102.9055	Rhodium	Rh	12.41	2.28					
46	106.42	Palladium	Pd	12	2.2					
47	107.8602	Silver	Ag	10.501	1.93					
48	112.411	Cadmium	Cd	8.65	42					
49	114.818	Indium	In	7.31	1.78					
50	118.71	Tin	Sn	7.287	1.96					
51	121.76	Antimony	Sb	6.685	2.05					
52	127.6	Tellurium	Te	6.232	2.1					
56	137.327	Barium	Ba	3.51	0.89					
57	138.90547	Lanthanum	La	6.145	1.1					
58	140.116	Cerium	Ce	6.77	1.12					
60	144.242	Neodymium	Nd	7.007	1.14					
72	178.49	Hafnium	Hf	13.31	1.3					
73	180.94788	Tantalum	Ta	16.4	1.5					
74	183.84	Tungsten	W	19.25	2.36					
75	186.207	Rhenium	Re	21.02	1.9					
76	190.23	Osmium	Os	22.59	2.2					
77	192.217	Iridium	Ir	22.562	2.2					
78	195.084	Platinum	Pt	21.5	2.28					
79	196.966569	Gold	Au	19.282	2.54					
80	200.592	Mercury	Hg	13.534	2					
81	204.38	Thallium	Tl	11.85	1.62					
82	207.2	Lead	Pb	11.342	1.87					
83	208.9804	Bismuth	Bi	9.807	2.02					
90	232.0377	Thorium	Th	11.72	1.3					
92	238.02891	Uranium	U	18.95	1.38					

					Count
8	8	8	8	8	8 Count
100	2205.754	1	264.8284	238.9633333	Sum
12.5	275.7192	0.125	33.10355	29.87041667	Arithmetic Mean
7.770165	161.9593	0.073426	22.53982	13.22768356	Geometric Mean
4.204717	66.64491	0.030214	14.57145	4.507175608	Harmonic Mean
				37.57896207	Population SD

Table One
Harmonic Density Computed for Zussmanite
Using MicroSoft® EXCEL

Computation of the Crystallographic Unit Cell Volume[11]

There are a finite number of possible spacial configurations of atoms in a crystal structure. These are called the fourteen (three-dimensional) Bravais Lattices, and they constitute seven fundamental Symmetry Systems: Cubic (sometimes known as Isometric), tetragonal, orthorhombic, trigonal, hexagonal, monoclinic and triclinic.

A Unit Cell is a nominally-fundamental element of constituent atom arrangement having a pattern of one of the lattice systems. Because atoms are very little, unit cells are inevitably so small as only to be perceptible using x-ray or electron beams. Intuitively, a unit cell could be visualised as a right or distorted cuboid or (in two cases) trigonal or hexagonal prism.

To assist scaling calculations it is helpful to use Avogadro's Number, the population of atoms or molecules in a mole-weight of substance. Avogadro's Number, N_A or L, is taken to be $6.022140857 \times 10^{+23}$ particles per (gram) mol^{-1}.

The three axial Edges of a Unit Cell a, b and c are so short that they are conveniently measured in ångströms. An ångström is 0.1 nanometers, i.e. 10^{-10} meters.

The Volume of a Unit Cell, V, depends not only upon the axial lengths a, b and c but also upon the corresponding Interaxial Angles, α, β, and γ which are recorded in degrees but are of course re-expressed in radians for computation.

Table Two represents the Edge and Interangle conformations that characterise the seven Symmetry Systems:-

Crystal Symmetry System						
Length x y z						
	Edge			Angle		
	a	b	c	α	β	γ
Cubic	x	x	x	$\pi/2$	$\pi/2$	$\pi/2$
Tetragonal	x	x	z	$\pi/2$	$\pi/2$	$\pi/2$
Orthorhombic	x	y	z	$\pi/2$	$\pi/2$	$\pi/2$
Trigonal	x	x	z	$\pi/2$	$\pi/2$	$2\pi/3$
Hexagonal	x	x	z	$\pi/2$	$\pi/2$	$2\pi/3$
Monoclinic	x	y	z	$\pi/2$	β	$\pi/2$
Triclinic	x	y	z	α	β	γ

x,y,z and α,β,χ all differ

Table Two
Unit Cell Geometries for the Symmetry Systems
(Angles for General Cosine Solutions)

To quantify V we use an assemblage of trigonometrical equations described below. The units of volume are in practice cubic ångströms.

The Triclinic Case

In a triclinic crystal a, b, c, α, β, γ are, or are capable of being, numerically different. The applicable formula is[12]:-

$$V - abc\sqrt{1 + 2Cos\alpha\, Cos\beta\, Cos\gamma - Cos^2\alpha - Cos^2\beta - Cos^2\gamma}$$
Equation 7

The Cubic, Tetragonal and Orthorhombic Cases

The commonality of these lattice systems is that $\alpha = \beta = \gamma = \pi/2$; that is all angles are orthogonal.

In the Cubic system a = b = c and accordingly:-

$$V = a^3$$
Equation 8

The Tetragonal system forms a square cuboid where a = b ≠ c. Therefore:-

$$V = a^2 c$$
Equation 9

The Orthorhombic system is a general cuboid cell like a house brick in which a ≠ b ≠ c. Therefore:-

$$V = abc$$
Equation 10

It is notable that Equation Seven is sufficiently general to apply to the cubic, Tetragonal and Orthorhombic systems because:-

$$Cos(\alpha) = Cos(\beta) = Cos(\gamma) = 0$$
Equation 11

The Hexagonal and Trigonal Cases

The cells are respectively hexagonal and triangular prisms. Accordingly:-

$$V = a^2 c.Sin\left(\frac{\pi}{3}\right)$$
Equation 12

The Monoclinic Case

In this final case a ≠ b ≠ c but $\alpha = \gamma = \pi/2$. The internal angle β may be other than orthogonal.

The required formula for Cell Volume is:-

$$V = abc.Sin(\beta)$$
Equation 13

The Theoretical Cell Density of a Crystal[11]

Having now the means of knowing Cell Volume and also constituent Atomic and Formula Weights, we may also incorporate the constant Avogadro's Number to establish the Theoretical Cell Mass Density, ρ_{theo}, as:-

$$\rho_{theo} = \frac{ZW_{cry}}{VL} \times k$$
Equation 14

where W_{cry} is the crystal's chemical Formula Weight; V is Cell Volume and L (or N_A) is Avogadro's Number. k is an arbitrary Scaling Constant used to bring ρ_{theo} to co-ordinate units. In my calculations k was ten.

Measured Density and Harmonic Density

Measured Mass Density, ρ_{meas}, is actually assessed using inertial (in practice gravimetric) laboratory methods. Traditionally, that was done by comparing the weight of mineral to that of water in a glass density jar (or more properly specific gravity jar).

Theoretical Mass Density, ρ_{theo}, is indirectly established from radiological determinations.

Our third approach, Harmonic Mass Density, ρ_H, is defined by:-

$$\rho_H = \frac{H(x_e)}{n_e} = \frac{H(\rho_e / f_e)}{n_e}$$
Equation 15

where n_e is the Number of Different Elements in the Mineral Formula; ρ_e is the Mass Density of Element e; f_e is the Fraction of the Formula Weight accounted for by Element e; x_e is the Density/Fraction Ratio for Element e; and $H(x_e)$ is the Harmonic Mean of the Density/Fraction Ratios of all the constituent elements.

We cannot of course expect ρ_H closely to resemble inertial density because of the action of the interatomic forces that condense solid matter. Notwithstanding, Harmonic Density has its uses as we shall see.

Lattice Strain

In our context, the phrase "Lattice Strain" denotes the apparent isometric volume contraction engendered by interatomic binding forces.

The Z of Equation Fourteen is the Number of Formula Units in the Crystallographic Unit Cell. It is almost always an even integer, often 2 or 4. Deviant odd Z are known, but very rare.

Alongside this parameter we may define a more flexible analog Ξ (big Xi) which is the ratio of Measured Density to Harmonic Density. That is:-

$$\Xi = \frac{\rho_{meas}}{\rho_H}$$

Equation 16

Ξ is apparent Cell Lattice Strain (not necessarily the number of formula weights in the repeating lattice element, which is of course Z).

Appendix D presents plots of Ξ for all one hundred and twenty-five minerals. In the plots Ξ is named "Precise Density Ratio".

Figure D1 represents Mineral Formula versus Ξ as a column chart *in an arbitrary order*. This clarifies that Ξ tends to cluster around one, near to two, and also to some extent near 1.7: And it is obvious that Ξ is near to, but not at, low integral values for many simple chemical formulae, especially heavy metal sulphides.

Figure D2 is a line chart of the same data but *in order of ascending Ξ*. The approximate quantisation of Ξ is now striking.

Inspection infers that:-

$$\Xi \approx \sqrt{n}$$

Equation 17

where n is a low positive integer. n = 1 for simple or complex heavy metal sulphides; n = 2 for some hydrated phyllosilicates; n = 3 for some simple silicates and also zussmanite; n = 4 for several light metal silicates and alkaline carbonates, nitrates and sulphates.

There is a level of $\Xi = 5^{1/2} \approx 2.236$ very near to the Ξ of the hydrated ferro-aluminium phosphate childrenite. I speculate that this may relate to the relation of mass density to the Phidian Ratio $[(1+5^{1/2})/2]$ proposed by some researchers[13,14].

It is of course possible statistically to assess the agreement of mineral Ξ values with our $\Xi = n^{1/2}$ theory. We can do this using simple or sophisticated methods. I prefer simple things. I can understand them.

So I adopted a simple criterion:-

$$\Xi_{test} - t < \Xi_{mineral} < \Xi_{test} + t$$

Criterion 1

where t is the Tolerance and Ξ_{test} is some exact Ξ for $\Xi = n^{1/2}$ where n = 1,2,3,4,5,...

Table Three shows how ninety-three of the available one hundred and twenty-five minerals group into ten Ξ classes when the tolerance is set to 0.092880799 (i.e. ±9.3%). This was set to agree with the histogram plotting tolerance.

Some 42.4% of minerals have Ξ = 1 or 2, whilst 12.8% have Ξ = $3^{1/2}$. Strikingly, thirteen minerals (10.4%) have Ξ = $5^{1/2}$; not only childrenite, which is almost exact, but also the very different though simpler spodumene and goethite.

More than 74.4% of minerals fall into some Ξ = $n^{1/2}$ category. These values cover 80% of the total range of Ξ.

Quandaries

The obvious corollary is that 25.6% of minerals do not conform to our Ξ = $n^{1/2}$ model.

This is not because the 25.6% are deviant: In fact they are almost typical. It is because they are defective, as are easily 25.6% of men.

Or else Ξ is merely a chimera, a trick of cognition, a fallacy of judgement.

Figure One is a histogram representing the frequencies of minerals in twenty-five subranges of the total range of Equation Sixteen Density Ratio Ξ. The quantisation of Ξ is dramatically pronounced but so too is a groundmass of non-conformable crystal constitutions: And it is difficult to say whether the peaks are culminations of statistical distributions, in particular Gaussian Distributions, or whether they are not.

Basis Integer	1	2	3	4	5	6	7	8	9	10
Test Ξ	1.0000	1.4142	1.7321	2.0000	2.2361	2.4495	2.6458	2.8284	3.0000	3.1623
0.092880799 Tolerance	0.0929	0.0929	0.0929	0.0929	0.0929	0.0929	0.0929	0.0929	0.0929	0.0929
LB	0.9071	1.3213	1.6392	1.9071	2.1432	2.3566	2.5529	2.7355	2.9071	3.0694
UB	1.0929	1.5071	1.8249	2.0929	2.3289	2.5424	2.7386	2.9213	3.0929	3.2552
93 Count	24	6	16	29	13	2	1	1	0	1
Mean	1.0237	1.3885	1.7560	2.0012	2.2531	2.3829	2.5563	2.8075	0	3.2184
Pop.SD	0.0360	0.0445	0.0561	0.0525	0.0659	0.0078	0	0	0	0.0000
Fractional SpDef(Ξ,Mean)	-0.0237	0.0182	-0.0138	-0.0006	-0.0076	0.0272	0.0338	0.0074	0	-0.0177
Prob(Ξ)	0.1920	0.0480	0.1280	0.2320	0.1040	0.0160	0.0080	0.0080	0	0.0080
Prob(Ξ=1,...,10)	0.7440									
F_R	0.8000									

Table Three
Mineral Ξ Groupings According to
The Model $\Xi = n^{1/2}$

Distribution of Ξ Values

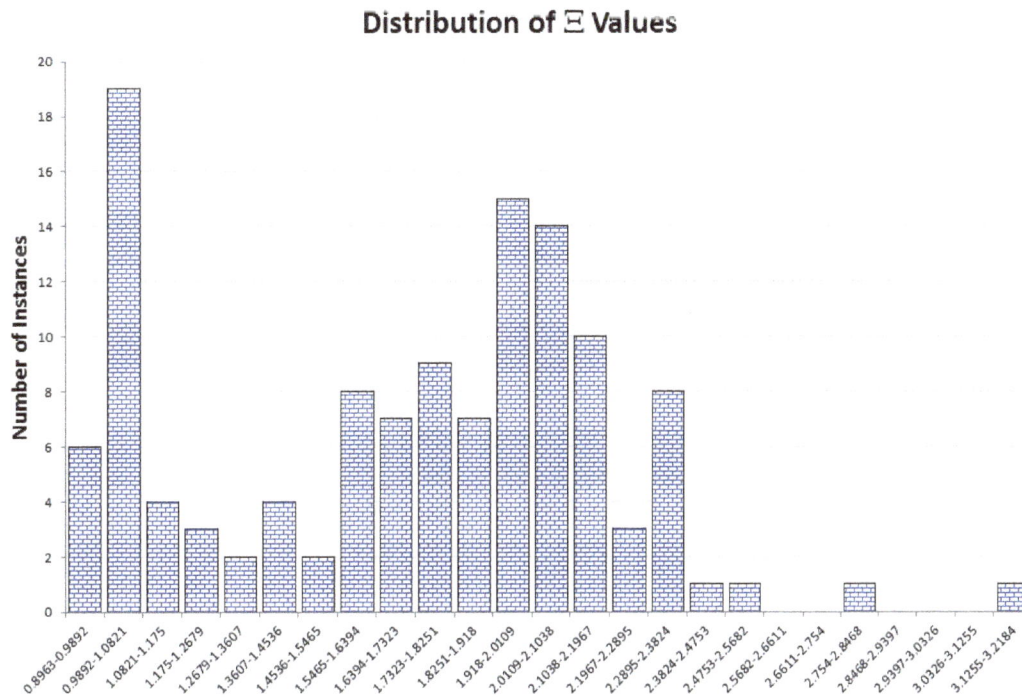

Figure One
Density Ratio (Ξ) Subranges for One hundred and Twenty-Five Crystalline Minerals

Allow that R is the Total Range of Ξ Density Ratios, and that the n values of subranges r_j represent the clusters of minerals conformable to $\Xi = n^{1/2}$.

Then, neglecting any end-of-range overlaps, we may write:-

$$F_R = \frac{2tn}{R_+ - R_-}$$

Equation 18

where F_R is the Fraction of the Total Range defined by $R_+ - R_-$, given that R_+ is the Upper Limit of R and R_- the Lower Limit. t is the Tolerance.

The data of Table Three show that for our assemblage of one hundred and twenty-five minerals $F_R = 0.8$ or four-fifths of the total range R is conformable to $\Xi = n^{1/2}$. (Where t varies from point to point, mean t is employed in Equation Eighteen).

Summary Equations for the Lattice Strain Density Ratio

Should we be willing to identify the experimentally-measured mass density with that determined by radiographic studies of the crystal lattice, i.e.:-

$$\rho_{meas} \equiv \rho_{theo}$$

Equation 19

then:-

$$\rho_{theo} = \frac{kZW_{cry}}{VL}$$

Equation 14

Also:-

$$\Xi = \frac{\rho_{meas}}{\rho_H}$$

Equation 16

So by incorporation of Equation Fifteen:-

$$\Xi = \frac{kZW_{cry}}{VL} \times \frac{n_e}{H(\rho_e / f_e)}$$

Equation 20

Or:-

$$\Xi = \frac{kZ\sum_{j=1}^{n_e} A_j}{VL} \cdot \sum_{j=1}^{n_e} \frac{1}{x_j} = \frac{kZ\sum_{j=1}^{n_e} A_j}{VL} \cdot \sum_{j=1}^{n_e} \frac{f_j}{\rho_j}$$

Equation 21

which allows us to write:-

$$\Xi = \frac{kZ}{VL} \sum_{j=1}^{n_e} A_j \sum_{j=1}^{n_e} \left[\frac{1}{\rho_j} \cdot \frac{A_j}{\sum_{j=1}^{n_e} A_j} \right]$$

Equation 22

Equation Twenty-Two may be re-arranged as:-

$$\Xi = \frac{kZ}{VL} \cdot \sum_{j=1}^{n_e} \left(\sum_{j=1}^{n_e} A_j \right) \cdot \left[\sum_{j=1}^{n_j} \left(\frac{1}{\rho_j} \cdot \frac{A_j}{\sum_{j=1}^{n_e} A_j} \right) \right]$$

Equation 23

from which simplification yields:

$$\Xi = \frac{kZ}{VL} \cdot \sum_{j=1}^{n_e} \frac{A_j}{\rho_j}$$

Equation 24

As we aforenoted Ξ is sometimes equal to $n^{\frac{1}{2}}$ where $n = 1,2,3,4,5,\ldots$?

$$\diamondsuit \left(\Xi = \frac{kZ}{VL} \cdot \sum_{j=1}^{n_e} \frac{A_j}{\rho_j} \right)$$

Equation 25

where $\diamondsuit$ is the Logical Possibility Operator.

Formula Weight

We know from Equation Sixteen that:-

$$\Xi = \frac{\rho_{meas}}{\rho_H}$$

Equation 16

and from Equation Fourteen that:-

$$\rho_{theo} = \frac{kZW_{cry}}{VL}$$

Equation 14

If we identify ρ_{meas} and ρ_{theo} as a Synthetic Mass Density, ρ_{syn}, it is possible to write:-

$$\rho_{syn} = \frac{1}{\sum_{j=1}^{n_e} \frac{1}{x_j}} \cdot K \cdot \sum_{j=1}^{n_e} \left(\frac{A_j}{\rho_j} \right)$$

Equation 26

Also:-

$$\rho = K.W_{cry}$$

Equation 27

where W_{cry} is the chemical Formula Weight of the mineral.

By elimination of the crystallographic construct $K = kZ/VL$ we may expand Formula Weight as:-

$$W_{cry} = \frac{\sum_{j=1}^{n_e}\left(\dfrac{A_j}{\rho_j}\right)}{\sum_{j=1}^{n_e}\dfrac{1}{x_j}} = \frac{\sum_{j=1}^{n_e}\left(\dfrac{A_j}{\rho_j}\right)}{\sum_{j=1}^{n_e}\dfrac{f_j}{\rho_j}}$$

Equation 28

which simplifies to:-

$$W_{cry} = \frac{1}{n_e}\sum_{j=1}^{n_e}\left(\frac{A_j}{f_j}\right)$$

Equation 29

PART TWO
THE QUANTISATIONS OF
APPARENT LATTICE STRAIN AND CELL VOLUME MULTIPLIER

We saw in Equations Sixteen and Fifteen of this disquisition that the Apparent Lattice Strain, Ξ, can be expressed as:-

$$\Xi = \frac{\rho_{meas}}{\rho_H} = \frac{n_e \cdot \rho_{meas}}{H(\rho_e / f_e)}$$
Equation 30

We also saw that the crystallographic Unit Cell Volume, V, was given by:-

$$V = abc\sqrt{1 + 2\,Cos\alpha\,Cos\beta\,Cos\gamma - Cos^2\alpha - Cos^2\beta - Cos^2\gamma}$$
Equation 7

which was general for every crystal lattice system, though the expression under the square root was degenerate for certain lattice structures.

Let us then define the Lattice Angle Constant, Ψ (big Psi), as:-

$$\Psi = \sqrt{1 + 2\,Cos\alpha\,Cos\beta\,Cos\gamma - Cos^2\alpha - Cos^2\beta - Cos^2\gamma}$$
Equation 31

so that:-

$$V = abc\Psi$$
Equation 32

We saw also that for the cubic (isometric), tetragonal and orthorhombic systems Ψ was unity; and that for the trigonal and hexagonal systems $\Psi = Sin(\pi/3)$. To that limited extent Ψ is quantised, but for the monoclinic and triclinic cases Ψ could be some apparently random value based upon the peculiar angular inclinations of the mineral's unit cell.

Table Four explicates the Ψ angular relations and general values for the seven lattice systems.

Figure Two shows the distribution of mineral Ψ values illustrating the expected pronounced peaks at $\Psi = Sin(\pi/3)$ and $\Psi = 1$. But it is also notable that other angular expressions are represented, in line with the natural proportions of monoclinic and triclinic mineral species.

Length	Crystal Symmetry System	x	y	z	α	β	γ	α	β	γ	Ψ (big Psi) Functional Value	Ψ (big Psi) Numerical Value	Ψ (big Psi) Numerical Value (by Formula)
			Edge			Angle			Angle				
		a	b	c	α	β	γ	α	β	γ			
	Cubic	x	x	x	$\pi/2$	$\pi/2$	$\pi/2$	1.570796	1.570796	1.570796	Unity	1	1
	Tetragonal	x	x	z	$\pi/2$	$\pi/2$	$\pi/2$	1.570796	1.570796	1.570796	Unity	1	1
	Orthorhombic	x	y	z	$\pi/2$	$\pi/2$	$\pi/2$	1.570796	1.570796	1.570796	Unity	1	1
	Trigonal	x	x	z	$\pi/2$	$\pi/2$	$2\pi/3$	1.570796	1.570796	2.094395	Sin($\pi/3$)	0.866025404	0.866025404
	Hexagonal	x	x	z	$\pi/2$	$\pi/2$	$2\pi/3$	1.570796	1.570796	2.094395	Sin($\pi/3$)	0.866025404	0.866025404
	Monoclinic	x	y	z	$\pi/2$	β	$\pi/2$	1.570796	β	1.570796	Sin(β)		
	Triclinic	x	y	z	α	β	γ	α	β	γ	Formula		
x,y,z and α,β,χ all differ													

Table Four
Lattice System Angular Relations and the
Lattice Angle Constant, Ψ

Distribution of Ψ Values

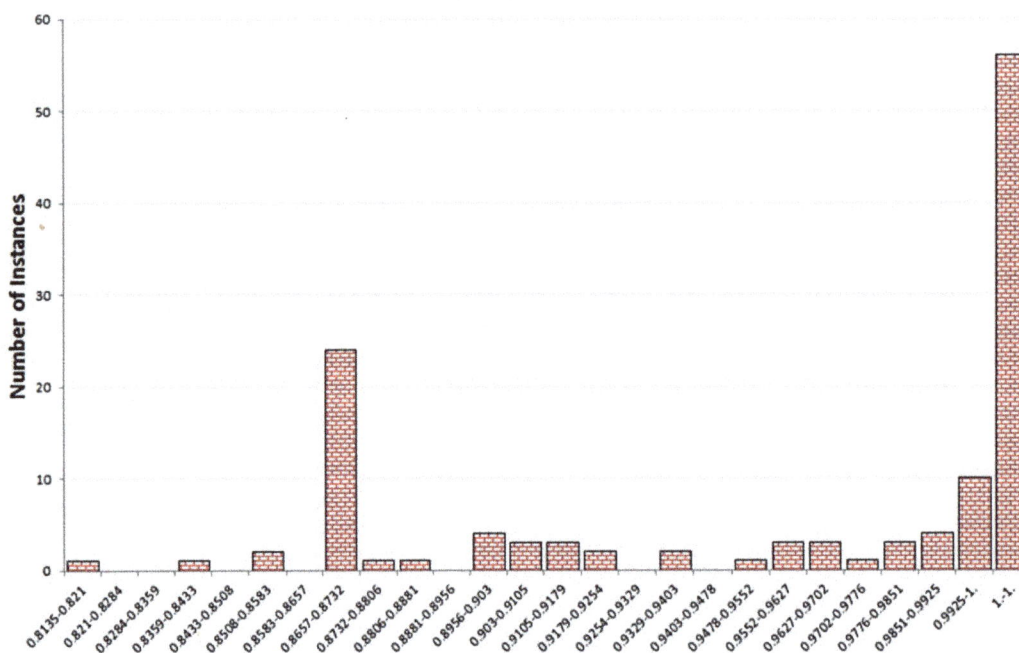

Figure Two
The Distribution of the Lattice Angle Constants Ψ

Another relationship that we may recollect from the first part of this study is:-

$$\rho_{theo} = \frac{ZW_{cry}}{VL} \times k$$

Equation 14

which we may now recast as:-

$$\rho_{theo} = \frac{ZW_{cry}}{abc\Psi L} \times k$$

Equation 33

And also:-

$$\Xi = \frac{kZ}{VL} \cdot \sum_{j=1}^{n_e} \frac{A_j}{\rho_j}$$

Equation 24

Or:-

$$\Xi = \frac{kZ}{abc\Psi L} \cdot \sum_{j=1}^{n_e} \frac{A_j}{\rho_j}$$

Equation 34

Accordingly:-

$$\Xi\Psi = \frac{kZ}{abcL} \cdot \sum_{j=1}^{n_e} \frac{A_j}{\rho_j}$$

Equation 35

Equation Thirty-Five yields what appears to be a random number, except where qualified by the two modes of Ψ. This property is illustrated by Figure Three, in which mineral crystals crowd the $\Psi = Sin(\pi/3)$ and $\Psi = 1$ parallels but the monoclinic and triclinic groups appear to form a random cloud mostly between those two lines. In terms of the red Pearson linear regression both the grade and the coefficient of determination should be zero for a perfect random scatter, whilst the intercept should be the mean of the y-values. We can see that Grade b = -0.00563245 and Coefficient of Determination r^2 is 0.00213317, indicating that a mere 0.2% of the variation in Ψ is "accounted for" by Ξ. The intercept is 0.96389601. There is no sensible relation between Apparent Lattice Strain and Lattice Angle Constant.

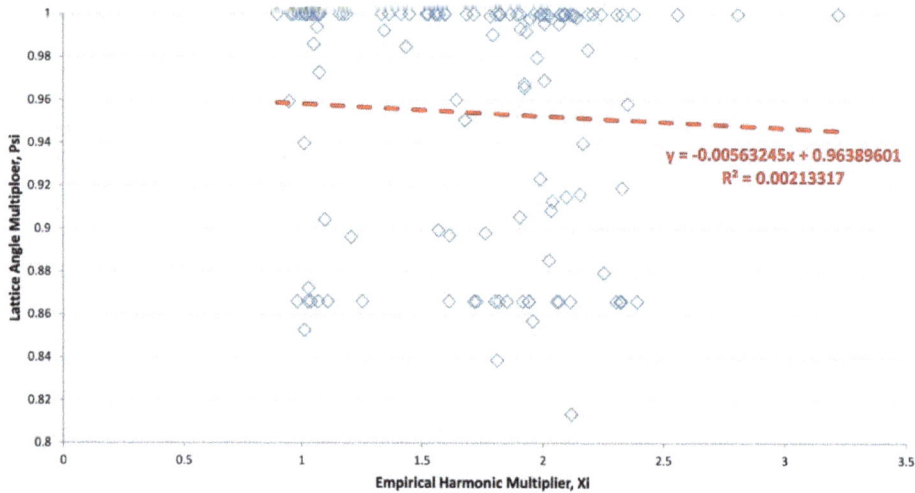

The Non-Relation of Empirical Harmonic Mutiplier Ξ
and
Lattice Angle Multiplier Ψ

$y = -0.00563245x + 0.96389601$
$R^2 = 0.00213317$

Figure Three
The Non-Relation of Ξ and Ψ for All 125 Minerals Studied

Figure Four presents the same data differentiated for Lattice System in order to explicate the contrast between orthohedral systems as squares at $\Psi = 1$; and trigonal and hexagonal forms at $Sin(\pi/3)$. Meanwhile, the monoclinic and triclinic systems are shown as circles.

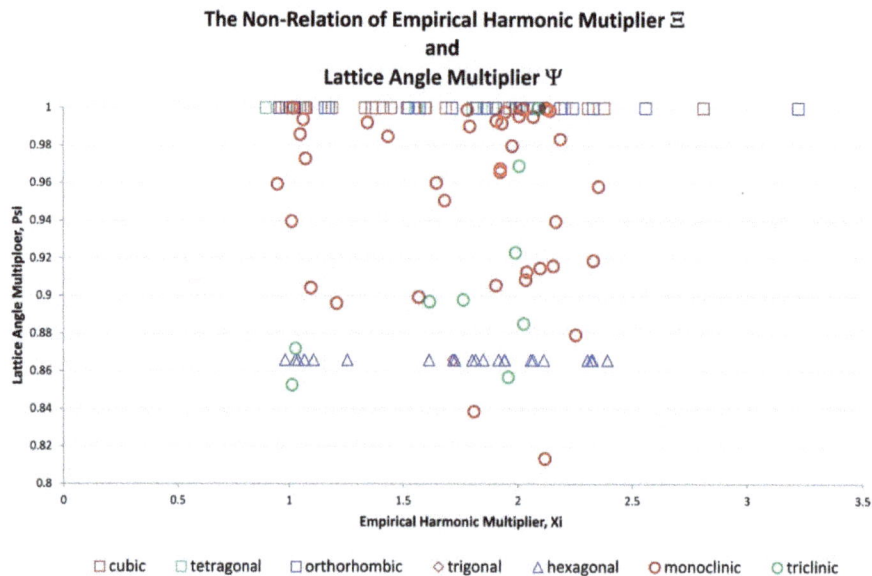

The Non-Relation of Empirical Harmonic Mutiplier Ξ
and
Lattice Angle Multiplier Ψ

□ cubic   □ tetragonal   □ orthorhombic   ◇ trigonal   △ hexagonal   ○ monoclinic   ○ triclinic

Figure Four
All 125 Minerals Studied Differentiated by Lattice System

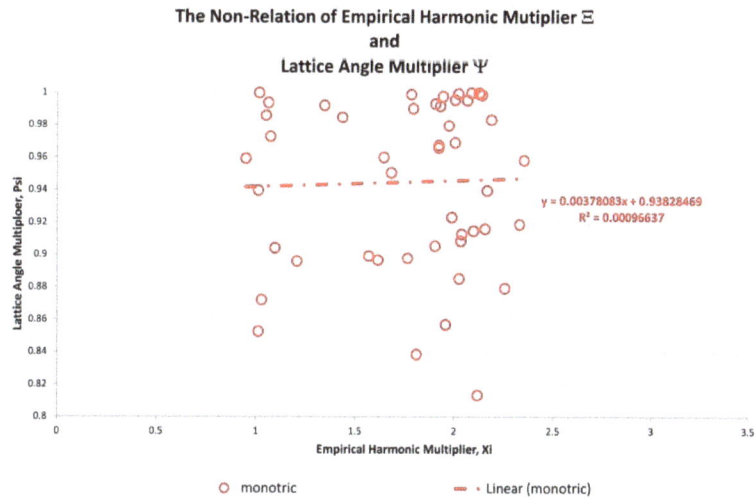

Figure Five
Demonstration of the Random Scatter of Monoclinic and Triclinic Crystals

Figure Five plots the minerals of "monotric" class (i.e. either monoclinic or triclinic) that populate the "random cloud" of Ξ versus Ψ. The fitted Pearson regression has an Intercept a = 0.93828469, near to the mean of Ψ. The Grade is 0.00378083 (ideally zero) and the Coefficient of Determination r^2 is 0.00096637 (also ideally zero). In formal terms, this demonstrates that Ξ has no influence upon Ψ for either the monoclinic or triclinic systems, or, to be pedantic, Ξ accounts for 0.096637% of the variation in Ψ.

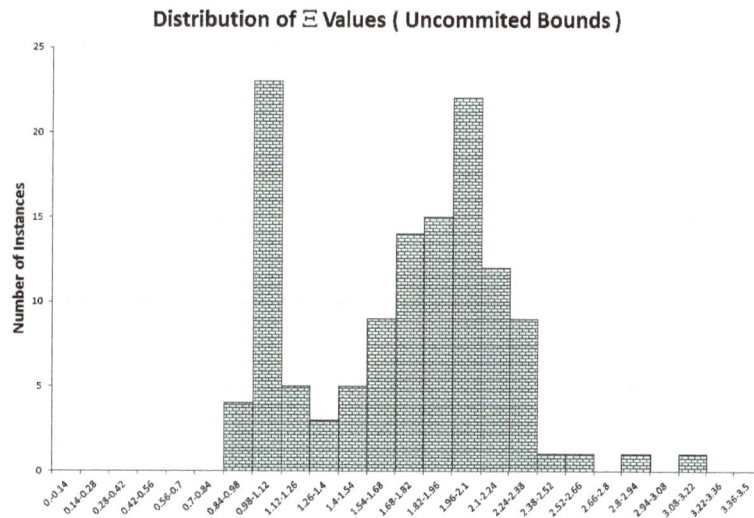

Figure Six
The Distribution of Apparent Lattice Strain Ξ for
125 Minerals plotted in Twenty-Five Arbitrary Bins between Zero and 3.5

Figure Six is a Ξ frequency histogram of arbitrarily-bounded bins on the range { 0, 3.5/25, 3.5 }. This bin choice tends to occlude the quantisation of Ξ, though of course the grouping at $\Xi = 1$ and $\Xi = 2$ is obvious enough.

The modalities of Ξ are much better discriminated in Figure Seven which is defined on { min(Ξ), (max(Ξ)-min(Ξ))/25, max(Ξ) } where max(Ξ) is the largest available Apparent Strain and min(Ξ) the smallest. For our one hundred and twenty-five minerals these are respectively 3.218357608 (for Obradovicite) and 0.896337622 (for Zaccariniite). The relevant statistics are numerically presented in Table Five.

Figure Eight also presents the one hundred and twenty-five values of Ξ, but without the color differentiation for lattice systems in order to assist the perception of the red verticals that position the integer square roots. The diagram clarifies that the modes of Apparent Lattice Strain Ξ co-ordinate with the roots of 1,2,3 and 4, but that the agreement with higher values is less definite.

Start	0.896337622	Serial	1	2	3	4	5	6
Finish	3.218357608	LB	0.8963	0.9892	1.0821	1.1750	1.2679	1.3607
Range	2.322019986	UB	0.9892	1.0821	1.1750	1.2679	1.3607	1.4536
Increment	0.092880799	Label	0.8963-0.9892	0.9892-1.0821	1.0821-1.175	1.175-1.2679	1.2679-1.3607	1.3607-1.4536
Count: In	125	Count	6	19	4	3	2	4
Count: Out	125							
		Serial	7	8	9	10	11	12
		LB	1.4536	1.5465	1.6394	1.7323	1.8251	1.9180
		UB	1.5465	1.6394	1.7323	1.8251	1.9180	2.0109
		Label	1.4536-1.5465	1.5465-1.6394	1.6394-1.7323	1.7323-1.8251	1.8251-1.918	1.918-2.0109
		Count	2	8	7	9	7	15
		Serial	13	14	15	16	17	18
		LB	2.0109	2.1038	2.1967	2.2895	2.3824	2.4753
		UB	2.1038	2.1967	2.2895	2.3824	2.4753	2.5682
		Label	2.0109-2.1038	2.1038-2.1967	2.1967-2.2895	2.2895-2.3824	2.3824-2.4753	2.4753-2.5682
		Count	14	10	3	8	1	1
		Serial	19	20	21	22	23	24
		LB	2.5682	2.6611	2.7540	2.8468	2.9397	3.0326
		UB	2.6611	2.7540	2.8468	2.9397	3.0326	3.1255
		Label	2.5682-2.6611	2.6611-2.754	2.754-2.8468	2.8468-2.9397	2.9397-3.0326	3.0326-3.1255
		Count	0	0	1	0	0	0
		Serial	25					
		LB	3.1255					
		UB	3.2184					
		Label	3.1255-3.2184					
		Count	1					

Table Five
Frequency Counts of the Distribution of Apparent Lattice Strain Ξ for One Hundred and Twenty-Five Minerals counted in Twenty-Five Bins Between the Lowest and Highest Sample Ξ Values

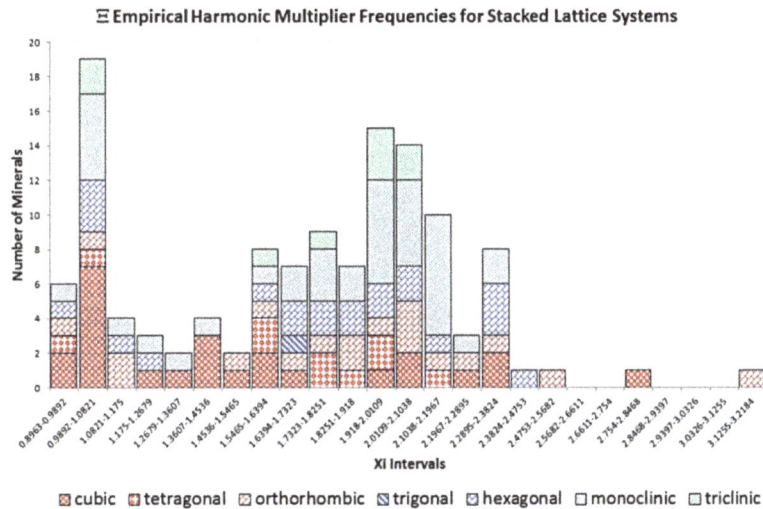

Ξ Empirical Harmonic Multiplier Frequencies for Stacked Lattice Systems

cubic ⊠ tetragonal ⊠ orthorhombic ◩ trigonal ⊞ hexagonal ☐ monoclinic ☐ triclinic

Figure Seven
The Distribution of Apparent Lattice Strain Ξ
for
One Hundred and Twenty-Five Minerals counted in Twenty-Five Bins
Between the Lowest and Highest Sample Ξ Values
Resolved for Lattice System Bin Populations

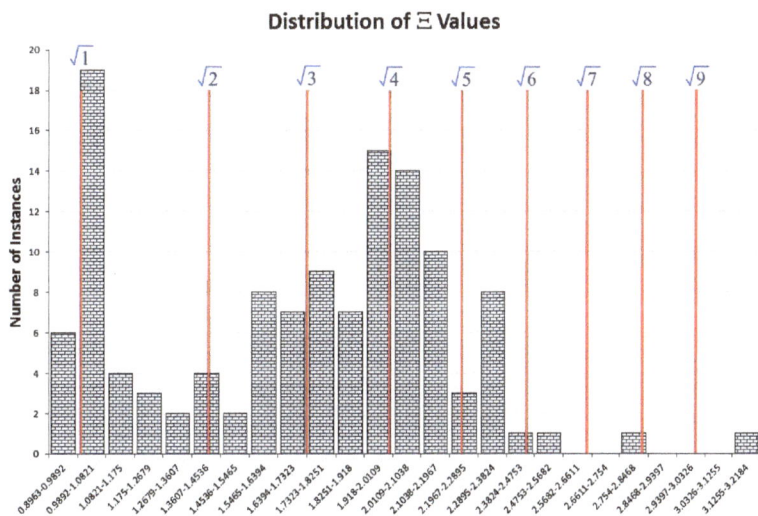

Distribution of Ξ Values

Figure Eight
The Distribution of Apparent Lattice Strain Ξ for
One Hundred and Twenty-Five Minerals counted in Twenty-Five Bins
Between the Lowest and Highest Sample Ξ Values
Showing the Positions of Root-Integer Values

PART THREE
AN EXAMINATION OF GROUPED FREQUENCY TRIADS

Further to accumulate evidence that the Apparent Lattice Strains, Ξ, really are the square roots of Natural or Countable Numbers, $\mathbb{N}$, or to confute that hypothesis, it is useful to examine grouped frequency triads.

These (single) Grouped Frequency Triads (GFTs) are each three adjacent count bins whose central member includes a natural-number square root.

But to facilitate analysis we shall initially review all twenty-one available bin triads ("moving triads") whether or not they include such roots.

The lower Submodal Bin has f_0 members and has Lower Bound l_0 and Upper Bound u_0. Accordingly, The Interval Mid-Point x_0 is, in general, defined by:-

$$x_j = \frac{(u_j + l_j)}{2}$$

Equation 36

or for this Submodal Bin by x_0.

Similarly, the central Modal Bin is defined by the parameters f_1, l_1, u_1 and x_1; whilst the upper Supermodal Bin is defined by f_2, l_2, u_2 and x_2.

In general:-

$$l_{j+1} = u_j \qquad \textbf{Equation 37a}$$
$$l_{j+2} = u_{j+1} \qquad \textbf{Equation 37a}$$

Also:-

$$w_j = u_j - l_j$$

Equation 38

where w_j is the Interval Width, which should be held constant throughout the range where practicable, or in other words equal intervals are desirable.

For the case of $\Xi = 4^{1/2} = 2$ Figure Nine illustrates these triad conformations:-

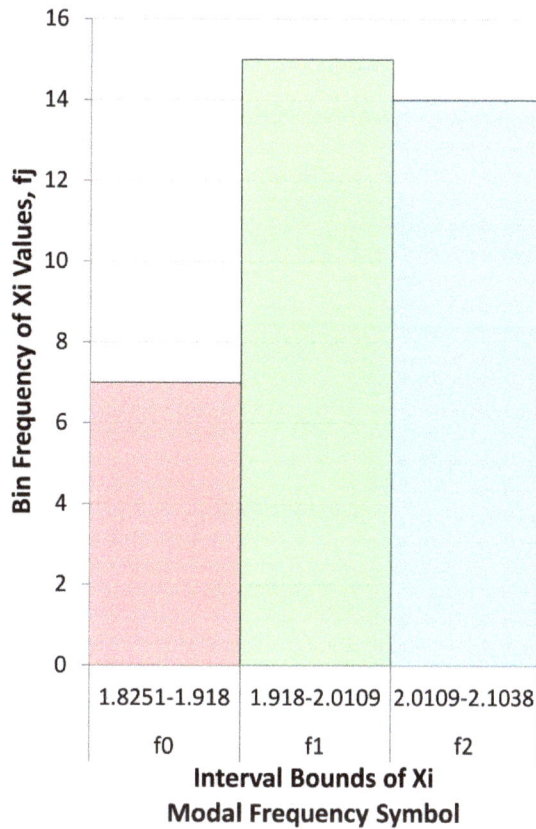

Figure Nine
The Grouped Frequency Bin Triad
for Ξ is Root Four

Grouped Frequency Moments

The Mean

For a complete batch of grouped frequencies as a triad of three consecutive bins the Arithmetic Mean is defined by:-

$$\mu = \frac{\sum fx}{\sum f}$$

Equation 39

Or for a triad:-

$$\mu = \frac{\sum fx}{\sum f} = \frac{f_0 x_0 + f_1 x_1 + f_2 x_2}{f_0 + f_1 + f_2}$$

Equation 40

The Sample Standard Deviation

The Sample Standard Deviation, s, for a grouped frequency distribution is:-

$$s = \sqrt{\frac{\sum fx^2}{\sum f} - \mu^2}$$

Equation 41

Or for a triad:-

$$s = \sqrt{\frac{f_0 x_0{}^2 + f_1 x_1{}^2 + f_2 x_2{}^2}{f_0 + f_1 + f_2} - \left(\frac{f_0 x_0 + f_1 x_1 + f_2 x_2}{f_0 + f_1 + f_2}\right)^2}$$

Equation 42

The Mode

The Mode, M, is defined by:-

$$M_j = l_j + w_j \left[\frac{f_j - f_{j-1}}{2f_j - f_{j-1} - f_{j+1}}\right]$$

Equation 43

The Specific Defect

As part of the assessment of the co-incidence of two real numbers it is convenient to employ the Percentage Specific Defect, %Spec.Def(theory:measurement) where:-

$$\%Spec.Def(t:m) = 100\left(\frac{m-t}{t}\right)$$

Equation 44

In particular, we shall examine:-

$$\%Spec.Def(root:\mu) = 100\left(\frac{\mu - root}{root}\right)$$

Equation 45

and:-

$$\%Spec.Def(\mu, M) = 100\left(\frac{M - \mu}{\mu}\right)$$

Equation 46

Specific Defect is closely related both to Standard Deviation and to Root Mean Square, but unlike those it preserves the sign of deviation.

Findings

Appendix E presents triad analysis for all twenty one triads of the twenty-five bins; and also the nine triads which include the square root of a natural number within the central Modal Bin.

Table E1 presents the Bin Parameters f, l, u, x_{mid} (i.e. x_j), Σf, Σfx and Σfx^2 for the twenty-one consecutive and overlapping triads and the derived statistics μ, s, M, %Spec.Def(Root:μ) and %Spec.Def(μ:M).

It is seen that, neglecting instances for which the relevant statistics are indeterminate, the mean %Spec.Def(root:m) is 4.612771 with s = 8.781095. Given the number of (moving) triads of common square root the large standard deviation is to be expected. On the other hand, %Spec.Def(m:M) collective standard deviation is 0.025454, showing that triad means and modes are almost co-incident.

Allow that the square root of $\mathbb{N}$, a Natural (or Counting) Number is defined by:-

$$\Omega = \sqrt{N}$$
Equation 47

Figure E1 presents the interesting first-order linearity of Ξ versus Ω. Our hypothesis of Part Two is of course that Ω and Ξ are equivalent:-

$$\Omega = \sqrt{N} = \frac{kZ}{VL} \sum_{j=1}^{n_e} \frac{A_j}{\rho_j}$$
Equation 24x

Perfect equivalence would be represented by the linear regression:-

$$\Xi = 0 + 1\Omega = c_0 + c_1 \Omega$$
Equation 48

The actual fitted line has Intercept c_0 = -0.21817078 and Grade c_1 = 1.07668741. The Coefficient of Determination r^2 should be unity at equivalence: It is actually computed at 0.98446498.

This is compelling evidence that Equation Twenty-Four x is valid.

But we should proceed to consider the Single Trials abstract in which the tabulated triads are *only* those which include the square root of a natural number in their modal bin.

These nine triads are isolated in Table E2.

The mean %Spec.Def(root:μ) is now reduced to -1.86657 with a sample standard deviation of 2.820412.

The sample standard deviation of %Spec.Def(μ:M) is 2.543668.

The nine Ω-inclusive triads' $\{\Xi,\Omega\}$ values are plotted in Figure E2.

The Figure E2 linear regression coefficients are identical to the moving-triads' case, and the same arguments of equivalence may be adduced.

Accordingly, by the expansion of Equation Twenty-Four followed by a minor transposition we may declare:-

$$\Omega\Psi = \sqrt{N} = \frac{kZ}{abcL}\sum_{j=1}^{n_e}\frac{A_j}{\rho_j}$$

Equation 49

The first seeming paradox of Equation Forty-Nine is that the central root of a natural number is quantised whilst the RHS is apparently continuous.

The second apparent paradox is that both the LHS and the RHS of Equation Forty-Nine are seen to be quasi-random numbers since we established that the product $\Xi\Psi$ is a quasi-random number according to Equation Thirty-Five:-

$$\Omega\Psi = \frac{kZ}{abcL}\sum_{j=1}^{n_e}\frac{A_j}{\rho_j}$$

Equation 35

PART FOUR
FURTHER STATISTICAL OBSERVATIONS

Some Aspects of the Interpretation of Linear Regression

As discussed in Parts Two and Three, a linear regression in which the Grade is zero is very suggestive of a lack of correlation between the Independent and the Dependent Variables.

Furthermore, a Coefficient of Determination, r^2, which is at or near zero is often felt to confirm such a non-relation.

Thus the ideal equation of such a system may be given as:-

$$y = c_0 + c_1 x = \mu_y + 0 \times x$$

Equation 36

where y is the Dependent Variable; x is the Independent Variable; c_0 is the Intercept, c_1 the Grade and μ_y is the Arithmetic Mean of the Dependent y-values.

This ideal state of affairs is illustrated in the case of (nearly) uniformly-distributed y-values as illustrated in Figure Ten:-

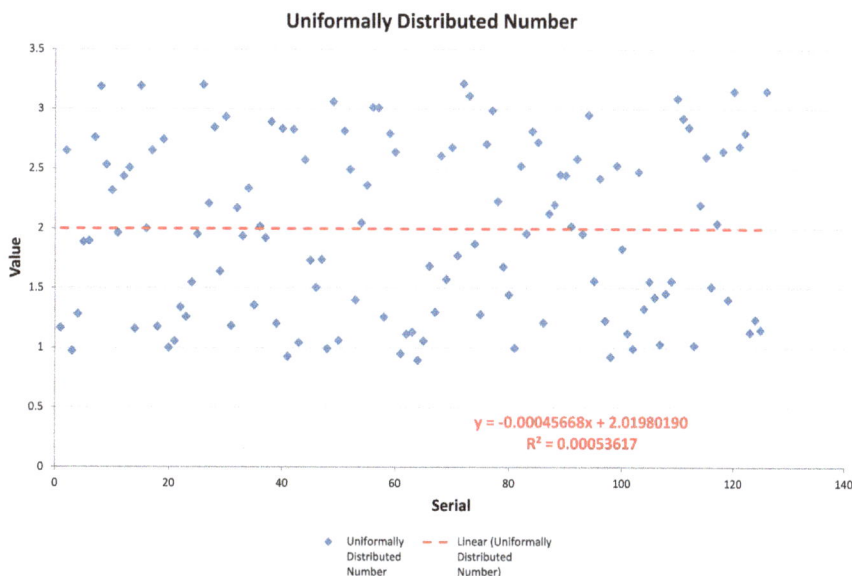

Figure Ten
A Linear Regression of
Uniformly Distributed Data

Excel® uniformly-distributed random numbers, as yielded by the intrinsic function RAND() lie on the interval {0,1}. So we shift and scale our rendition of RAND() according to:-

$$y = \min(\Xi) + RAND() * (\max(\Xi) - \min(\Xi))$$

Equation 37

Thus, for uniformly-distributed y, c_0 is 2.01980190, c_1 is -0.00045668 and crucially r^2 is 0.00053617, indicating that only about 0.054% of the variation in y is "accounted for" by x-values.

Now great care is needed. In the context of Figure Eleven, treating of three complete cycles of a sine wave, the x-y relation is very manifestly deterministic according to the relation:-

$$y = \max(\Xi) + [\max(\Xi) - \min(\Xi)] \times \sin(\omega x + \phi)$$
Equation 38

in which ϕ is the Phase Lag and the Static Angular Frequency (i.e. the angular frequency in which kinematic parameters, in this case velocity, do not figure), ω, is specified as:-

$$\omega = \frac{2\pi}{42}$$
Equation 39

The value 42 is the dividend of 126/3, employed to engineer exactly three cycles with 126 data points.

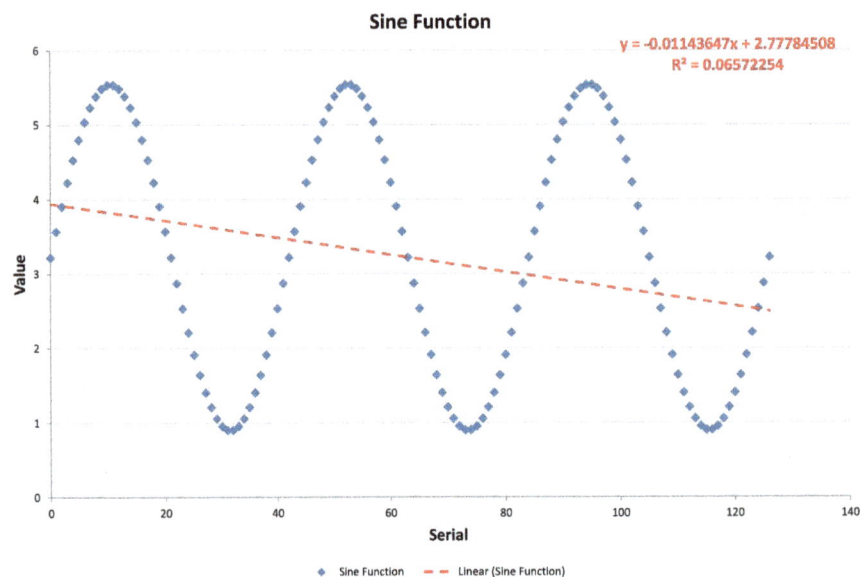

Sine Function

$y = -0.01143647x + 2.77784508$
$R^2 = 0.06572254$

**Figure Eleven
A Linear Regression of
Three Cycles of a Sine Function**

Clearly, this system is not only deterministic but evenly-balanced, so we might reasonably expect a horizontal regression line of zero grade. What we get is a Grade of -0.01143647 *with these particular numerical inputs*. In fact both the Intercept and the Grade of this system are very sensitive to the scaling of the Independent Variable, x.

The only parameter reliably conserved under scaling is the Coefficient of Determination, r^2, which for this sine system is 0.06572254, indicating somewhat misleadingly that the Independent Variable accounts for some 6.5% of the variation in the dependent.

When we use Pearson regression to throw light upon functional dependencies we must at all times remember that it is a species of *linear* regression, designed on the assumption of a *linear* relationship, and that it is never valid for non-linear hypotheses without the appropriate transformations.

This latter principle is strikingly illustrated by Figures Twelve and Thirteen which employ exactly the same set of data couples. And with regard to those couples it is manifest that Pearson correlation processes only the paired data *values* and in no way the data *conformation*.

Both regressions employ five lots of 25 y-values that are respectively square roots of 1, 2, 3, 4 and 5 (the final batch of $5^{\frac{1}{2}}$ is 26 in order to make up a total of 126 data pairs).

So in both instances the data is quantised. In both instances, the quantisation is readily apparent. Otherwise, both plots appear dissimilar.

Figure Twelve presents the data sorted by y-value to yield a highly-determinate stepwise conformation.

Figure Thirteen is exactly the same data in a random order.

For both conformations the linear regression coefficients are c_0 = +0.94139010, c_1 = +0.01165327 and r^2 = 0.94790221.

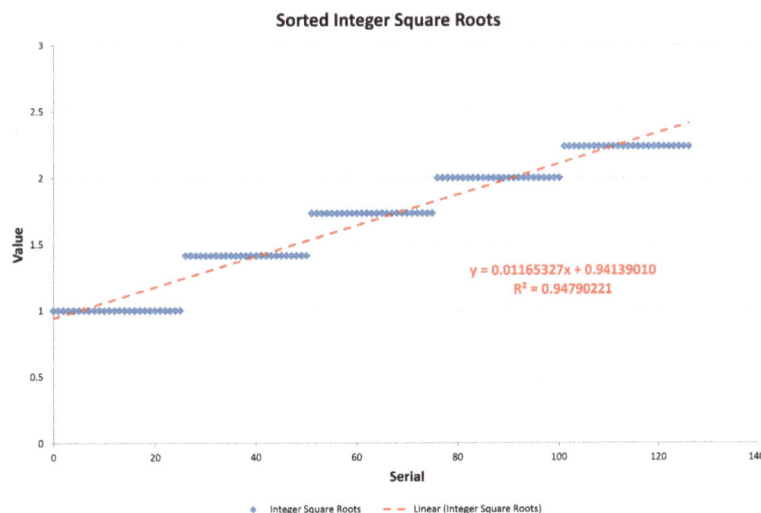

Sorted Integer Square Roots

$y = 0.01165327x + 0.94139010$
$R^2 = 0.94790221$

Figure Twelve
A Linear Regression of
Sorted Integer Square Roots

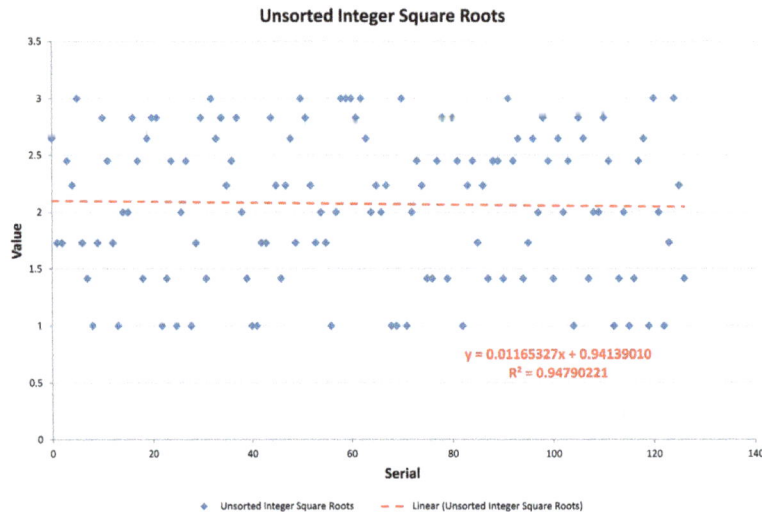

Unsorted Integer Square Roots

$y = 0.01165327x + 0.94139010$
$R^2 = 0.94790221$

♦ Unsorted Integer Square Roots   – – Linear (Unsorted Integer Square Roots)

Figure Thirteen
A Linear Regression of
Unsorted Integer Square Roots

Chi-Squared Tests[15]

      Chi-squared tests attempt to correlate two parallel grouped frequency distributions. They are "non-parametric", or in other words do not assume that the underlying distributions are Gaussian. This gives them great latitude and robustness, but they are not suited to all circumstances, and in particular if data are both few and integral Fisher's Exact Test is often superior.

      The Chi-squared test hinges upon the comparison of an empirical Observed Frequency, O_i, with a corresponding Expected Frequency, E_i, to form an Observed Chi-Squared Value, X.

      X is defined by:-

$$X = \sum \frac{(O_i - E_i)^2}{E_i}$$

Equation 40

      where i is a Bin Label.

      O_i and E_i are the counts in the bins' two separate but parallel grouped frequency distributions. The Chi-Squared Method compares the two distributions to see where they differ at a pre-determined Confidence (or Significance) Level, α_{crit}, which is expressed as a fractional probability, often 0.95 or a chance of 1 in 20. In our trials, we will use the much more stringent criterion of 0.001 (one chance in a thousand). Notwithstanding that, we shall see that the Probabilities $P(X,\nu)$ and its complement $Q(X,\nu)$ are natural fallouts of the process and can be used to assess the quality of similitude without worry about arbitrary confidence levels.

P and Q have introduced us to a second criterion active in Chi-Squared analyses: The Degrees of Freedom, ν. Broadly, The Degrees of Freedom is a positive non-zero integer (a natural number) defined as below:-

$$\nu = m - M - 1$$
Equation 41

where in context ν is the Degrees of Freedom; m is the Number of Bin Pairs and M is the Number of Parameters already Fixed by the Data Processing.

For example if we processed 125 mineral data initially to 125 Z-scores which implicitly fixed the parameters mean and standard deviation, and we gathered the Z-scores into twenty-five pairs of bins, then ν would be 25 bin-pairs less 2 parameters minus one would be $\nu = 22$.

Now we are gathering to four pairs of bins, with zero parameters pre-calculated, and accordingly we will use n = 3.

χ^2 is actually a statistic that represents a probability of difference between the two distributions. In particular, χ^2 has a value that identifies with the Confidence Level α_{crit} and with ν and can thus be tabulated against both.

It is then possible to establish the criterion:-

if X>= χ^2 the Distributions Differ; otherwise the Distributions are the Same
Criterion 2

Now if one distribution is, say, a flat uniform distribution, perhaps generated on purpose using RAND() or some other appropriate computer intrinsic function, or alternatively specified by:-

$$E_j = \frac{n}{m}$$
Equation 42

Then we can ask the question "Is our data random (uniformly-distributed) or is it generated by some deterministic 'natural law' type effect?".

If empirical X equals or exceeds theoretical χ^2, then the latter.

To complete this introductory discussion of chi-squared principles, let us return to the Probabilities of Association P(X,ν) and Q(X,ν). They are defined below[16,17]:-

$$P(X,\nu) = \left[2^{\frac{\nu}{2}}.\Gamma\left(\frac{\nu}{2}\right) \right]^{-1} . \int_0^{cq} t^{\frac{\nu}{2}-1}.e^{-\frac{t}{2}}.dt$$
Equation 43

$$Q(X,\nu) = \left[2^{\frac{\nu}{2}}.\Gamma\left(\frac{\nu}{2}\right) \right]^{-1} . \int_{cq}^{\infty} t^{\frac{\nu}{2}-1}.e^{-\frac{t}{2}}.dt$$
Equation 44

Thus:-

$$P(X,v) + Q(X,v) = 1$$
Equation 45

The boundary limit cq can either be X or χ^2 depending upon whether you wish to examine the confidence probability α_{crit} or the empirical probability p = Q(X,v).

Equation Forty-Five does of course spare us the computation of both probabilities, and P may be computationally the more desirable, though our interest centers upon Q.

The tabulated theoretical χ^2 is, in contrast to empirical X, exceedingly difficult to compute. I tried a number of elementary approaches including methods as basic as regula falsi, which proved divergent. I do not fancy paying \$35 or even \$15 to read the definitive paper, much less re-join the IEEE or the ACM, and am too idle in my old age to go and request the relevant offprint at Bloxwich Library.

But to test our mineral data for fortuity we do not really need to mither about the actual values of χ^2, since Q(X,v) tells us all we really need to know.

Nevertheless, I exploited the EXCEL® intrinsic CHISQ.INV.RT to compute critical χ^2 in the following analyses.

Simplifications for Three Degrees of Freedom

As mentioned above, we have four sets of bins and therefore three degrees of Freedom, v. Accordingly the Q of Equation Forty-Four is susceptible of simplification.

Firstly, we shall examine the composite constant C_1 defined by:-

$$Q(X,v) = C_1 . \int_{cq}^{\infty} t^{\frac{v}{2}-1} . e^{-\frac{t}{2}} . dt$$
Equation 46

where:-

$$C_1 = \left[2^{\frac{v}{2}} . \Gamma\left(\frac{v}{2}\right) \right]^{-1}$$
Equation 47

We perceive at once that when v = 3:-

$$2^{\frac{v}{2}} = 2^{\frac{3}{2}} = \sqrt{8}$$
Equation 48

Now with regard to the Gamma Function of Half-Integers[18] we may note that:-

$$\Gamma\left(\frac{v}{2}\right) = \frac{\sqrt{\pi}.(v-2)!!}{2^{\frac{v-1}{2}}}$$

Equation 49

$(v-2)!!$ is a Double Factorial[19] of $v-2$ where for odd arguments:-

$$n!! = \prod_{k=1}^{\frac{n+1}{2}} (2k-1)$$

Equation 50a

and for even arguments:-

$$n!! = \prod_{k=1}^{\frac{n}{2}} (2k)$$

Equation 50b

Therefore:-

$$(v-2)!! = (3-2)!! = 1!! = \prod_{k=1}^{1} (2k-1) = 1$$

Equation 51

So, by recollection of Equation Forty-Nine:-

$$\Gamma\left(\frac{3}{2}\right) = \frac{\sqrt{\pi}.(3-2)!!}{2^{\frac{3-1}{2}}} = \frac{\sqrt{\pi}.1}{2^1} = \frac{\sqrt{\pi}}{2}$$

Equation 49

Accordingly:-

$$C_1 = \left[2^{\frac{v}{2}}.\Gamma\left(\frac{v}{2}\right)\right]^{-1} = \left[2^{\frac{3}{2}}.\Gamma\left(\frac{3}{2}\right)\right]^{-1} = \left[2^{\frac{3}{2}}.\frac{\sqrt{\pi}}{2}\right]^{-1} = \frac{1}{\sqrt{2\pi}}$$

Equation 52

Turning to the integral, we may chose the form for $P(X,v)$ in anticipation of extended campaigns of simplification. Therefore:-

$$P(X,v) = \frac{1}{\sqrt{2\pi}} \int_0^{cq} t^{\frac{v}{2}-1} . e^{-\frac{t}{2}} . dt = \frac{1}{\sqrt{2\pi}} \int_0^{cq} t^{\frac{3}{2}-1} . e^{-\frac{3}{2}} . dt = \frac{1}{\sqrt{2\pi}} \int_0^{cq} \sqrt{t} . e^{-\frac{3}{2}} . dt$$

Equation 53

Clearly, it is therefore the case that:-

$$Q(X,v) = 1 - \frac{1}{\sqrt{2\pi}} \int_0^{cq} \sqrt{t} . e^{-\frac{3}{2}} . dt$$

Equation 54

By finding the antiderivative using Integral Calculator®, followed by re-arrangements and a bit of trial-and-error, I was able to establish that:-

$$p_{cq} = \sqrt{2\pi} \left[\left\{ erf\left(\frac{\sqrt{cq+4.5}}{\sqrt{2}} \right) - erf\left(\frac{\sqrt{cq}}{\sqrt{2}} \right) \right\} - \frac{1}{\pi} \left\{ \sqrt{cq+4.5} . e^{-\frac{cq+4.5}{2}} - \sqrt{cq} . e^{-\frac{cq}{2}} \right\} \right]$$

Equation 55

where p_{cq} is essentially $Q(X,3)$, the probability that the distribution is random.

This expression is okay for $p_{cq} < 0.1$ where error can vary from around 6% to 12%, but breaks down completely for probabilities in the upper ranges.

Notwithstanding Equation Fifty-Five further application of black arts enabled:-

$$p_{cq} = \sqrt{cq} . e^{-\frac{cq}{2}} - \sqrt{cq+4.5} . e^{-\frac{cq+4.5}{2}}$$

Equation 56

which was an even better estimate of probability for lower values but also broke down in excess of 0.1.

I recommend Equation Fifty-Four for full-range accuracy. The integral part of Equation Fifty-Four may expeditiously be solved using some standard numerical integration method such as Romberg[20] or a Newton-Coates formula: Abramowitz and Stegun offers a number of candidates.

Often enough the old tricks are the best tricks, and frequently the cheapest too.

Chi-Squared Results for the Mineral Crystal Characteristics

Appendix F presents four tabulations of Chi-Squared analysis outcomes for:-

(i)   The hypothesis that RAND() generates a uniform distribution
(ii)   The hypothesis that Ξ is distributed uniformly (i.e. is "random")

(iii)   The hypothesis that Ψ is random

(iv)   The hypothesis that the product $\Xi.\Psi$ is random

The common hypothesis is that the objective distribution is uniformly random: So its negation infers determinacy.

All judgments are at the 0.001 α_{crit} confidence level, or the *critical* odds are about 1000 to 1 that the distribution is uniform (or not as the case may be).

The first test is to confirm that the EXCEL® intrinsic function RAND() does indeed generate a uniformly-distributed output. The probability p that it does is computed, for 125 suitably scaled and shifted values, as 0.9990182. Accordingly the verdict is that the data conforms to the model. This implies that we can rely upon EXCEL® to provide a valid set of random numbers for comparison.

The second test is to confirm that Ξ is uniformly random. We hope for confutation. The probability p that Ξ indeed forms a random distribution is 3.586279×10^{-10}. Or the odds that Ξ is *not* random are about 2.8 milliard to 1. The Table rubricates that "The data does not conform to model".

Thirdly, we test the hypothesis that Ψ is random, and again desire contradiction. Table F3 gives p equal to 6.598912×10^{-25} or the odds that Ψ is determinate are something like 1.5 billion billion to one. (I am using British billions, where a British billion is 10^{-12}). As we might expect Lattice Angle Multiplier is even more deterministic than Apparent Lattice Strain.

Lastly, is the product of Apparent Lattice Strain and Lattice Angle Multiplier also non-random?: Common sense would suggest so. The odds against randomness are around $7.43639E^{+13}$, intermediate between the odds for Ξ and those for Ψ.

The key results are summarised in Table Six:-

Test Serial	Subject Date	Chi-Squared	Degrees of Freedom	α	p	CHISQ. INV.RT	Verdict
0 RAND()		0.024	3	0.001	0.999018227	16.26624	ACCEPT
1 Ξ		46.936	3	0.001	3.58628E-10	16.26624	REJECT
2 Ψ		115.6717	3	0.001	6.59891E-25	16.26624	REJECT
3 Product $\Xi.\Psi$		67.672	3	0.001	1.34474E-14	16.26624	REJECT

One-Hundred and Twenty-Five Data Tried for being Uniformally Distributed

Table Six
Summary of the Chi-Squared Test Results

Given that we have only one hundred and twenty-five data, this is as close as we may get to proof categorical that Apparent Lattice Strain Ξ is no chimera or trick of cognition, but is an actual feature of the natural world.

References

Element Properties Data

1   https://en.wikipedia.org/wiki/list_of_chemical_elements
2   http://periodictable.com
3   http://www.knowledgedoor.com/2/elements_handbook
4   http://www.espimetals.com

Minerals Properties Data

5   http://www.handbookofmineralogy.org
6   http://webmineral.com
7   http://som.web.cmu.edu/structures/
8   http://www.spec2000.net/05-mineralprops.htm
9   http://database.iem.ac.ru/mincryst/
10   http://www.minweb.co.uk/

Theoretical Cell Volume

11   http://webmineral.com/help/
           CellDimensions.shtml#WZg2nE1K0r9

General Equation of Cell Volume

12   Lattice constant. (2017, July 31).
           In *Wikipedia, The Free Encyclopedia.*
           Retrieved 08:19, August 29, 2017
           from
           https://en.wikipedia.org/w/index.php?
               title=Lattice_constant&oldid=793291329

The Crystal Lattice and the Golden Mean (The Ratio of Phidias)

13   "X-Ray Multiple-Wave Diffraction: Theory and Application"
           Shih-Lin Chang
           Springer-Verlag of Berlin   2004 Edition
           ISBN 978-3540211969
           pp 452

14   "The Golden Ratio, ionic and atomic radii and bond lengths"
           Raji Heyrovska
           Molecular Physics; V103; No.6-8; 20 March – 20 April 2005;
            877-882
           ISSN 0026 8976
           Taylor and Francis Group Ltd

15   Chi-Squared Distribution
        http://mathworld.wolfram.com/Chi-SquaredDistribution.html

Statistical Mathematics

16   "Handbook of Mathematical Functions:
        with Formulas, Graphs and Mathematical Tables"
        Milton Abramowitz and Irene A Stegun
        Dover Publications of New York   1965
        (National Bureau of Standards   1964)
        SBN: 486-61272-4
        pp 1046

        Gamma Function
        Eqn.:   6.1.1
        Page   255

17   "Handbook of Mathematical Functions:
        with Formulas, Graphs and Mathematical Tables"
        Milton Abramowitz and Irene A Stegun
        Dover Publications of New York   1965
        (National Bureau of Standards   1964)
        SBN: 486-61272-4
        pp 1046

        Chi-Squared Function
        Eqn.:   26.4.1
        Page   940

18   Particular values of the Gamma function. (2017, October 14).
        In *Wikipedia, The Free Encyclopedia*.
        Retrieved 13:20, October 16, 2017, from
        https://en.wikipedia.org/w/index.php?
                title=Particular_values_of_the_Gamma_function
                &oldid=805266221

19   Double factorial. (2017, October 7).
        In *Wikipedia, The Free Encyclopedia*.
        Retrieved 13:22, October 16, 2017, from
        https://en.wikipedia.org/w/index.php?
                title=Double_factorial&oldid=804230791

20   "Basic Numerical Methods"
           "An Introduction to Numerical Mathematics
           on a Microcomputer"
           RE Scraton
           Edward Arnold of London   1984
           ISBN 0-7131-3521-2
           pp 92
           Romberg Integration pp61-64

APPENDIX A
DATA FOR THE
PRINCIPAL ROCK-FORMING ELEMENTS

Atomic Number Z	Atomic Weight	Element Name	Element Symbol	Density ρ_l (grams/cc) (STP)	Density Reference Temperature (°K)	Coefficient of Linear Expansion α (10^{-6}/°K)	Bulk Modulus K (Gpa)	Electro negativity χ	Atomic Radius R (pm)
1	1.008	Hydrogen	H	0.088	4		0.2	2.2	53
3	6.997	Lithium	Li	0.534	298.15	46	11.6	0.98	152
4	9.0121831	Beryllium	Be	1.85	273.15	11.3	130	1.57	112
5	10.811	Boron	B	2.46	293.15	6	320	2.04	87
6	12.011	Carbon	C	2.2	293.15	25	33	2.55	67
7	14.007	Nitrogen	N	1.03	20		1.2	3.04	56
8	15.999	Oxygen	O	1.19	80			3.44	48
9	18.9984032	Flourine	F	1.706	53.48			3.98	42
11	22.98977	Sodium	Na	0.968	293.15	71	6.3	0.93	190
12	24.305	Magnesium	Mg	1.738	293.15	8.2	45	1.31	145
13	26.9815385	Aluminium	Al	2.698	293.15	23.1	76	1.61	118
14	28.085	Silicon	Si	2.3296	293.15	2.6	100	1.9	111
15	30.973761998	Phosphorous	P	1.82	293.15	124.5	11	2.19	98
16	32.06	Sulfur	S	2.067	293.15	64	7.7	2.58	88
17	35.453	Chlorine	Cl	2.03	93		1.1	3.16	79
19	39.0983	Potassium	K	0.856	293.15	83.3	3.1	0.82	243
20	40.078	Calcium	Ca	1.55	293.15	22.3	17	1	194
22	47.867	Titanium	Ti	4.506	293.15	8.6	110	1.54	176
23	50.9415	Vanadium	V	6	293.15	8.4	160	1.63	171
24	51.9961	Chromium	Cr	7.15	293.15	4.9	160	1.66	166
25	54.938044	Manganese	Mn	7.3	293.15	21.7	120	1.55	161
26	55.845	Iron	Fe	7.874	293.15	11.8	170	1.83	156
27	58.9332	Cobalt	Co	8.9	298.15	13	180	1.88	152
28	58.6934	Nickel	Ni	8.912	293.15	13.4	180	1.91	149
29	63.546	Copper	Cu	8.96	293.15	16.5	140	1.9	145
30	65.38	Zinc	Zn	7.134	293.15	30.2	70	1.65	142
31	69.723	Gallium	Ga	5.904	298.15	120	22	1.81	136
33	74.9216	Arsenic	As	5.778	299.15	5.6	22	2.18	114
34	78.971	Selenium	Se	4.809	293.15	37	8.3	2.55	103
35	79.904	Bromine	Br	3.122	293.15		1.9	2.96	94
38	87.62	Strontium	Sr	2.63	298.15	22.5	11.6	0.95	219
41	92.90637	Niobium	Nb	8.61	298.15	7.3	170.2	1.6	198
40	91.224	Zirconium	Zr	6.52	298.15	5.7	91.1	1.33	160
42	95.95	Molybdenum	Mo	10.2	293.15	4.8	230	2.16	190
44	101.07	Ruthenium	Ru	12.1	298.15	6.4	320.8	2.2	178
45	102.9055	Rhodium	Rh	12.41	273.15	8.2	380	2.28	173
46	106.42	Palladium	Pd	12	298.15	11.8	180.8	2.2	169
47	107.8602	Silver	Ag	10.501	293.15	18.9	100	1.93	165
48	112.411	Cadmium	Cd	8.65	298.15	30.8	46.7	42	161
49	114.818	Indium	In	7.31	298.15	32.1	41.1	1.78	156
50	118.71	Tin	Sn	7.287	293.15	22	58	1.96	145
51	121.76	Antimony	Sb	6.685	293.15	11	42	2.05	133
52	127.6	Tellurium	Te	6.232	293.15	18	65	2.1	123
56	137.327	Barium	Ba	3.51	298.15	20.6	10.3	0.89	253
57	138.90547	Lanthanum	La	6.145	273.15	12.1	28	1.1	207
58	140.116	Cerium	Ce	6.77	273.15	6.3	22	1.12	204
60	144.242	Neodymium	Nd	7.007	273.15	9.6	32	1.14	206
72	178.49	Hafnium	Hf	13.31	273.15	5.9	110	1.3	208
73	180.94788	Tantalum	Ta	16.4	298.15	6.3	200	1.5	200
74	183.84	Tungsten	W	19.25	293.15	4.5	310	2.36	193
75	186.207	Rhenium	Re	21.02	273.15	6.2	370	1.9	188
76	190.23	Osmium	Os	22.59	298.15	5.1	418	2.2	185
77	192.217	Iridium	Ir	22.562	298.15	6.4	355	2.2	180
78	195.084	Platinum	Pt	21.5	298.15	8.8	278.3	2.28	177
79	196.966569	Gold	Au	19.282	293.15	14.2	220	2.54	174
80	200.592	Mercury	Hg	13.534	298.15	60.4	38.2	2	151
81	204.38	Thallium	Tl	11.85	273.15	29.9	43	1.62	156
82	207.2	Lead	Pb	11.342	293.15	28.9	46	1.87	154
83	208.9804	Bismuth	Bi	9.807	273.15	13.4	31	2.02	143
90	232.0377	Thorium	Th	11.72	273.15	11	54	1.3	206
92	238.02891	Uranium	U	18.95	273.15	13.9	100	1.38	196

Table A1
Selected Element Data

APPENDIX B
DATA FOR THE ARBITRARILY-CHOSEN
ROCK-FORMING MINERALS

Table B1a
Selected Mineral Data

Serial	Mineral	Chemical Formula	Handbook of mineralogy (Density g/cc)				Bulk Electron Density	Webmineral Com Measured Density	Calculated Formula Weight	Lattice System	Space Group*	a	b	c	α	β	γ	Z
			LB	UB	mean	calc												
1	ARGENTOPYRITE	$AgFe_2S_3$	4.25	4.25	4.25	4.27	3.98	4.25	315.7302	orthorhombic	Pmmn	6.639	11.463	6.452	1.570796	1.570796	1.570796	4
2	ARGENTOTENNANTITE	$(Ag,Cu)_{10}(Zn,Fe)_2(As,Sb)_4S_{13}$	4.71	4.71	4.71	5.05	4.71	4.71	1788.399	cubic	I43m	10.584	10.584	10.584	1.570796	1.570796	1.570796	2
3	BARITE	$BaSO_4$	4.5	4.5	4.5	4.47	3.99	4.5	233.383	orthorhombic	Pnma	8.884	5.457	7.157	1.570796	1.570796	1.570796	4
4	CASSITERITE	SnO_2	6.98	7.01	6.995	6.993	6.26	6.995	150.708	tetragonal	P4$_2$/mnm	4.7382	4.7382	3.1871	1.570796	1.570796	1.570796	2
5	CERUSSITE	$PbCO_3$	6.552	6.552	6.552	6.577	5.52	6.552	267.208	orthorhombic	Pmcn	5.179	8.492	6.141	1.570796	1.570796	1.570796	4
6	CHALCOPYRITE	$CuFeS_2$	4.1	4.3	4.2	4.283	3.98	4.2	183.511	tetragonal	I42d	5.281	5.281	10.401	1.570796	1.570796	1.570796	4
7	CHILDRENITE	$Fe^{2+}Al(PO_4)(OH)_2 \cdot H_2O$	3.11	3.19	3.15	3.135	3.14	3.15	228.8173	orthorhombic	Bba2	10.395	13.394	6.918	1.570796	1.570796	1.570796	8
8	CUPRITE	$Cu_2^{1+}O$	6.14	6.14	6.14	6.15	5.64	6.14	143.091	cubic	Pn3m	4.2685	4.2685	4.2685	1.570796	1.570796	1.570796	2
9	EDGARITE	$FeNb_3S_6$	4.98	4.98	4.98	4.99	3.85	4.98	526.9241	hexagonal	P6$_3$22	5.771	5.771	12.19	1.570796	1.570796	2.094395	2
10	GALENA	PbS	7.58	7.58	7.58	7.57	6.23	7.58	239.26	cubic	Fm3m	5.936	5.936	5.936	1.570796	1.570796	1.570796	4
11	SMITHSONITE	$ZnCO_3$	4.431	4.431	4.431	4.43	4.24	4.431	125.388	hexagonal	R3c	4.6526	4.6526	15.9257	1.570796	1.570796	2.094395	6
12	SPHALERITE	$(Zn,Fe)S$	3.9	4.1	4	4.096	3.85	4	92.6725	cubic	F43m	5.406	5.406	5.406	1.570796	1.570796	1.570796	4
13	ZUSSMANITE	$K(Fe^{2+},Mg,Mn^{2+})_{13}(AlSi)_{18}O_{42}(OH)_{14}$	3.146	3.146	3.146	3.14	3.09	3.146	2205.754	hexagonal	R3	11.66	11.66	28.69	1.570796	1.570796	2.094395	3
14	TENNANTITE	$(Cu,Ag,Sn,Fe)_{12}As_4S_{12}$	4.72	4.72	4.72	4.62	4.3	4.72	1722.29	cubic	I43m	10.19	10.19	10.19	1.570796	1.570796	1.570796	2
15	WOLLASTONITE 2M	$CaSiO_3$	2.86	3.09	2.975	2.9	2.91	2.84	116.16	monoclinic	2/m	15.409	7.322	7.063	1.570796	1.663299	1.570796	12
16	WOLLASTONITE 1A	$CaSiO_3$	2.86	3.09	2.975	2.9	2.91	2.84	116.16	triclinic	P1	7.94	7.32	7.063	1.567306	1.479969	1.339366	6
17	PYRITE	FeS_2	5.018	5.018	5.018	5.013	4.84	5.018	119.98	cubic	Pa3	5.4179	5.4179	5.4179	1.570796	1.570796	1.570796	4
18	CALCITE	$CaCO_3$	2.7102	2.7102	2.7102	2.711	2.71	2.7102	100.086	hexagonal	R3c	4.9896	4.9896	17.061	1.570796	1.570796	2.094395	6
19	STRONTIANITE	$SrCO_3$	3.76	3.76	3.76	3.78	3.48	3.76	147.628	orthorhombic	Pmcn	5.1059	8.4207	6.0319	1.570796	1.570796	1.570796	4
20	QUARTZ-α	SiO_2	2.65	2.65	2.65	2.66	2.65	2.65	60.083	trigonal	P3$_1$21	4.9135	4.9135	5.405	1.570796	1.570796	2.094395	3
21	QUARTZ-β	SiO_2	2.65	2.65	2.65	2.66	2.53	2.65	60.083	hexagonal	P3$_2$21	4.9965	4.9965	5.4546	1.570796	1.570796	2.094395	6
22	CORUNDUM	Al_2O_3	3.98	4.1	4.04	3.997	3.93	4.04	101.9601	hexagonal	R3c	4.754	4.754	12.982	1.570796	1.570796	2.094395	6
23	ACANTHITE	Ag_2S	7.2	7.22	7.21	7.24	6.43	7.21	247.7804	monoclinic	P2$_1$/n	4.229	6.931	7.862	1.570796	1.738522	1.570796	4
24	ARGENTITE	Ag_2S	7.2	7.4	7.3	7.04	6.25	7.3	247.7804	cubic	Im3m	4.88	4.88	4.88	1.570796	1.570796	1.570796	2
25	PYRARGYRITE	Ag_3SbS_3	5.82	5.82	5.82	5.855	5.19	5.82	541.5206	hexagonal	R3c	11.047	11.047	8.719	1.570796	1.570796	2.094395	6

Table B1b
Selected Mineral Data

Serial	Mineral	Chemical Formula	Density (g/cc) Handbook of mineralogy LB	UB	mean	calc	Bulk Electron Density	Webmineral Com Measured Density	Calculated Formula Weight	Lattice System	Space Group*	a	b	c	α	β	γ	Z
26	HALITE	NaCl	2.168	2.168	2.168	2.165	2.07	2.16	58.44	cubic	Fm3m	5.6404	5.6404	5.6404	1.570796	1.570796	1.570796	4
27	SYLVITE	KCl	1.993	1.993	1.993	1.987	1.92	1.99	74.55	cubic	Fm3m	6.2931	6.2931	6.2931	1.570796	1.570796	1.570796	4
28	NATRON	$Na_2CO_3 \cdot 10H_2O$	1.478	1.478	1.478	1.458	1.55	1.46	286.14	monoclinic	Cc	12.83	9.026	13.44	1.570796	2.146755	1.570796	4
29	NITRATITE	$NaNO_3$	2.24	2.29	2.265	2.25	2.23	2.26	84.99	hexagonal	R3c	5.07	5.07	16.829	1.570796	1.570796	2.094395	6
30	SPODUMENE	LiAlSi2O6	3.03	3.23	3.13	3.184	3.11	3.15	186.09	monoclinic	C2/c	9.45	8.39	5.215	1.570796	1.919862	1.570796	4
31	GYPSUM	$CaSO_4 \cdot 2H_2O$	2.317	2.317	2.317	2.31	2.36	2.31	172.17	monoclinic	I2/a	5.679	15.202	6.522	1.570796	2.066993	1.570796	4
32	ANHYDRITE	$CaSO_4$	2.98	2.98	2.98	2.95	2.97	2.96	136.14	orthorhombic	Amma	6.993	6.995	6.245	1.570796	1.570796	1.570796	4
33	BIOTITE	$K(Mg,Fe^{2+})_3(Al,Fe^{3+})Si_3O_{10}(OH,F)_2$	2.7	3.3	3	3.25	3.07	2.89	433.53	monoclinic	C2/m	5.3	9.2	10.2	1.570796	1.745329	1.570796	2
34	PENNANTITE	$Mn_5{}^{2+}Al(Si_3Al)O_{10}(OH)_8$	2.89	3.07	2.98	3.18	2.98	3.2	708.06	monoclinic	C1	5.45	9.5	14.4	1.570796	1.698205	1.570796	2
35	MUSCOVITE	$KAl_2(AlSi_3O_{10})(F,OH)_2$	2.77	2.88	2.825	2.83	2.81	2.83	398.71	monoclinic	C2/c	5.199	9.027	20.106	1.570796	1.666789	1.570796	4
36	CINNABAR	HgS	8.176	8.176	8.176	8.2	6.76	8.19	232.66	hexagonal	P3₁21	4.145	4.145	9.496	1.570796	1.570796	2.094395	3
37	GOETHITE	$\alpha\text{-}Fe^{3+}O(OH)$	4.28	4.28	4.28	4.18	4.13	4.27	88.85	orthorhombic	Pbnm	4.608	9.956	3.0215	1.570796	1.570796	1.570796	4
38	CHLORARGYRITE	AgCl	5.556	5.556	5.556	5.57	4.96	5.55	143.32	cubic	Fm3m	5.554	5.554	5.554	1.570796	1.570796	1.570796	4
39	MAGNETITE	$Fe^{2+}Fe^{3+}{}_2O_4$	5.175	5.175	5.175	5.2	4.89	5.21	231.54	cubic	Fd3m	8.397	8.397	8.397	1.570796	1.570796	1.570796	8
40	HEAMATITE	$\alpha\text{-}Fe_2O_3$	5.26	5.26	5.26	5.255	5.04	5.28	159.69	hexagonal	R3c	5.038	5.038	13.772	1.570796	1.570796	2.094395	6
41	OLIVINE	$(Mg,Fe)_2SiO_4$	3.27	3.37	3.32	3.3	3.25	3.3	153.31	orthorhombic	Pbnm	4.78	10.25	6.3	1.570796	1.570796	1.570796	4
42	MAGNESIOHORNBLENDE	$Ca_2[Mg_4(Al,Fe^{++})]Si_7AlO_{22}(OH)_2$	3	3.47	3.235	2.96	3.22	2.96	821.16	monoclinic	C2/m	9.887	18.174	5.308	1.570796	1.832596	1.570796	2
43	VERMICULITE	$(MgFe,Al)_3(Al,Si)_4O_{10}(OH)_2 \cdot 4H_2O$	2.2	2.6	2.4	2.26	2.51	2.32	504.19	monoclinic	C2/c	5.349	9.255	28.89	1.570796	1.695006	1.570796	4
44	MAGNESITE	$MgCO_3$	3	3	3	3.01	2.97	2.98	84.31	hexagonal	R3c	4.6632	4.6632	15.015	1.570796	1.570796	2.094395	6
45	CHALCOCITE	Cu_2S	5.5	5.8	5.65	5.8	6.01	6.46	159.16	monoclinic	P2₁/c	15.246	11.884	13.494	1.570796	2.030691	1.570796	48
46	COVELLITE	CuS	4.6	4.76	4.68	4.602	4.41	4.68	95.61	hexagonal	P63/mmc	3.7938	3.7938	16.341	1.570796	1.570796	2.094395	6
47	NATIVE PALLADIUM	Pd	11.9	11.9	11.9	12.04	10.4	12.19	107.44	cubic	Fm3m	3.8898	3.8898	3.8898	1.570796	1.570796	1.570796	4
48	SILLIMANITE	Al_2SiO_5	3.23	3.24	3.235	3.24	3.2	3.25	162.05	orthorhombic	Pbnm	7.4883	7.6808	5.7774	1.570796	1.570796	1.570796	4
49	TITANITE	$CaTiSiO_5$	3.48	3.6	3.54	3.53	3.47	3.55	197.76	monoclinic	P2₁/a	7.057	8.707	6.555	1.570796	1.986359	1.570796	4
50	ZIRCON	$ZrSiO_4$	4.6	4.7	4.65	4.714	4.52	4.85	190.31	tetragonal	I4₁/amd	6.607	6.607	5.982	1.570796	1.570796	1.570796	4

Table B1c
Selected Mineral Data

Serial	Mineral	Chemical Formula	LB	UB	mean	calc	Bulk Electron Density	Measured Density	Calculated Formula Weight	Lattice System	Space Group*	a	b	c	α	β	γ	Z
			Handbook of mineralogy					Webmineral Com										
51	ALLARGENTUM	Ag_6Sb	10	10	10	10.12	8.72	10.01	108.01	hexagonal	nd	2.952	2.952	4.773	1.570796	1.570796	2.094395	2
52	FLUORITE	CaF_2	3.175	3.184	3.1795	3.18	3.1	3.18	78.07	cubic	Fm3m	5.4626	5.4626	5.4626	1.570796	1.570796	1.570796	4
53	ABELSONITE	$C_{31}H_{32}N_4Ni$	1.33	1.48	1.405	1.45	1.48	1.45	519.31	triclinic	P1	8.508	11.185	7.299	1.585632	1.992002	1.395973	1
54	ABERNATHYITE	$K_2(UO_2)_2(AsO_4)_2 \cdot 6H_2O$	3.32	3.32	3.32	3.572	3.3	3.44	520.11	tetragonal	P4/ncc	7.176	7.176	18.126	1.570796	1.570796	1.570796	4
55	AMMONIOALUNITE	$(NH_4)Al_3(SO_4)_2(OH)_6$	2.4	2.4	2.4	2.58	2.61	2.4	393.15	hexagonal	R3m	7.013	7.013	17.855	1.570796	1.570796	1.570796	5
56	BABEFPHITE	$BaBe(PO_4)F$	4.31	4.31	4.31	4.325	3.94	4.31	259.56	triclinic	P1	6.889	16.814	6.902	1.570971	1.570622	1.576381	8
57	BABINGTONITE	$Ca_2(Fe^{2+},Mn)Fe^{3+}Si_5O_{14}(OH)$	3.34	3.37	3.355	3.26	3.48	3.4	573.05	triclinic	P1	7.5	12.18	6.68	1.506219	1.633454	1.958085	2
58	CABRIITE	Pd_2SnCu	11.1	11.1	11.1	10.7	9.29	10.7	395.1	orthorhombic	Pmmm	7.88	7.88	3.94	1.570796	1.570796	1.570796	4
59	CACOXENITE	$Fe^{3+}_{24}AlO_6(PO_4)_{17}(OH)_{12} \cdot 75H_2O$	2.2	2.6	2.4	2.217	2.89	2.84	4481.51	hexagonal	P6$_3$/m	27.559	27.559	10.55	1.570796	1.570796	1.570796	2
60	DACHIARDITE	$(Ca,Na_2,K_2)_5Al_{10}Si_{38}O_{96} \cdot 25H_2O$	2.165	2.206	2.1855	2.1395	2.18	2.16	1808.84	monoclinic	C2/m	18.625	7.508	10.247	1.570796	1.835933	1.570796	1
61	DADSONITE	$Pb_{23}Sb_{25}S_{60}Cl$	5.68	5.68	5.68	5.51	4.67	5.72	8950.53	monoclinic	P1	17.33	4.11	19.05	1.570796	1.680752	1.577778	2
62	DIAMOND	C	3.511	3.511	3.511	3.515	3.52	3.51	12.01	cubic	Fd3m	3.5595	3.5595	3.5595	1.570796	1.570796	1.570796	8
63	EAKERITE	$Ca_2Sn^{4+}Al_2Si_6O_{18}(OH)_2 \cdot 2H_2O$	2.93	2.93	2.93	2.931	2.83	2.93	779.38	monoclinic	P2$_1$/a	15.829	7.721	7.438	1.570796	1.772673	1.570796	2
64	EARLANDITE	$Ca_3(C_6H_5O_7)_2 \cdot 4H_2O$	1.8	1.95	1.875	1.96	2.02	1.95	570.5	monoclinic	NoDet	30.94	5.93	10.56	1.570796	1.635955	1.570796	4
65	FABIANITE	$CaB_3O_5(OH)$	2.77	2.796	2.783	2.788	2.77	2.77	169.51	monoclinic	P2$_1$/a	6.593	10.488	6.365	1.570796	1.943276	1.570796	4
66	GABRIELITE	$Tl_2AgCu_2As_3S_7$	5.36	5.36	5.36	5.1	4.69	5.36	1082.85	triclinic	P1	12.138	12.196	15.944	1.370729	1.478556	1.055401	6
67	GABRIELSONITE	$PbFe^{2+}(AsO_4)(OH)$	6.67	6.67	6.67	6.69	5.97	6.67	418.97	orthorhombic	P2$_1$ma	7.86	5.98	8.62	1.570796	1.570796	1.570796	4
68	GRAPHITE	C	2.09	2.23	2.16	2.26	2.25	2.16	12.01	hexagonal	P6$_3$/mmc	2.464	2.464	6.711	1.570796	1.570796	1.570796	4
69	HAFNON	$HfSiO_4$	6.16	6.16	6.16	6.97	6.16	6.97	244.28	tetragonal	I4$_1$/amd	6.5725	6.5725	5.9632	1.570796	1.570796	1.570796	4
70	HAGENDORFITE	$NaCaMn^{2+}(Fe^{2+},Fe^{3+},Mg)_2(PO_4)_3$	3.71	3.71	3.71	3.84	3.58	3.71	501.07	monoclinic	I2/a	10.933	12.594	6.515	1.570796	1.770074	1.570796	4
71	IANGREYITE	$Ca_2Al_7(PO_4)_2(PO_3OH)_2(OH,F)_{15} \cdot 8H_2O$	2.46	2.46	2.46	2.451				hexagonal	P321	6.988	6.988	16.707	1.570796	1.570796	1.570796	1
72	ICE	H_2O	0.9167	0.9167	0.9167	0.93	1.02	0.9167	18.02	hexagonal	P6$_3$/mmc	4.498	4.498	7.338	1.570796	1.570796	1.570796	4
73	JACHYMOVITE	$(UO_2)_8(SO_4)(OH)_{14} \cdot 13H_2O$	4.79	4.79	4.79	4.79	4.1	4.79	2728.59	monoclinic	P2$_1$	18.553	9.276	13.532	1.570796	2.191435	1.570796	2
74	JACOBSITE	$(Mn^{2+},Fe^{2+},Mg)(Fe^{3+},Mn^{3+})_2O_4$	4.76	4.76	4.76	5.03	4.49	4.75	227.38	cubic	Fd3m	8.499	8.499	8.499	1.570796	1.570796	1.570796	8
75	KAATIALAITE	$Fe^{3+}(H_2AsO_4)_3 \cdot 3\text{-}5H_2O$	2.64	2.64	2.64	2.64	2.58	2.64	577.74	monoclinic	P2$_1$	15.3065	19.722	4.736	1.570796	1.602823	1.570796	4

CABRITE: n is about 75

Table B1d
Selected Mineral Data

Serial	Mineral	Chemical Formula	Density (g/cc) Handbook of mineralogy LB	UB	mean	calc	Bulk Electron Density	Webmineral Com Measured Density	Calculated Formula Weight	Lattice System	Space Group*	a	b	c	α	β	γ	Z
76	KADYRELITE	$Hg^{1+}_6H(Br,Cl)_3O_2$	8.93	8.93	8.93	8.79	7.27	8.93	955.94	cubic	[Ia3d]	16.22	16.22	16.22	1.570796	1.570796	1.570796	16
77	LABRADORITE	$Na_{0.5-0.3}Ca_{0.5-0.7}Al_{1.5-1.7}Si_{2.5-2.3}O_8$	2.68	2.72	2.7	2.6975	2.67	2.69	271.81	triclinic	C1(low)	8.1648	12.8585	7.0973	1.580687	2.026327	1.567C15	4
78	LANARKITE	$Pb_2O(SO_4)$	6.92	6.92	6.92	7.08	5.87	6.92	526.46	monoclinic	C2/m	13.769	5.698	7.079	1.570796	2.023418	1.570796	4
79	MACAULAYITE	$(Fe^{3+})_{24}Si_4O_{43}(OH)_2$	4.41	4.41	4.41	4.41	4.22	4.41	2116.93	monoclinic	C-centered	5.038	8.726	36.342	1.570796	1.605703	1.570796	2
80	MACDONALDITE	$BaCa_4Si_{16}O_{36}(OH)_2 \cdot 10H_2O$	2.27	2.27	2.27	2.27	2.35	2.27	1537.15	orthorhombic	Bmmb	14.06	23.52	13.08	1.570796	1.570796	1.570796	4
81	NABALAMPROPHYLLITE	$Ba(Na,Ba)(Na_3Ti(Ti_2O_2Si_4O_{14})(OH,F)_2)$	3.62	3.62	3.62	3.58	3.37	3.58	810.56	monoclinic	P2/m	19.741	7.105	5.408	1.570796	1.68721	1.570796	2
82	NABAPHITE	$NaBaPO_4 \cdot 9H_2O$	2.3	2.3	2.3	2.26	2.21	2.3	417.43	cubic	P2₁3	10.711	10.711	10.711	1.570796	1.570796	1.570796	4
83	OBOYERITE	$H_6(Pb,Ca)_6(Te^{4+}O_3)_3(Te^{6+}O_6)_2 \cdot 2H_2O$	6.4	6.4	6.4	6.66	5.59	6.4	2259.27	triclinic	P1	12.249	15.113	6.868	2.032436	1.720546	1.497842	2
84	OBRADOVICITE	$H_4(K,Na)CuFe^{3+}_2(AsO_4)(MoO_4)_5 \cdot 12H_2O$	3.55	3.55	3.55	3.68	3.17	3.55	1257.18	orthorhombic	Pcnm	15.046	14.848	11.056	1.570796	1.570796	1.570796	4
85	PAAKKONENITE	Sb_2AsS_2	5.21	5.21	5.21	5.21	4.61	5.21	382.55	monoclinic	C2/m	10.75	3.959	12.49	1.570796	2.012016	1.570796	4
86	PAARITE	$Cu_{1.7}Pb_{1.7}Bi_{6.3}S_{12}$	6.96	6.96	6.96	6.948	5.82	6.96	2166.36	orthorhombic	Pmcn	4.007	55.998	11.512	1.570796	1.570796	1.570796	5
87	PEROVSKITE	$CaTiO_3$	3.98	4.26	4.12	4.02	3.91	4	135.96	orthorhombic	Pnma	5.447	7.654	5.388	1.570796	1.570796	1.570796	4
88	QUANDILITE	$(Mg,Fe^{2+})_2(Ti,Fe^{3+},Al)O_4$	4.03	4.08	4.055	4.04				cubic	Fd3m	8.4376	8.4376	8.4376	1.570796	1.570796	1.570796	8
89	QILIANSHANITE	$H_3Na(HCO_3)(BO_3) \cdot 2H_2O$	1.706	1.706	1.706	1.639	1.68	1.706	181.87	monoclinic	C2	16.119	6.928	6.73	1.570796	1.753358	1.570796	4
90	RABBITTITE	$Ca_3Mg_3(UO_2)_2(CO_3)_6(OH)_4 \cdot 18H_2O$	2.57	2.57	2.57	2.69	2.46	2.58	1485.56	monoclinic	NoDet	32.6	23.8	9.45	1.570796	1.570796	1.570796	8
91	RABERITE	$Tl_5Ag_4As_6SbS_{15}$	5.65	5.65	5.65	5.654				triclinic	P1	8.92	9.429	20.062	1.390329	1.550551	1.094571	2
92	REALGAR	AsS	3.56	3.56	3.56	3.59	3.3	3.56	106.99	monoclinic	P2₁/n	9.325	13.571	6.587	1.570796	1.856739	1.570796	16
93	SABATIERITE	Cu_6TlSe_4	7.34	7.34	7.34	6.78	6.31	7.34	695.45	orthorhombic		3.986	5.624	9.778	1.570796	1.570796	1.570796	1
94	TACHARANITE	$Ca_{12}Al_2Si_{18}O_{51} \cdot 18H_2O$	2.33	2.36	2.345	2.28	2.32	2.28	2180.68	monoclinic	A-centered	17.07	3.65	27.9	1.570796	1.991421	1.570796	1
95	TACHYHYDRITE	$CaMg_2Cl_6 \cdot 12H_2O$	1.667	1.667	1.667	1.673	1.72	1.66	517.59	hexagonal	R3	10.136	10.136	17.318	1.570796	1.570796	1.570796	3
96	UCHUCCHACUAITE	$AgPb_3MnSb_5S_{12}$	5.16	5.16	5.16	5.61	4.74	5.16	1777.95	monoclinic	Pmmm	12.67	19.32	4.38	1.570796	1.570796	1.570796	2
97	UEDAITE-(CE)	$Mn^{2+}CeAl_1Fe(Si_2O_7)(SiO_4)O(OH)$	4.1	4.1	4.1	4.19	3.88	4.1	584.87	monoclinic	P2₁/m	8.939	5.742	10.187	1.570796	2.008874	1.57C796	2
98	UKLONSKOVITE	$NaMg(SO_4)F \cdot 2H_2O$	2.42	2.42	2.42	2.414	2.43	2.46	198.39	monoclinic	P2₁/m	7.202	7.214	5.734	1.570796	1.976236	1.57C796	2
99	VAESITE	NiS_2	4.45	4.45	4.45	4.47	4.35	4.45	122.82	cubic	Pa3	5.66787	5.66787	5.66787	1.570796	1.570796	1.57C796	4
100	VALENTINITE	Sb_2O_3	5.76	5.76	5.76	5.828	5.04	5.69	291.5	orthorhombic	Pccn	4.911	12.464	5.412	1.570796	1.570796	1.57C796	4

Table B1e
Selected Mineral Data

Serial	Mineral	Chemical Formula	Density (g/cc) Handbook of mineralogy LB	UB	mean	calc	Bulk Electron Density	Webmineral Com Measured Density	Calculated Formula Weight	Lattice System	Space Group*	a	b	c	α	β	γ	Z
101	WADEITE	$K_2ZrSi_3O_9$	3.1	3.13	3.115	3.16	3.01	3.11	397.67	hexagonal	P6₃/m	6.893	6.893	10.172	1.570796	1.570796	1.570796	2
102	WADSLEYITE	$\beta\text{-}(Mg,Fe^{2+})_2SiO_4$	3.77	3.77	3.77	3.84	3.77	3.84	156.46	orthorhombic	Imma	5.7	11.71	8.24	1.570796	1.570796	1.570796	8
103	XANTHIOSITE	$Ni_3(AsO_4)_2$	4.98	5.42	5.2	5.388	5.08	5.42	453.91	monoclinic	P2₁/a	10.174	9.548	5.766	1.570796	1.52272	1.570796	4
104	XANTHOCONITE	Ag_3AsS_3	5.54	5.54	5.54	5.53	4.97	5.55	494.72	monoclinic	C2/c	12	6.26	17.08	1.570796	1.949862	1.570796	8
105	YAFSOANITE	$Ca_3Zn_3(Te^{6+}O_6)_2$	5.4	5.55	5.475	5.54	4.61	5.55	763.6	cubic	Ia3d	12.632	12.632	12.632	1.570796	1.570796	1.570796	8
106	YAGIITE	$(Na,K)_{1.5}Mg_4(Al,Mg)_3(Si,Al)_{12}O_{30}$	2.7	2.7	2.7	2.7	2.6	2.7	1991.71	hexagonal	P6/mcc	10.09	10.09	14.29	1.570796	1.570796	1.570796	2
107	ZABUYELITE	Li_2CO_3	2.09	2.09	2.09	2.1	2.05	2.09	73.89	monoclinic	C2/c	8.361	4.976	6.193	1.570796	2.001718	1.570796	4
108	ZACCARINIITE	RhNiAs	10.09	10.09	10.09	10.19				tetragonal	P4/nmm	3.5498	3.5498	6.1573	1.570796	1.570796	1.570796	2
109	ACTINOLITE	$Ca_2(Mg,Fe^{2+})_5Si_8O_{22}(OH)_2$	3.03	3.24	3.135	3.07	3.02	3.04	853.16	monoclinic	C2/m	9.891	18.2	5.305	1.570796	1.826313	1.570796	2
110	ALMANDINE	$Fe^{2+}_3Al_2(SiO_4)_3$	4.318	4.318	4.318	4.313	4.08	4.19	497.75	cubic	Ia3d	11.526	11.526	11.526	1.570796	1.570796	1.570796	8
111	AZURITE	$Cu_3(CO_3)_2(OH)_2$	3.773	3.773	3.773	3.78	3.67	3.83	344.67	monoclinic	P2₁/c	5.0109	5.8485	10.345	1.570796	1.63208	1.570796	2
112	BERYL	$Be_3Al_2Si_6O_{18}$	2.63	2.97	2.8	2.64	2.74	2.76	537.5	hexagonal	P6/mcc	9.227	9.227	9.218	1.570796	1.570796	1.570796	2
113	BISMUTHITE	$Bi_2O_2(CO_3)$	6.1	7.7	6.9	8.25	5.82	7	509.97	tetragonal	I4/mmm	3.8685	3.8585	13.6915	1.570796	1.570796	1.570796	2
114	BORAX	$Na_2B_4O_5(OH)_4 \cdot 8H_2O$	1.715	1.715	1.715	1.7	1.78	1.71	381.37	monoclinic	C2/c	11.879	10.644	12.2012	1.570796	1.860818	1.570796	4
115	GOLD	Au	19.3	19.3	19.3	19.302	15.5	17.64	196.97	cubic	Fm3m	4.0786	4.0786	4.0786	1.570796	1.570796	1.570796	4
116	MONAZITE-(CE)	$(Ce,La,Nd,Th)PO_4$	4.98	5.43	5.205	5.26	4.58	5.15	240.21	monoclinic	P2₁/n	6.7902	7.0203	6.4674	1.570796	1.864321	1.570796	4
117	PYROPE	$Mg_3Al_2(SiO_4)_3$	3.582	3.582	3.582	3.563	3.53	3.74	403.13	cubic	Ia3d	11.459	11.459	11.459	1.570796	1.570796	1.570796	8
118	RUTILE	TiO_2	4.23	4.23	4.23	4.25	4.04	4.25	79.88	tetragonal	P4₂/mnm	4.5937	4.5937	2.9587	1.570796	1.570796	1.570796	2
119	SCHEELITE	$CaWO_4$	6.1	6.1	6.1	6.09	5.26	6.01	287.93	tetragonal	I4₁/a	5.2429	5.2429	11.3737	1.570796	1.570796	1.570796	4
120	SCOTLANDITE	$PbS^{4+}O_3$	6.37	6.37	6.37	6.418	5.49	6.37	287.26	monoclinic	P2₁/m	4.505	5.333	6.405	1.570796	1.854238	1.570796	2
121	SPINEL	$MgAl_2O_4$	3.6	4.1	3.85	3.578	3.59	3.64	142.27	cubic	Fd3m	8.0898	8.0898	8.0898	1.570796	1.570796	1.570796	8
122	TALC	$Mg_3Si_4O_{10}(OH)_2$	2.58	2.83	2.705	2.78	2.76	2.75	379.27	triclinic	P1	5.291	9.46	5.29	1.722291	2.09265	1.488242	2
123	TETRAHEDRITE	$(Cu,Fe,Ag,Zn)_{12}Sb_4S_{13}$	4.97	4.97	4.97	4.99	4.52	4.9	1643.31	cubic	I43m	10.39	10.39	10.39	1.570796	1.570796	1.570796	2
124	TOPAZ	$Al_2SiO_4(F,OH)_2$	3.49	3.57	3.53	3.55	3.51	3.55	182.25	orthorhombic	Pbnm	4.6499	8.7968	8.3909	1.570796	1.570796	1.570796	4
125	WULFENITE	$PbMoO_4$	6.5	7.5	7	7.18	5.8	6.75	367.14	tetragonal	I4₁/a	5.436	5.436	12.068	1.570796	1.570796	1.570796	4

* Rotoinversion Axes Neglected

Serial	Mineral	Chemical Formula	Formula Weight	a	b	c	α	β	γ	Z	V	V by General Equation
1	ARGENTOPYRITE	$AgFe_2S_3$	315.7302	6.639	11.463	6.452	1.570796	1.570796	1.570796	4	491.0156	491.0156
2	ARGENTOTENNANTITE	$(Ag,Cu)_{10}(Zn,Fe)_2(As,Sb)_4S_{13}$	1788.3992	10.584	10.584	10.584	1.570796	1.570796	1.570796	2	1185.631	1185.631
3	BARITE	$BaSO_4$	233.383	8.884	5.457	7.157	1.570796	1.570796	1.570796	4	346.9713	346.9713
4	CASSITERITE	SnO_2	150.708	4.7382	4.7382	3.1871	1.570796	1.570796	1.570796	2	71.55211	71.55211
5	CERUSSITE	$PbCO_3$	267.208	5.179	8.492	6.141	1.570796	1.570796	1.570796	4	270.0816	270.0816
6	CHALCOPYRITE	$CuFeS_2$	183.511	5.281	5.281	10.401	1.570796	1.570796	1.570796	4	290.0731	290.0731
7	CHILDRENITE	$Fe^{2+}Al(PO_4)(OH)_2 \cdot H_2O$	228.8173	10.395	13.394	6.918	1.570796	1.570796	1.570796	8	963.1975	963.1975
8	CUPRITE	$Cu_2^{1+}O$	143.091	4.2685	4.2685	4.2685	1.570796	1.570796	1.570796	2	77.77246	77.77246
9	EDGARITE	$FeNb_3S_6$	526.92411	5.771	5.771	12.19	1.570796	1.570796	2.094395	2	351.59	351.59
10	GALENA	PbS	239.26	5.936	5.936	5.936	1.570796	1.570796	1.570796	4	209.1615	209.1615
11	SMITHSONITE	$ZnCO_3$	125.388	4.6526	4.6526	15.9257	1.570796	1.570796	2.094395	6	298.5524	298.5524
12	SPHALERITE	$(Zn,Fe)S$	92.6725	5.406	5.406	5.406	1.570796	1.570796	1.570796	4	157.9895	157.9895
13	ZUSSMANITE	$K(Fe^{2+},Mg,Mn^{2+})_{13}(AlSi)_{18}O_{42}(OH)_{14}$	2205.7535	11.66	11.66	28.69	1.570796	1.570796	2.094395	3	3377.989	3377.989
14	TENNANTITE	$(Cu,Ag,Sn,Fe)_{12}As_4S_{12}$	1722.29	10.19	10.19	10.19	1.570796	1.570796	1.570796	2	1058.09	1058.09
15	WOLLASTONITE 2M	$CaSiO_3$	116.16	15.409	7.322	7.063	1.570796	1.663299	1.570796	12	793.4739	793.4739
17	WOLLASTONITE 1A	$CaSiO_3$	116.16	7.74	7.32	7.063	1.567306	1.47969	1.339366	6	375.8138	387.8205
18	PYRITE	FeS_2	119.965	5.4179	5.4179	5.4179	1.570796	1.570796	1.570796	4	159.0351	159.0351
19	CALCITE	$CaCO_3$	100.086	4.9896	4.9896	17.061	1.570796	1.570796	2.094395	6	367.8465	367.8465
20	STRONTIANITE	$SrCO_3$	147.628	5.1059	8.4207	6.0319	1.570796	1.570796	1.570796	4	259.3431	259.3431
21	QUARTZ-a	SiO_2	60.083	4.9135	4.9135	5.405	1.570796	1.570796	2.094395	3	113.0078	113.0078
22	QUARTZ-b	SiO_2	60.083	4.9965	4.9965	5.4546	1.570796	1.570796	2.094395	6	117.9303	117.9303
23	CORUNDUM	Al_2O_3	101.96008	4.754	4.754	12.982	1.570796	1.570796	2.094395	6	254.0918	254.0918
24	ACANTHITE	Ag_2S	247.7804	4.229	6.931	7.862	1.570796	1.738522	1.570796	4	227.2108	227.2108
25	ARGENTITE	Ag_2S	247.7804	4.88	4.88	4.88	1.570796	1.570796	1.570796	2	116.2143	116.2143
26	PYRARGYRITE	Ag_3SbS_3	541.5206	11.047	11.047	8.719	1.570796	1.570796	2.094395	6	921.4802	921.4802
27	HALITE	$NaCl$	58.44277	5.6404	5.6404	5.6404	1.570796	1.570796	1.570796	4	179.4443	179.4443
28	SYLVITE	KCl	74.5513	6.2931	6.2931	6.2931	1.570796	1.570796	1.570796	4	249.2263	249.2263
29	NATRON	$Na_2CO_3 \cdot 10H_2O$	276.05754	12.83	9.026	13.44	1.570796	2.146755	1.570796	8	1305.307	1305.307
30	NITRATITE	$NaNO_3$	84.99377	5.07	5.07	16.829	1.570796	1.570796	2.094395	6	374.632	374.632
31	SPODUMENE	$LiAlSi2O6$	186.14254	9.45	8.39	5.215	1.570796	1.919862	1.570796	4	388.5384	388.5384
32	GYPSUM	$CaSO_4 \cdot 2H_2O$	172.164	5.679	15.202	6.522	1.570796	2.066993	1.570796	4	495.1532	495.1532
33	ANHYDRITE	$CaSO_4$	136.134	6.993	6.995	6.245	1.570796	1.570796	1.570796	4	305.4806	305.4806
34	BIOTITE	$K(Mg,Fe^{2+})_3(Al,Fe^{3+})Si_3O_{10}(OH,F)_2$	566.61734	5.3	9.2	10.2	1.570796	1.745329	1.570796	2	489.7961	489.7961
35	PENNANTITE	$Mn_5^{2+}Al(Si_3Al)O_{10}(OH)_8$	708.9543	5.45	9.5	14.4	1.570796	1.698205	1.570796	2	739.5168	739.5168
36	MUSCOVITE	$KAl_2(AlSi_3O_{10})(F,OH)_2$	436.29872	5.199	9.027	20.106	1.570796	1.666789	1.570796	4	939.258	939.258
37	CINNABAR	HgS	232.652	4.145	4.145	9.496	1.570796	1.570796	2.094395	3	141.2929	141.2929
38	GOETHITE	$\alpha\text{-}Fe^{3+}O(OH)$	88.851	4.608	9.956	3.0215	1.570796	1.570796	1.570796	4	138.6181	138.6181
39	CHLORARGYRITE	$AgCl$	143.3132	5.554	5.554	5.554	1.570796	1.570796	1.570796	4	171.3238	171.3238
40	MAGNETITE	$Fe^{2+}Fe^{3+}_2O_4$	175.686	8.397	8.397	8.397	1.570796	1.570796	1.570796	8	592.0692	592.0692
41	HEAMATITE	$\alpha\text{-}Fe_2O_3$	159.687	5.038	5.038	13.772	1.570796	1.570796	2.094395	6	302.722	302.722
42	OLIVINE	$(Mg,Fe)_2SiO_4$	144.146	4.78	10.25	6.3	1.570796	1.570796	1.570796	4	308.6685	308.6685
43	MAGNESIOHORNBLENDE	$Ca_2[Mg_4(Al,Fe^{+})]Si_7AlO_{22}(OH)_2$	869.77108	9.887	18.174	5.308	1.570796	1.832596	1.570796	2	921.276	921.276
44	VERMICULITE	$(MgFe,Al)_3(Al,Si)_4O_{10}(OH)_2 \cdot 4H_2O$	803.69277	5.349	9.255	28.89	1.570796	1.695006	1.570796	4	1419.181	1419.181
45	MAGNESITE	$MgCO_3$	84.313	4.6632	4.6632	15.015	1.570796	1.570796	2.094395	6	282.764	282.764
46	CHALCOCITE	Cu_2S	159.152	15.246	11.884	13.494	1.570796	2.030691	1.570796	48	2190.864	2190.864
47	COVELLITE	CuS	95.606	3.7938	3.7938	16.341	1.570796	1.570796	2.094395	6	203.6846	203.6846
48	NATIVE PALLADIUM	Pd	106.42	3.8898	3.8898	3.8898	1.570796	1.570796	1.570796	4	58.85479	58.85479
49	SILLIMANITE	Al_2SiO_5	162.04308	7.4883	7.6808	5.7774	1.570796	1.570796	1.570796	4	332.2937	332.2937
50	TITANITE	$CaTiSiO_5$	196.025	7.057	8.707	6.555	1.570796	1.986359	1.570796	4	368.4935	368.4935
51	ZIRCON	$ZrSiO_4$	183.305	6.607	6.607	5.982	1.570796	1.570796	1.570796	4	261.1289	261.1289
52	ALLARGENTUM	Ag_6Sb	768.9212	2.952	2.952	4.773	1.570796	1.570796	2.094395	2	36.02092	36.02092
53	FLUORITE	CaF_2	78.074806	5.4626	5.4626	5.4626	1.570796	1.570796	1.570796	4	163.004	163.004
54	ABELSONITE	$C_{31}H_{32}N_4Ni$	519.31	8.508	11.185	7.299	1.585632	1.992002	1.395973	1	694.5873	622.9906
55	ABERNATHYITE	$K_2(UO_2)_2(AsO_4)_2 \cdot 6H_2O$	520.11	7.176	7.176	18.126	1.570796	1.570796	1.570796	4	933.3979	933.3979
56	AMMONIOALUNITE	$(NH_4)Al_3(SO_4)_2(OH)_6$	393.15	7.013	7.013	17.855	1.570796	1.570796	2.094395	5	878.1476	760.4982
57	BABEFPHITE	$BaBe(PO_4)F$	259.56	6.889	16.814	6.902	1.570971	1.570622	1.576381	8	799.47	799.4575
58	BABINGTONITE	$Ca_2(Fe^{2+},Mn)Fe^{3+}Si_5O_{14}(OH)$	573.05	7.5	12.18	6.68	1.506219	1.633454	1.958085	2	610.218	563.3622
59	CABRIITE	Pd_2SnCu	395.1	7.88	7.88	3.94	1.570796	1.570796	1.570796	4	244.6519	244.6519
60	CACOXENITE	$Fe^{3+}_{24}AlO_6(PO_4)_{17}(OH)_{12} \cdot 75H_2O$	4481.51	27.559	27.559	10.55	1.570796	1.570796	2.094395	2	8012.709	6939.21
61	DACHIARDITE	$(Ca,Na_2,K_2)_5Al_{10}Si_{38}O_{96} \cdot 25H_2O$	1808.84	18.625	7.508	10.247	1.570796	1.885933	1.570796	1	1432.905	1362.34
62	DADSONITE	$Pb_{21}Sb_{23}S_{55}Cl$	8950.53	17.33	4.11	19.05	1.570796	1.680752	1.577778	2	1356.861	1348.634
63	DIAMOND	C	12.01	3.5595	3.5595	3.5595	1.570796	1.570796	1.570796	8	45.09901	45.09901

Table C1a
Given and Computed Data for the Rock-Forming Mineral Species

Serial	Mineral	Chemical Formula	Formula Weight	a	b	c	α	β	γ	Z	V	V by General Equation
64	EAKERITE	$Ca_2Sn^{4+}Al_2Si_6O_{18}(OH)_2.2H_2O$	779.38	15.829	7.721	7.438	1.570796	1.772673	1.570796	2	909.0404	890.5797
65	EARLANDITE	$Ca_3(C_6H_5O_7)_2.4H_2O$	570.5	30.94	5.93	10.56	1.570796	1.635955	1.570796	4	1937.488	1933.376
66	FABIANITE	$CaB_3O_5(OH)$	169.51	6.593	10.488	6.365	1.570796	1.983276	1.570796	4	440.1231	403.2098
67	GABRIELITE	$Tl_2AgCu_2As_3S_7$	1082.85	12.138	12.196	15.944	1.370729	1.478556	1.055401	6	2360.271	2012.655
68	GABRIELSONITE	$PbFe^{2+}(AsO_4)(OH)$	418.97	7.86	5.98	8.62	1.570796	1.570796	1.570796	4	405.1641	405.1641
69	GRAPHITE	C	12.01	2.464	2.464	6.711	1.570796	1.570796	2.094395	2	40.74447	35.28574
70	HAFNON	$HfSiO_4$	244.28	6.5725	6.5725	5.9632	1.570796	1.570796	1.570796	4	257.5969	257.5969
71	HAGENDORFITE	$NaCaMn^{2+}(Fe^{2+},Fe^{3+},Mg)_2(PO_4)_3$	501.07	10.933	12.594	6.515	1.570796	1.710074	1.570796	4	897.0517	888.3651
72	IANGREYITE	$Ca_2Al_7(PO_4)_2(PO_3OH)_2(OH,F)_{15}.8H_2O$	1335.123	6.988	6.988	16.707	1.570796	1.570796	2.094395	1	815.8386	706.537
73	ICE	H_2O	18.02	4.498	4.498	7.338	1.570796	1.570796	2.094395	4	148.4624	128.5722
74	JACHYMOVITE	$(UO_2)_8(SO_4)(OH)_{14}.13H_2O$	2728.59	18.553	9.276	13.532	1.570796	2.191435	1.570796	2	2328.825	1894.515
75	JACOBSITE	$(Mn^{2+},Fe^{2+},Mg)(Fe^{3+},Mn^{3+})_2O_4$	227.38	8.499	8.499	8.499	1.570796	1.570796	1.570796	8	613.9083	613.9083
76	KAATIALAITE	$Fe^{3+}(H_2AsO_4)_3.3-5H_2O$	577.74	15.3065	19.722	4.736	1.570796	1.602823	1.570796	4	1429.679	1428.946
77	KADYRELITE	$Hg^{1+}_6H(Br,Cl)_3O_2$	955.94	16.22	16.22	16.22	1.570796	1.570796	1.570796	16	4267.294	4267.294
78	LABRADORITE	$Na_{0.5-0.3}Ca_{0.5-0.7}Al_{1.5-1.7}Si_{2.5-2.3}O_8$	271.81	8.1648	12.8585	7.0973	1.580687	2.026327	1.567015	4	745.1248	669.1098
79	LANARKITE	$Pb_2O(SO_4)$	526.46	13.769	5.698	7.079	1.570796	2.023418	1.570796	4	555.3883	499.4627
80	MACAULAYITE	$(Fe^{3+},Al)_{24}Si_4O_{43}(OH)_2$	2116.93	5.038	8.726	36.342	1.570796	1.605703	1.570796	2	1597.652	1596.679
81	MACDONALDITE	$BaCa_4Si_{16}O_{36}(OH)_2.10H_2O$	1537.15	14.06	23.52	13.08	1.570796	1.570796	1.570796	4	4325.441	4325.441
82	NABALAMPROPHYLLITE	$Ba(Na,Ba)\{Na_3Ti[Ti_2O_2Si_4O_{14}](OH,F)_2\}$	810.56	19.741	7.105	5.408	1.570796	1.68721	1.570796	2	758.525	753.391
83	NABAPHITE	$NaBaPO_4.9H_2O$	417.43	10.711	10.711	10.711	1.570796	1.570796	1.570796	4	1228.825	1228.825
84	OBOYERITE	$H_6(Pb,Ca)_6(Te^{4+}O_3)_3(Te^{6+}O_6)_2.2H_2O$	2259.27	12.249	15.113	6.868	2.032436	1.720546	1.497842	2	1271.398	1125.543
85	OBRADOVICITE	$H_4(K,Na)CuFe^{3+}_2(AsO_4)(MoO_4)_5.12H_2O$	1257.18	15.046	14.848	11.056	1.570796	1.570796	1.570796	4	2469.944	2469.944
86	PAAKKONENITE	Sb_2AsS_2	382.55	10.75	3.959	12.49	1.570796	2.012016	1.570796	4	531.565	480.6579
87	PAARITE	$Cu_{1.7}Pb_{1.7}Bi_{6.3}S_{12}$	2166.36	4.007	55.998	11.512	1.570796	1.570796	1.570796	5	2583.108	2583.108
88	PEROVSKITE	$CaTiO_3$	135.96	5.447	7.654	5.388	1.570796	1.570796	1.570796	4	224.6329	224.6329
89	QUANDILITE	$(Mg,Fe^{2+})_2(Ti,Fe^{3+},Al)O_4$	354.9895	8.4376	8.4376	8.4376	1.570796	1.570796	1.570796	8	600.6988	600.6988
90	QILIANSHANITE	$H_3Na(HCO_3)(BO_3).2H_2O$	181.87	16.119	6.928	6.73	1.570796	1.753358	1.570796	4	751.5555	739.066
91	RABBITTITE	$Ca_3Mg_3(UO_2)_2(CO_3)_6(OH)_4.18H_2O$	1485.56	32.6	23.8	9.45	1.570796	1.570796	1.570796	8	7332.066	7332.066
92	RABERITE	$Tl_5Ag_4As_6SbS_{15}$	2505.53	8.92	9.429	20.062	1.390329	1.550551	1.094671	2	1687.348	1471.605
93	REALGAR	AsS	106.99	9.325	13.571	6.587	1.570796	1.856739	1.570796	16	833.5821	799.7353
94	SABATIERITE	Cu_6TiSe_4	695.45	3.986	5.624	9.778	1.570796	1.570796	1.570796	1	219.196	219.196
95	TACHARANITE	$Ca_{12}Al_2Si_{18}O_{51}.18H_2O$	2180.68	17.07	3.65	27.9	1.570796	1.991421	1.570796	1	1738.323	1586.801
96	TACHYHYDRITE	$CaMg_2Cl_6.12H_2O$	517.59	10.136	10.136	17.318	1.570796	1.570796	2.094395	3	1779.225	1540.854
97	UCHUCCHACUAITE	$AgPb_3MnSb_5S_{12}$	1777.95	12.67	19.32	4.38	1.570796	1.570796	1.570796	2	1072.156	1072.156
98	UEDAITE-(CE)	$Mn^{2+}CeAl_2Fe(Si_2O_7)(SiO_4)O(OH)$	584.87	8.939	5.742	10.187	1.570796	2.008874	1.570796	2	522.8757	473.4999
99	UKLONSKOVITE	$NaMg(SO_4)F.2H_2O$	198.39	7.202	7.214	5.734	1.570796	1.976236	1.570796	2	297.9113	273.7593
100	VAESITE	NiS_2	122.82	5.66787	5.66787	5.66787	1.570796	1.570796	1.570796	4	182.0789	182.0789
101	VALENTINITE	Sb_2O_3	291.5	4.911	12.464	5.412	1.570796	1.570796	1.570796	4	331.2723	331.2723
102	WADEITE	$K_2ZrSi_3O_9$	397.67	6.893	6.893	10.172	1.570796	1.570796	2.094395	2	483.3068	418.556
103	WADSLEYITE	$\beta-(Mg,Fe^{2+})_2SiO_4$	156.46	5.7	11.71	8.24	1.570796	1.570796	1.570796	8	549.9953	549.9953
104	XANTHIOSITE	$Ni_3(AsO_4)_2$	453.91	10.174	9.548	5.766	1.570796	1.62272	1.570796	4	560.117	559.3622
105	XANTHOCONITE	Ag_3AsS_3	494.72	12	6.26	17.08	1.570796	1.919862	1.570796	8	1283.05	1205.672
106	YAFSOANITE	$Ca_3Zn_3(Te^{6+}O_6)_2$	763.6	12.632	12.632	12.632	1.570796	1.570796	1.570796	8	2015.656	2015.656
107	YAGIITE	$(Na,K)_{1.5}Mg_2(Al,Mg)_3(Si,Al)_{12}O_{30}$	1991.71	10.09	10.09	14.29	1.570796	1.570796	2.094395	2	1454.838	1259.926
108	ZABUYELITE	Li_2CO_3	73.89	8.361	4.976	6.193	1.570796	2.001718	1.570796	4	257.6557	234.1011
109	ZACCARINIITE	$RhNiAs$	319.822	3.5498	3.5498	3.5498	1.6573	1.570796	1.570796	2	77.58863	77.58863
110	ACTINOLITE	$Ca_2(Mg,Fe^{2+})_5Si_8O_{22}(OH)_2$	853.16	9.891	18.2	5.305	1.570796	1.826313	1.570796	2	954.9859	923.9804
111	ALMANDINE	$Fe^{2+}_3Al_2(SiO_4)_3$	497.75	11.526	11.526	11.526	1.570796	1.570796	1.570796	8	1531.214	1531.214
112	AZURITE	$Cu_3(CO_3)_2(OH)_2$	344.67	5.0109	5.8485	10.345	1.570796	1.613208	1.570796	2	303.1731	302.9005
113	BERYL	$Be_3Al_2Si_6O_{18}$	537.5	9.227	9.227	9.218	1.570796	1.570796	2.094395	2	784.7977	679.6548
114	BISMUTHITE	$Bi_2O_2(CO_3)$	509.97	3.8685	3.8685	13.6915	1.570796	1.570796	1.570796	2	204.8973	204.8973
115	BORAX	$Na_2B_4O_5(OH)_4.8H_2O$	381.37	11.879	10.644	12.2012	1.570796	1.860818	1.570796	4	1542.721	1478.293
116	GOLD	Au	196.97	4.0786	4.0786	4.0786	1.570796	1.570796	1.570796	4	67.84742	67.84742
117	MONAZITE-(CE)	$(Ce,La,Nd,Th)PO_4$	240.21	6.7902	7.0203	6.4674	1.570796	1.804321	1.570796	4	308.296	299.9279
118	PYROPE	$Mg_3Al_2(SiO_4)_3$	403.13	11.459	11.459	11.459	1.570796	1.570796	1.570796	8	1504.666	1504.666
119	RUTILE	TiO_2	79.88	4.5937	4.5937	2.9587	1.570796	1.570796	1.570796	2	62.43472	62.43472
120	SCHEELITE	$CaWO_4$	287.93	5.2429	5.2429	11.3737	1.570796	1.570796	1.570796	4	312.6403	312.6403
121	SCOTLANDITE	$PbS^{4+}O_3$	287.26	4.505	5.333	6.405	1.570796	1.854238	1.570796	2	153.3812	147.7411
122	SPINEL	$MgAl_2O_4$	142.27	8.0898	8.0898	8.0898	1.570796	1.570796	1.570796	8	529.4359	529.4359
123	TALC	$Mg_3Si_4O_{10}(OH)_2$	379.27	5.291	9.46	5.29	1.722291	2.09265	1.488242	2	264.7796	226.8996
124	TETRAHEDRITE	$(Cu,Fe,Ag,Zn)_{12}Sb_4S_{13}$	1643.31	10.39	10.39	10.39	1.570796	1.570796	1.570796	2	1121.622	1121.622
125	TOPAZ	$Al_2SiO_4(F,OH)_2$	182.25	4.6499	8.7968	8.3909	1.570796	1.570796	1.570796	4	343.2234	343.2234
126	WULFENITE	$PbMoO_4$	367.14	5.436	5.436	12.068	1.570796	1.570796	1.570796	4	356.6106	356.6106

Table C1b
Given and Computed Data for the Rock-Forming Mineral Species

Serial	Mineral	Chemical Formula	Lattice System	Space Group	Theoretical ρ	Measured Density	Harmonic Density	Fractional Defect for Measured Density	Fractional Defect for Harmonic Density	Precise Harmonic ρ Multiplier (Empirical Ξ)	Lattice Angle Multiplier Ψ
1	ARGENTOPYRITE	$AgFe_2S_3$	orthorhombic	Pmmn	4.2710031	4.25	4.447701	0.0049176	-0.041371464	0.955549898	1
2	ARGENTOTENNANTITE	$(Ag,Cu)_{10}(Zn,Fe)_2(As,Sb)_4S_{13}$	cubic	I43m	5.00949629	4.71	4.872075	0.05978571	0.027432224	0.966733952	1
3	BARITE	$BaSO_4$	orthorhombic	Pnma	4.46770798	4.5	2.15272	-0.00722787	0.51815999	2.090378237	1
4	CASSITERITE	SnO_2	tetragonal	P4$_2$/mnm	6.99508386	6.995	3.490249	1.1989E-05	0.501042614	2.00415514	1
5	CERUSSITE	$PbCO_3$	orthorhombic	Pmcn	6.57148536	6.552	4.171114	0.00296514	0.365270732	1.570803354	1
6	CHALCOPYRITE	$CuFeS_2$	tetragonal	I42d	4.20207454	4.2	4.0595	0.0004937	0.033929538	1.034610149	1
7	CHILDRENITE	$Fe^{3+}Al(PO_4)(OH)_2 \cdot H_2O$	orthorhombic	Bba2	3.15582265	3.15	1.407354	0.00184505	0.554045194	2.238242387	1
8	CUPRITE	$Cu_2^{1+}O$	cubic	Pn3m	6.11034241	6.14	5.179031	-0.00485367	0.152415561	1.185549928	1
9	EDGARITE	$FeNb_3S_6$	hexagonal	P6$_3$22	4.97726393	4.98	3.975997	-0.00054971	0.20116821	1.252516143	0.866025404
10	GALENA	PbS	cubic	Fm3m	7.59796876	7.58	7.083144	0.00236494	0.06775819	1.07014623	1
11	SMITHSONITE	$ZnCO_3$	hexagonal	R3c	4.18442436	4.431	2.281536	-0.05892702	0.454755211	1.942113045	0.866025404
12	SPHALERITE	$(Zn,Fe)S$	cubic	F43m	3.89611562	4	3.920348	-0.02666358	-0.006219503	1.020317707	1
13	ZUSSMANITE	$K(Fe^{2+},Mg,Mn^{2+})_{13}(AlSi)_{18}O_{42}(OH)_{14}$	hexagonal	R3	3.25288751	3.146	1.821431	0.03285927	0.440057111	1.727213166	0.866025404
14	TENNANTITE	$(Cu,Ag,Sn,Fe)_{12}As_4S_{12}$	cubic	I43m	5.40583535	4.72	4.781066	0.12686945	0.115573195	0.987227599	1
15	WOLLASTONITE 2M	$CaSiO_3$	monoclinic	P2$_1$/a	2.91711986	2.975	1.484547	0.01904154	0.491091624	2.00397868	0.995724698
17	WOLLASTONITE 1A	$CaSiO_3$	triclinic	P2$_1$/a	3.0795288	2.975	1.484547	0.03394312	0.517930558	2.00397868	0.969146729
18	PYRITE	FeS_2	cubic	Pa3	5.01038024	5.018	3.147603	-0.0015208	0.371783657	1.594229132	1
19	CALCITE	$CaCO_3$	hexagonal	R3c	2.71086005	2.7102	1.396875	0.00024348	0.484711504	1.940187925	0.866025404
20	STRONTIANITE	$SrCO_3$	orthorhombic	Pmcn	3.7809696	3.76	1.86614	0.00554609	0.506438769	2.014854184	1
21	QUARTZ-a	SiO_2	trigonal	P3$_1$21	2.64858371	2.65	1.542774	-0.00053473	0.417509963	1.717685575	0.866025404
22	QUARTZ-b	SiO_2	hexagonal	P3$_1$21	5.0760586	2.65	1.542774	0.47794141	0.696068594	1.717685575	0.866025404
23	CORUNDUM	Al_2O_3	hexagonal	R3c	3.99797344	4.04	1.689906	-0.01051197	0.577309288	2.39066518	0.866025404
24	ACANTHITE	Ag_2S	monoclinic	P2$_1$/n	7.24347667	7.21	6.872624	0.00462163	0.05119815	1.049089827	0.985966915
25	ARGENTITE	Ag_2S	cubic	Im3m	7.08086917	7.3	6.872624	-0.03094688	0.029409539	1.062185262	1
26	PYRARGYRITE	Ag_3SbS_3	hexagonal	R3c	5.85503244	5.82	5.666849	0.0059833	0.032140437	1.027025752	0.866025404
27	HALITE	$NaCl$	cubic	Fm3m	2.16326718	2.16	1.418022	0.0015103	0.344499902	1.523248743	1
28	SYLVITE	KCl	cubic	Fm3m	1.98687435	1.99	1.180728	-0.00157315	0.405735975	1.68540094	1
29	NATRON	$Na_2CO_3 \cdot 10H_2O$	monoclinic	Cc	1.40474029	1.46	0.806517	-0.03933803	0.425860285	1.810252797	0.838670568
30	NITRATITE	$NaNO_3$	hexagonal	R3c	2.26038582	2.26	1.094119	0.00017069	0.515959326	2.065589457	0.866025404
31	SPODUMENE	$LiAlSi2O6$	monoclinic	C2/c	3.18215123	3.15	1.455578	0.01010362	0.542580558	2.164089004	0.939692621
32	GYPSUM	$CaSO_4 \cdot 2H_2O$	monoclinic	I2/a	2.30946754	2.31	1.025686	-0.00023056	0.555877833	2.252151844	0.879399416
33	ANHYDRITE	$CaSO_4$	orthorhombic	Amma	2.96000195	2.96	1.430801	6.5994E-07	0.516621706	2.06877171	1
34	BIOTITE	$K(Mg,Fe^{2+})_3(Al,Fe^{3+})Si_3O_{10}(OH,F)_2$	monoclinic	C2/m	3.84196694	2.89	2.016051	0.24778114	0.475255427	1.433495267	0.984807753
35	PENNANTITE	$Mn_5^{2+}Al(Si_3Al)O_{10}(OH)_8$	monoclinic	C1	3.18382559	3.2	1.658624	-0.00508018	0.47904675	1.929309743	0.991894443
36	MUSCOVITE	$KAl_2(AlSi_3O_{10})(F,OH)_2$	monoclinic	C2/c	3.08537607	2.83	1.370455	0.08276984	0.555822323	2.065007342	0.995396198
37	CINNABAR	HgS	hexagonal	P3$_1$21	8.20269811	8.19	7.670249	0.00154804	0.064911466	1.067761953	0.866025404
38	GOETHITE	$\alpha\text{-}Fe^{3+}O(OH)$	orthorhombic	Pbnm	4.25746852	4.27	1.955522	-0.00294341	0.540684333	2.183560204	1
39	CHLORARGYRITE	$AgCl$	cubic	Fm3m	5.55619717	5.55	5.167055	0.00111536	0.070037494	1.074112807	1
40	MAGNETITE	$Fe^{2+}Fe^{3+}_2O_4$	cubic	Fd3m	3.94188339	5.21	2.585031	-0.32170323	0.344214126	2.015449371	1
41	HEAMATITE	$\alpha\text{-}Fe_2O_3$	hexagonal	R3c	5.25564389	5.28	2.929055	-0.00463428	0.44268395	1.802629364	0.866025404
42	OLIVINE	$(Mg,Fe)_2SiO_4$	orthorhombic	Pbnm	3.10183981	3.3	1.925671	-0.06388473	0.379184242	1.713688341	1
43	MAGNESIOHORNBLENDE	$Ca_2[Mg_4(Al,Fe^{3+})]Si_7AlO_{22}(OH)_2$	monoclinic	C2/m	3.1354097	2.96	1.5402	0.05594475	0.508772179	1.921827741	0.965925826
44	VERMICULITE	$(MgFe,Al)_3(Al,Si)_4O_{10}(OH)_2 \cdot 4H_2O$	monoclinic	C2/c	3.76150259	2.32	1.727268	0.3832252	0.540803754	1.343161672	0.992295941
45	MAGNESITE	$MgCO_3$	hexagonal	R3c	2.97078232	2.98	1.410444	-0.00310278	0.525228037	2.112809637	0.866025404
46	CHALCOCITE	Cu_2S	monoclinic	P2$_1$/c	5.79011194	6.46	5.359596	-0.11569518	0.07435368	1.205314771	0.896099437
47	COVELLITE	CuS	hexagonal	P63/mmc	4.67656918	4.68	4.229869	-0.00073362	0.095518692	1.106417136	0.866025404
48	NATIVE PALLADIUM	Pd	cubic	Fm3m	12.0102078	12.19	12	-0.01496995	0.000849925	1.015833333	1
49	SILLIMANITE	Al_2SiO_5	orthorhombic	Pbnm	3.23904838	3.25	1.63219	-0.00338112	0.496089673	1.99118984	1
50	TITANITE	$CaTiSiO_5$	monoclinic	P2$_1$/a	3.53338267	3.55	1.693402	-0.00470295	0.520742088	2.096372172	0.914889222
51	ZIRCON	$ZrSiO_4$	tetragonal	I4$_1$/amd	4.66260227	4.85	2.296328	-0.04019166	0.507500845	2.112067909	1
52	ALLARGENTUM	Ag_6Sb	hexagonal	nd	70.8934561	10.01	9.630482	0.8588022	0.864155558	1.039408002	0.866025404
53	FLUORITE	CaF_2	cubic	Fm3m	3.18142579	3.18	1.622191	0.00044816	0.490105623	1.960311559	1
54	ABELSONITE	$C_{31}H_{32}N_4Ni$	triclinic	P1	1.24150634	1.405	0.87021	-0.13168975	0.29066889	1.61455203	0.896921921
55	ABERNATHYITE	$K_2(UO_2)_2(AsO_4)_2 \cdot 6H_2O$	tetragonal	P4/ncc	3.70115651	3.32	1.82602	0.10298308	0.506635311	1.818161973	1
56	AMMONIOALUNITE	$(NH_4)Al_3(SO_4)_2(OH)_6$	hexagonal	R3m	3.71714791	2.4	1.041726	0.35434369	0.7197512	2.308868236	0.866025404
57	BABEFPHITE	$BaBe(PO_4)F$	triclinic	P1	4.31295235	4.31	2.067098	0.00068453	0.52072318	2.085048611	0.999984373
58	BABINGTONITE	$Ca_2(Fe^{2+},Mn)Fe^{3+}Si_5O_{14}(OH)$	triclinic	P1	3.11879327	3.355	1.687138	-0.07573658	0.459041443	1.988574842	0.923214733
59	CABRIITE	Pd_2SnCu	orthorhombic	Pmmm	10.7267326	11.1	9.608481	-0.03479786	0.104249085	1.155229481	1
60	CACOXENITE	$Fe^{3+}_{24}AlO_6(PO_4)_{17}(OH)_{12} \cdot 75H_2O$	hexagonal	P6$_3$/m	1.85747975	2.4	1.031898	-0.29207331	0.444463628	2.325812265	0.866025404
61	DACHIARDITE	$(Ca,Na_2,K_2)_5Al_{10}Si_{38}O_{96} \cdot 25H_2O$	monoclinic	C2/m	2.09619636	2.1855	1.300858	-0.04260271	0.379419635	1.680044626	0.950754034
62	DADSONITE	$Pb_{23}Sb_{25}S_{60}Cl$	monoclinic	P1	21.9074825	5.68	5.349634	0.74072785	0.755807914	1.061754905	0.993936438
63	DIAMOND	C	cubic	Fd3m	3.53765187	3.511	2.2	0.00753377	0.378118571	1.595909091	1

Table C1c
Given and Computed Data for the Rock-Forming Mineral Species

Serial	Mineral	Chemical Formula	Lattice System	Space Group	Theoretical ρ	Measured Density	Harmonic Density	Fractional Defect for Measured Density	Fractional Defect for Harmonic Density	Precise Harmonic ρ Multiplier (Empirical Ξ)	Lattice Angle Multiplier Ψ
64 EAKERITE	$Ca_2Sn^{4+}Al_2Si_6O_{18}(OH)_2.2H_2O$		monoclinic	P2$_1$/a	2.84737808	2.93	1.484938	-0.02901684	0.478489223	1.97314588	0.979692066
65 EARLANDITE	$Ca_3(C_6H_5O_7)_2.4H_2O$		monoclinic	NoDet	1.95580616	1.875	0.964861	0.04131604	0.506668208	1.943284372	0.997877906
66 FABIANITE	$CaB_3O_5(OH)$		monoclinic	P2$_1$/a	2.55817496	2.783	1.292387	-0.08788494	0.494801047	2.153379235	0.916129661
67 GABRIELITE	$Tl_2AgCu_2As_3S_7$		triclinic	P1	4.57095354	5.36	5.294448	-0.17262185	-0.15828082	1.012381306	0.852722104
68 GABRIELSONITE	$PbFe^{2+}(AsO_4)(OH)$		orthorhombic	P2$_1$ma	6.86848593	6.67	3.580794	0.02889806	0.478663196	1.862715104	1
69 GRAPHITE	C		hexagonal	P6$_3$/mmc (2H)	1.95786816	2.16	2.2	-0.10324078	-0.123671167	0.981818182	0.866025404
70 HAFNON	$HfSiO_4$		tetragonal	I4$_1$/amd	6.29877983	6.16	3.4144	0.02203281	0.457926779	1.804123779	1
71 HAGENDORFITE	$NaCaMn^{2+}(Fe^{2+},Fe^{3+},Mg)_2(PO_4)_3$		monoclinic	I2$_1$/a	3.71013768	3.71	2.071146	3.7109E-05	0.441760238	1.791278513	0.990316589
72 IANGREYITE	$Ca_2Al_7(PO_4)_2(PO_3OH)_2(OH,F)_{15}.8H_2O$		hexagonal	P321	2.71747841	2.46	1.060318	0.09474902	0.609815484	2.320058699	0.866025404
73 ICE	H_2O		hexagonal	P6$_3$/mmc	0.80620829	0.9167	0.495549	-0.13705107	0.385333931	1.849867968	0.866025404
74 JACHYMOVITE	$(UO_2)_8(SO_4)(OH)_{14}.13H_2O$		monoclinic	P2$_1$	3.89117261	4.79	2.262355	-0.2309914	0.418593036	2.117262902	0.813506961
75 JACOBSITE	$(Mn^{2+},Fe^{2+},Mg)(Fe^{3+},Mn^{3+})_2O_4$		cubic	Fd3m	4.92025773	4.76	3.572088	0.032571	0.274003901	1.332553989	1
76 KAATIALAITE	$Fe^{3+}(H_2AsO_4)_3.3-5H_2O$		monoclinic	P2$_1$	2.68412647	2.64	1.306635	0.01643979	0.513199098	2.02045684	0.999487186
77 KADYRELITE	$Hg^{1+}_6H(Br,Cl)_3O_2$		cubic	[Ia3d]	5.95178395	8.93	6.171374	-0.50039048	-0.036894752	1.447003643	1
78 LABRADORITE	$Na_{0.5-0.3}Ca_{0.5-0.7}Al_{1.5-1.7}Si_{2.5-2.3}O_8$		triclinic	C1(low)	2.42295579	2.7	1.531277	-0.11434142	0.368012845	1.763234292	0.897983475
79 LANARKITE	$Pb_2O(SO_4)$		monoclinic	C2/m	6.29618825	6.92	4.413982	-0.09907768	0.29894377	1.567745406	0.899303506
80 MACAULAYITE	$(Fe^{3+},Al)_{24}Si_4O_{43}(OH)_2$		monoclinic	C-centered cell	4.40051381	4.41	2.074584	-0.0021557	0.528558604	2.125726994	0.999390827
81 MACDONALDITE	$BaCa_4Si_{16}O_{36}(OH)_2.10H_2O$		orthorhombic	Bmmb	2.36045078	2.27	1.245664	0.03831928	0.472277252	1.822321891	1
82 NABALAMPROPHYLLITE	$Ba(Na,Ba)\{Na_3Ti[Ti_2O_2Si_4O_{14}](OH,F)_2\}$		monoclinic	P2/m	3.54890473	3.62	1.900627	-0.02003302	0.464446931	1.904634812	0.993231602
83 NABAPHITE	$NaBaPO_4.9H_2O$		cubic	P2$_1$3	2.25633033	2.3	0.819224	-0.01935429	0.6369221	2.807536032	1
84 OBOYERITE	$H_6(Pb,Ca)_6(Te^{4+}O_3)_3(Te^{6+}O_6)_2.2H_2O$		triclinic	P1	5.90154358	6.4	3.160626	-0.08446204	0.46444084	2.024915495	0.885279995
85 OBRADOVICITE	$H_4(K,Na)CuFe^{3+}_2(AsO_4)(MoO_4)_5.12H_2O$		orthorhombic	Pcnm	3.38080015	3.55	1.103047	-0.05004728	0.673731944	3.218357608	1
86 PAAKKONENITE	Sb_2AsS_2		monoclinic	C2/m	4.78014294	5.21	4.757322	-0.08992557	0.004774135	1.095153978	0.904231671
87 PAARITE	$Cu_{1.7}Pb_{1.7}Bi_{6.3}S_{12}$		orthorhombic	Pmcn	6.96317137	6.96	5.946755	0.00045545	0.145970354	1.170386246	1
88 PEROVSKITE	$CaTiO_3$		orthorhombic	Pnma	4.02019221	4.12	1.769771	-0.02482662	0.559779546	2.327985019	1
89 QUANDILITE	$(Mg,Fe^{2+})_2(Ti,Fe^{3+},Al)O_4$		cubic	Fd3m	7.85050846	4.055	2.870978	0.48347295	0.634293966	1.412410528	1
90 QILIANSHANITE	$H_3Na(HCO_3)_2(BO_3).2H_2O$		monoclinic	C2	1.60734502	1.706	0.78123	-0.0613776	0.513962441	2.183735756	0.983381892
91 RABBITTITE	$Ca_3Mg_3(UO_2)_2(CO_3)_6(OH)_4.18H_2O$		monoclinic	NoDet	2.69155283	2.57	1.210574	0.04516086	0.550232158	2.122960011	1
92 RABERITE	$Tl_5Ag_4As_6SbS_{15}$		triclinic	P1	4.93144254	5.65	5.494679	-0.14570938	-0.114213368	1.028267491	0.872140921
93 REALGAR	AsS		monoclinic	P2$_1$/n	3.41007481	3.56	3.756759	-0.04396537	-0.101664561	0.947625443	0.959396064
94 SABATIERITE	Cu_6TlSe_4		orthorhombic		5.26844393	7.34	7.184365	-0.39320074	-0.363659743	1.021663028	1
95 TACHARANITE	$Ca_{12}Al_2Si_{18}O_{51}.18H_2O$		monoclinic	See Below	2.08310155	2.345	1.150654	-0.12572524	0.447624466	2.037970863	0.912834177
96 TACHYHYDRITE	$CaMg_2Cl_6.12H_2O$		hexagonal	R3	1.44918985	1.667	0.870102	-0.15029787	0.399593937	1.915866517	0.866025404
97 UCHUCCHACUAITE	$AgPb_3MnSb_5S_{12}$		monoclinic	Pmmm	5.50732597	5.16	5.082719	0.06306617	0.077098638	1.01520473	1
98 UEDAITE-(CE)	$Mn^{2+}CeAl_2Fe^{2+}(Si_2O_7)(SiO_4)O(OH)$		monoclinic	P2$_1$/m	3.71483903	4.1	2.153679	-0.10368174	0.420249828	1.903719554	0.905568799
99 UKLONSKOVITE	$NaMg(SO_4)F.2H_2O$		monoclinic	P2$_1$/m	2.21162719	2.42	1.039362	-0.09421697	0.530046399	2.32835107	0.918928946
100 VAESITE	NiS_2		cubic	Pa3	4.480418	4.45	3.265733	0.0067891	0.271109747	1.362634358	1
101 VALENTINITE	Sb_2O_3		orthorhombic	Pccn	5.84470342	5.76	3.797701	0.01449234	0.350232012	1.516707008	1
102 WADEITE	$K_2ZrSi_3O_9$		hexagonal	P6$_3$/m	2.73261852	3.115	1.514859	-0.13993226	0.445638183	2.056296491	0.866025404
103 WADSLEYITE	$\beta-(Mg,Fe^{2+})_2SiO_4$		orthorhombic	Imma	3.77905699	3.77	1.981702	0.00239663	0.475609496	1.902405489	1
104 XANTHIOSITE	$Ni_3(AsO_4)_2$		monoclinic	P2$_1$/a	5.38269846	5.2	2.432052	0.0339418	0.548172381	2.138112329	0.998652276
105 XANTHOCONITE	Ag_3AsS_3		monoclinic	C2/c	5.12218311	5.54	5.47747	-0.08157008	-0.069362376	1.011415872	0.939692621
106 YAFSOANITE	$Ca_3Zn_3(Te^{6+}O_6)_2$		cubic	Ia3d	5.03255635	5.475	2.484353	-0.08791628	0.506343684	2.203792899	1
107 YAGIITE	$(Na,K)_{1.5}(Mg,Al)_3(Al,Si)_{12}O_{30}$		hexagonal	P6/mcc	4.54664063	2.7	1.674527	0.40615496	0.631700219	1.61239586	0.866025404
108 ZABUYELITE	Li_2CO_3		monoclinic	C2/c	1.90482496	2.09	1.027818	-0.09721368	0.460413586	2.033434589	0.908581095
109 ZACCARINIITE	RhNiAs		tetragonal	P4/nmm	13.6895552	10.09	11.25692	0.26294172	0.177700163	0.896337622	1
110 ACTINOLITE	$Ca_2(Mg,Fe^{2+})_5Si_8O_{22}(OH)_2$		monoclinic	C2/m	2.96696617	3.135	1.630958	-0.0566349	0.450294492	1.92218358	0.967532957
111 ALMANDINE	$Fe^{2+}_3Al_2(SiO_4)_3$		cubic	Ia3d	4.31831681	4.318	2.08452	7.3364E-05	0.517284197	2.07146033	1
112 AZURITE	$Cu_3(CO_3)_2(OH)_2$		monoclinic	P2$_1$/c	3.77565101	3.773	2.118934	0.00070213	0.4387898	1.780612444	0.999100767
113 BERYL	$Be_3Al_2Si_6O_{18}$		hexagonal	P6/mcc	2.27457263	2.8	1.540305	-0.23100048	0.322815748	1.817822068	0.866025404
114 BISMUTHITE	$Bi_2O_2(CO_3)$		tetragonal	I4/mmm	8.26584927	6.9	4.422923	0.16524004	0.464916082	1.560054285	1
115 BORAX	$Na_2B_4O_5(OH)_4.8H_2O$		monoclinic	C2/c	1.64198171	1.715	0.729572	-0.04446961	0.555675966	2.350693488	0.958237767
116 GOLD	Au		cubic	Fm3m	19.2830542	19.3	19.282	-0.00087879	5.46702E-05	1.000933513	1
117 MONAZITE-(CE)	$(Ce,La,Nd,Th)PO_4$		monoclinic	P2$_1$/n	5.17526038	5.205	4.856695	-0.0057465	0.061555441	1.071716477	0.972856714
118 PYROPE	$Mg_3Al_2(SiO_4)_3$		cubic	Ia3d	3.55913153	3.582	1.553716	-0.0064253	0.563456515	2.305441105	1
119 RUTILE	TiO_2		tetragonal	P4$_2$/mnm	4.2490413	4.23	2.129051	0.00448132	0.498933855	1.986800933	1
120 SCHEELITE	$CaWO_4$		tetragonal	I4$_1$/a	6.11717749	6.1	3.228276	0.00280807	0.472260483	1.889553261	1
121 SCOTLANDITE	$PbS^{4+}O_3$		monoclinic	P2$_1$/m	6.19967223	6.37	3.875964	-0.02747367	0.374811436	1.64346204	0.960098679
122 SPINEL	$MgAl_2O_4$		cubic	Fd3m	3.56976029	3.85	1.620989	-0.07850379	0.545910989	2.375093349	1
123 TALC	$Mg_3Si_4O_{10}(OH)_2$		triclinic	P1	4.75710796	2.705	1.382039	0.43137721	0.709479069	1.957252397	0.856937352
124 TETRAHEDRITE	$(Cu,Fe,Ag,Zn)_{12}Sb_4S_{13}$		cubic	I43m	4.86577409	4.97	4.746776	-0.02142021	0.024456172	1.04702647	1
125 TOPAZ	$Al_2SiO_4(F,OH)_2$		orthorhombic	Pbnm	3.52695357	3.53	1.380919	-0.00086376	0.60846692	2.556268701	1
126 WULFENITE	$PbMoO_4$		tetragonal	I4$_1$/a	6.83827541	7	4.507436	-0.02364991	0.340851889	1.552989217	1

TACHARANITE: A-centered monoclinic pseudocell   * Rotoinversion Axes Neglected

Table C1d
Given and Computed Data for the Rock-Forming Mineral Species

APPENDIX D
PLOTS OF THE
APPARENT LATTICE STRAIN

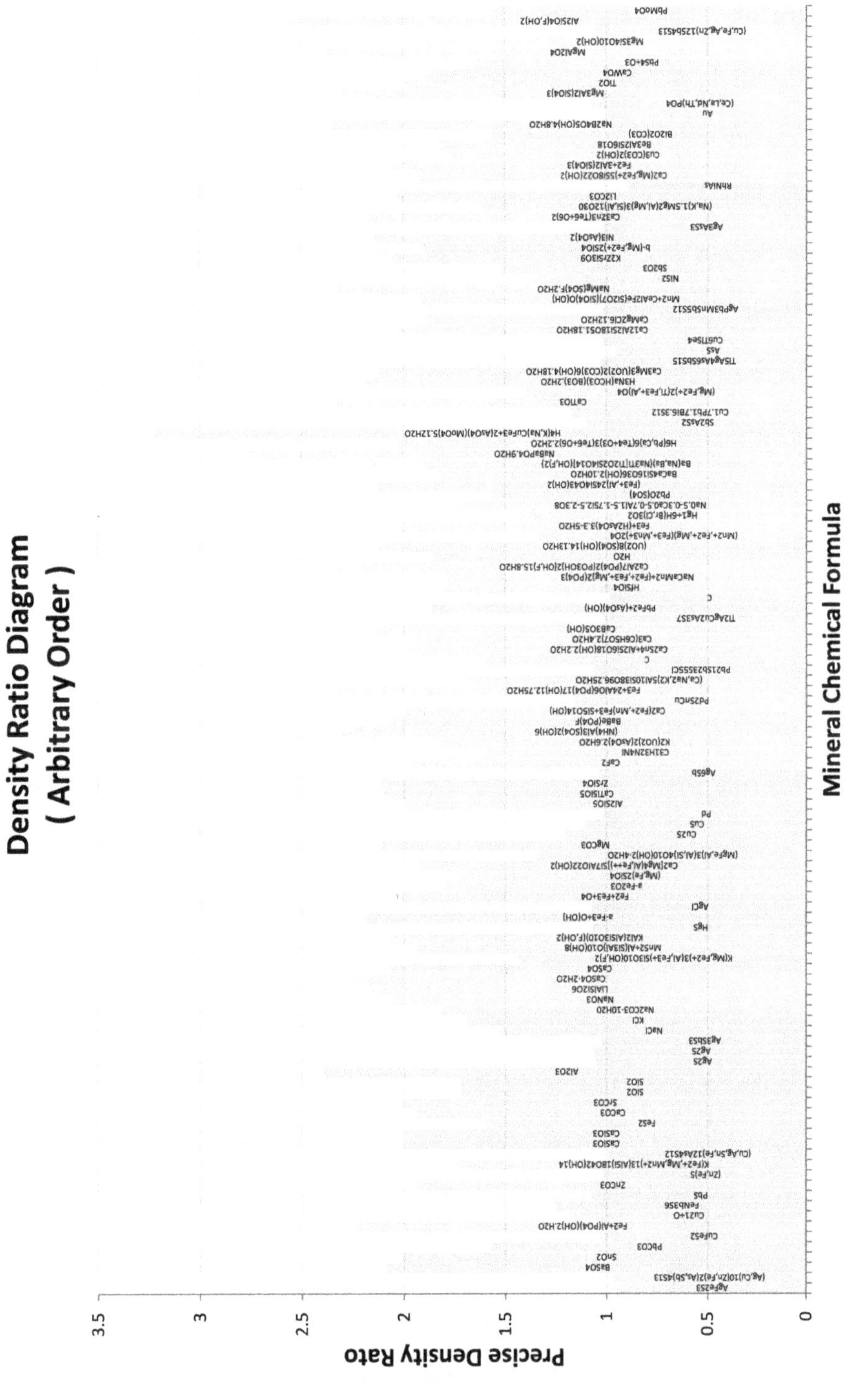

Figure D1
Lattice Strain Data for Each Mineral in Arbitrary Order of Record

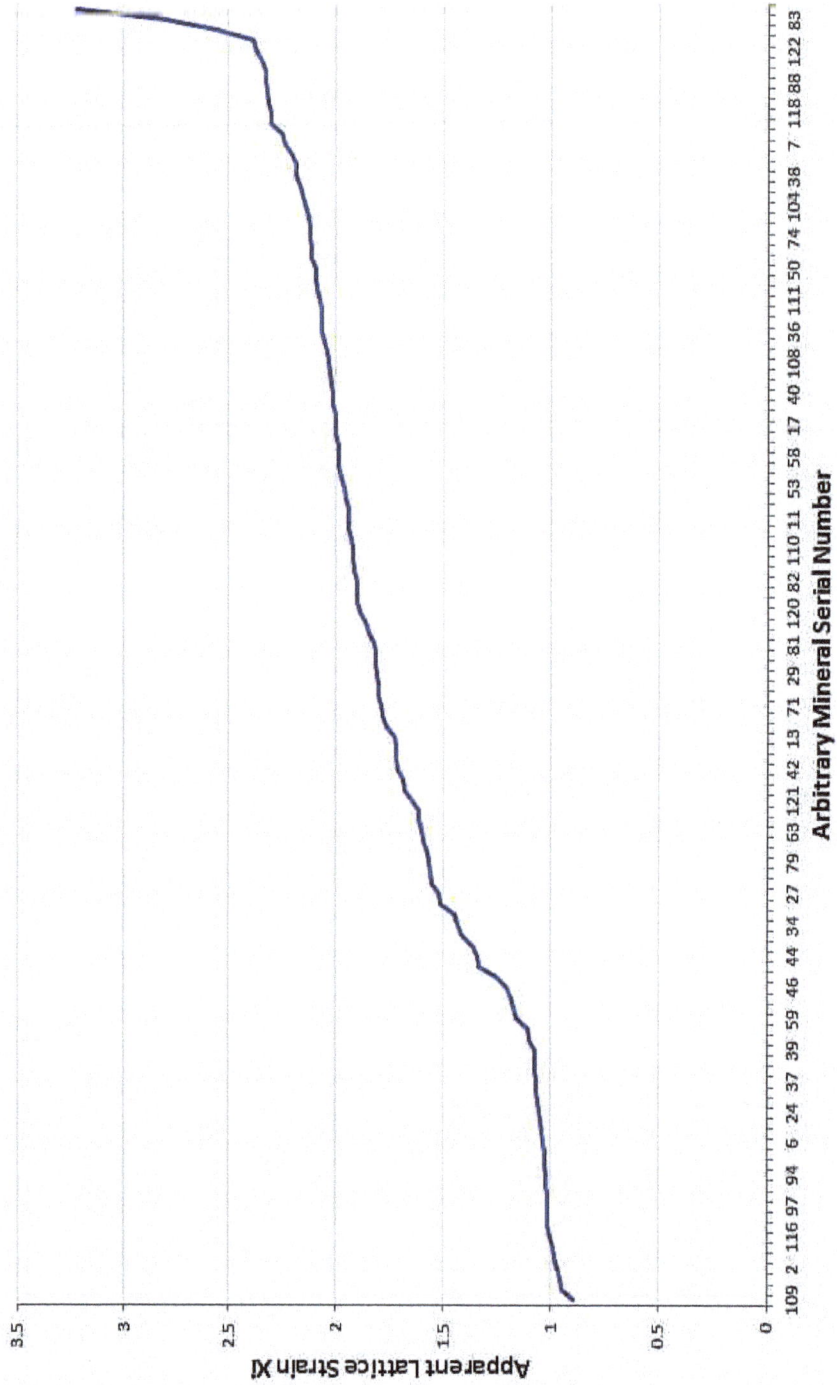

Figure D2
Lattice Strain Data for Each Mineral in Ascending Order of Lattice Strain

APPENDIX E
TABULATIONS AND PLOTS
OF THE
APPARENT LATTICE STRAIN
AND ITS RELATION TO THE
SQUARE ROOTS OF NATURAL NUMBERS

Table E1
Moving Triads Table
For All Twenty-One Possible Triads

Test Integer	Test Root	Column	f_0 Serial	f_0 LB	f_0 UB	f_0 f	f_1 Serial	f_1 LB	f_1 UB	f_1 f	f_2 Serial	f_2 LB	f_2 UB	f_2 f	x_{root} f_0	x_{root} f_1	x_{root} f_2	x_{mod}	ΣI	Σfx	Σfx^2	μ	s	M	%Sp.Def. (Root·µ)	%Sp.Def. (µ·M)	Graphical Ξ	Graphical Ω
1	1	J K	1	0.896338	0.989218	6	2	0.989218	1.082099	19	3	1.082099	1.17498	4	0.942778	1.035659	1.12854	1.17498	29	29.84824	30.80658	1.029253	0.054164	1.032342	2.925325	0.300062	1.0356588	1
1		K L	2	0.989218	1.082099	19	3	1.082099	1.17498	4	4	1.17498	1.267861	4	1.035659	1.12854	1.2142	1.267861	26	27.85594	29.9492	1.071382	0.063503	1.181614	7.138221	10.28878	1.1285396	1
1		L M	3	1.082099	1.17498	4	4	1.17498	1.267861	3	5	1.267861	1.360742	2	1.12854	1.2142	1.314301	1.314301		10.80702	13.02479	1.20078	0.072974		20.07802		1.2114204	1
1		M N	4	1.17498	1.267861	4	5	1.267861	1.360742	2	6	1.360742	1.453622	2	1.2142	1.314301	1.407182	1.407182	9	11.92159	15.85102	1.324621	0.08126	1.298821	32.46213	-1.94774	1.3143012	1
2	1.414214	N O	5	1.267861	1.360742	2	6	1.360742	1.453622	4	7	1.453622	1.546503	2	1.314301	1.407182	1.500063	1.500063		11.25746	15.8758	1.407182	0.065677	1.407182	-0.49721	0	1.407182	1.4142136
2	1.414214	O P	6	1.360742	1.453622	4	7	1.453622	1.546503	2	8	1.546503	1.639384	8	1.407182	1.500063	1.592944	1.592944	14	21.3744	32.7078	1.5256	0.081794	1.476843	7.946935	-3.25937	1.5000628	1.4142136
2	1.414214	P Q	7	1.453622	1.546503	2	8	1.546503	1.639384	4	9	1.639384	1.732265	7	1.500063	1.592944	1.685824	1.685824	17	27.54445	44.69416	1.620061	0.061813	1.626115	14.56979	0.361289	1.5929436	1.4142136
3	1.732051	Q R	8	1.546503	1.639384	4	9	1.639384	1.732265	8	10	1.732265	1.825146	9	1.592944	1.685824	1.778705	1.778705	24	40.5267	68.66791	1.689694	0.078075	1.670344	-2.4545	-1.14519	1.6858244	1.7320508
3	1.732051	R S	9	1.639384	1.732265	8	10	1.732265	1.825146	7	11	1.825146	1.918026	7	1.685824	1.778705	1.871586	1.871586	23	40.91022	72.888	1.778705	0.072465	1.778705	2.693594	1.25E-14	1.7787052	1.7320508
3	1.732051	S T	10	1.732265	1.825146	7	11	1.825146	1.918026	9	12	1.918026	2.010907	15	1.778705	1.871586	1.964467	1.964467	31	58.57645	110.8809	1.889563	0.079722	1.843722	9.093967	-2.42602	1.871586	1.7320508
4	2	T U	11	1.825146	1.918026	9	12	1.918026	2.010907	7	13	2.010907	2.103788	15	1.871586	1.964467	2.057348	2.057348	36	71.30567	141.6643	1.982527	0.068601	2.000587	-0.87365	0.910966	1.9644668	2
4	2	U V	12	1.918026	2.010907	7	13	2.010907	2.103788	15	14	2.103788	2.196669	14	1.964467	2.057348	2.150228	2.150228	39	79.77215	163.3793	2.04544	0.073405	1.979947	2.271991	-3.2019	2.0573476	2
4	2	V W	13	2.010907	2.103788	15	14	2.103788	2.196669	14	15	2.196669	2.28955	10	2.057348	2.150228	2.243109	2.243109	27	57.03448	120.5869	2.112388	0.063244	1.979947	5.619404	-6.26974	2.1502284	2
5	2.236068	W X	14	2.103788	2.196669	14	15	2.196669	2.28955	10	16	2.28955	2.38243	3	2.150228	2.243109	2.33599	2.33599	21	46.91953	104.9842	2.234263	0.085535	2.250849	-0.0807	0.742341	2.2431092	2.236068
5	2.236068	X Y	15	2.196669	2.28955	10	16	2.28955	2.38243	3	17	2.38243	2.475311	8	2.243109	2.33599	2.42871	2.42871	27	64.84612	64.64882	2.32051	0.051342	2.32825	3.776357	0.33355	2.33599	2.236068
6	2.44949	Y Z	16	2.28955	2.38243	3	17	2.38243	2.475311	8	18	2.475311	2.568192	1	2.33599	2.42871	2.521752	2.521752	10	23.63854	55.91344	2.363854	0.059473	2.475311	-3.49605	4.715052	2.4288708	2.4494897
6	2.44949	Z AA	17	2.38243	2.475311	8	18	2.475311	2.568192	1	19	2.568192	2.661073	1	2.42871	2.521752	2.614632	2.614632	2	4.950622	12.25864	2.475311	0.04694	2.475311	1.054157	0	2.5217516	2.4494897
7	2.645751	AA AB	18	2.475311	2.568192	1	19	2.568192	2.661073	1	20	2.661073	2.753954	0	2.521752	2.614632	2.707513	2.707513	1	2.521752	6.359231	2.521752	0	2.661073	-4.68675	5.524779	2.6146324	2.6457513
7	2.645751	AB AC	19	2.568192	2.661073	1	20	2.661073	2.753954	0	21	2.753954	2.846834	1	2.614632	2.707513	2.800394	2.800394	1	2.800354	7.842207	2.800394	0	2.661073	5.844945	-4.97506	2.7075132	2.6457513
8	2.828427	AC AD	20	2.661073	2.753954	0	21	2.753954	2.846834	1	22	2.846834	2.939715	0	2.707513	2.800394	2.893275	2.893275	1	2.800354	7.842207	2.800394	0	2.800394	-0.99112	0	2.800394	2.8284271
9	3 AC	AD AE	21	2.753954	2.846834	1	22	2.846834	2.939715	0	23	2.939715	3.032596	0	2.800394	2.893275	2.986156	2.986156	1	2.800354	7.842207	2.800394	0	2.939715	-6.65353	4.975067	2.8932748	3
9	3 AD	AE AF	22	2.846834	2.939715	0	23	2.939715	3.032596	0	24	3.032596	3.125477	0	2.893275	2.986156	3.079036	3.079036	0	0	0		0				2.9861556	3
9	3 AE	AF AG	23	2.939715	3.032596	0	24	3.032596	3.125477	0	25	3.125477	3.218358	1	2.986156	3.079036	3.171917	3.171917	1	3.171917	10.06106	3.171917	0	3.032596	5.730574	-4.39233	3.0790364	3

			23			23	23			23	23			23	23	23	23	23	23	23	23	23	23	23	23	21	23	23	23 Count
102 45.92984	0 0 0	0 ### ### ###	276	44.11461	46.25087	124	299	46.25087	48.38712	118	322	48.38712	50.52338	100	0	45.18274	47.319	49.45525	342	606.2738	1138.742	44.16719	1.159487	42.90074	101.481	0.534533	47.318995	45.929838 Sum
4.434783 1.996949		### ### ###	12	1.918026	2.010907	5.4	13	2.010907	2.103788	5.1	14	2.103788	2.196669	4.3	##	1.964467	2.057348	2.150228	15	26.35933	49.51051	2.007599	0.052704	2.042892	4.612771	0.025454	2.0573476	1.9969495 Mean
2.74402 0.683558		### ### ###	6.782	0.629948	0.629948	5.3	6.782	0.629948	0.629948	5.4	6.782	0.629948	0.629948	4.5	##	0.629948	0.629948	0.629948	13	23.68983	48.4471	0.612331	0.030901	0.595543	8.781095	3.874816	0.6299482	0.6835879 Sample SD

10:32 Tuesday, 19 July 2022
James R Warren

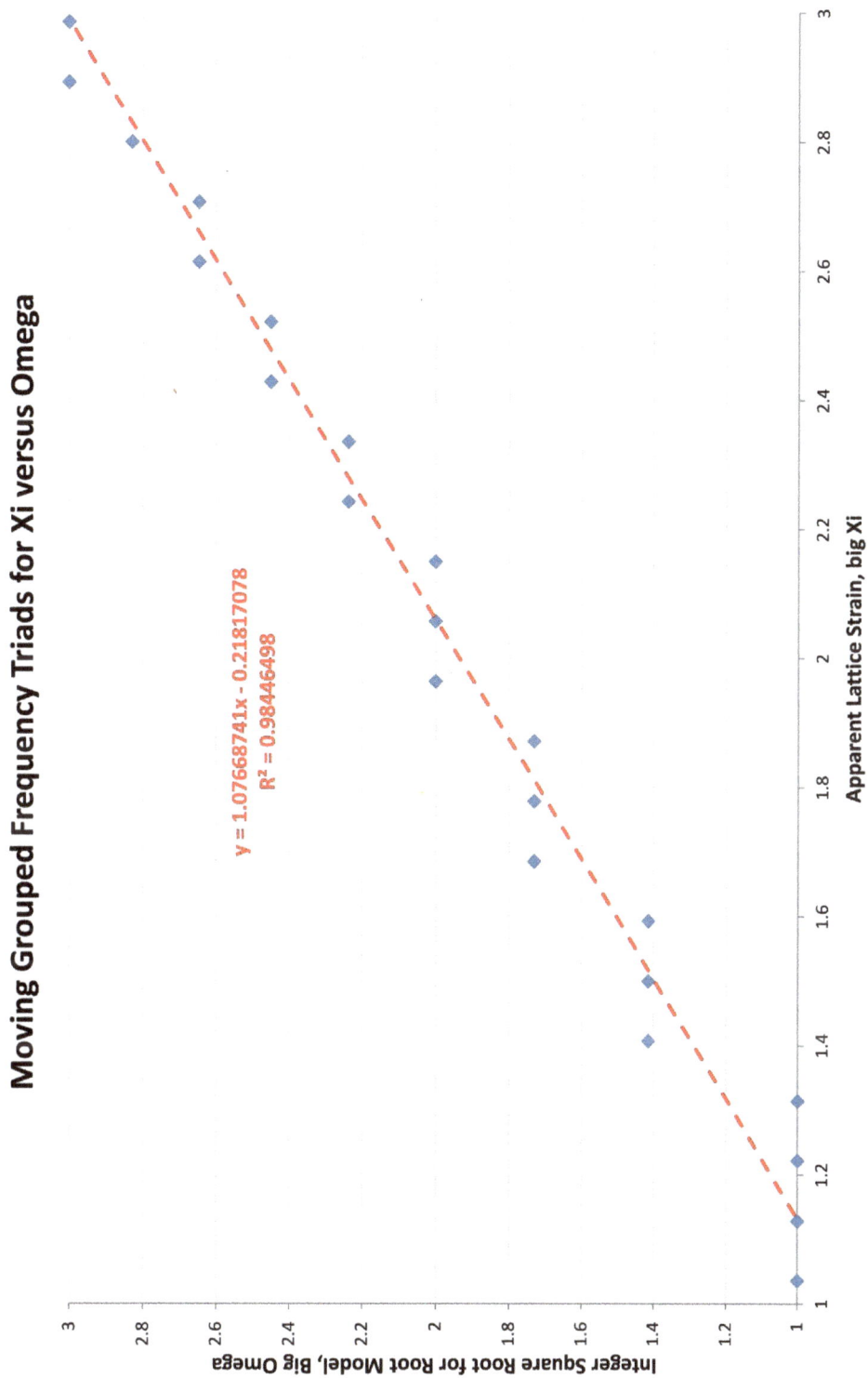

Figure E1
Plot of the Moving Triads
For All Twenty-One Possible Triads

Test Integer	Test Root	Column	f₀ Serial	f₀ LB	f₀ UB	f₀ f	f₁ Serial	f₁ LB	f₁ UB	f₁ f	f₂ Serial	f₂ LB	f₂ UB	f₂ f	f₀ x_{mid}	f₁ x_{mid}	f₂ x_{mid}	Σf	Σfx	Σfx²	μ	s	M	%Sp.Def. (Root±μ)	%Sp.Def. (μ-M)	Graphical Ξ	Graphical Ω
1	1	J K	1	0.896338	0.989218	6	2	0.989218	1.082099	19	3	1.082099	1.17498	4	0.942778	1.035659	1.12854	29	29.84834	30.80658	1.029253	0.054164	1.032342	2.925325	0.300062	1.0356388	1
2	1.414214	M N O	5	1.267861	1.360742	2	6	1.360742	1.453622	4	7	1.453622	1.546503	2	1.314301	1.407182	1.500063	8	11.25746	15.8758	1.407182	0.065677	1.407182	-0.49721	0	1.40782	1.4142136
3	1.732051	P Q R	8	1.546503	1.639384	8	9	1.639384	1.732265	7	10	1.732265	1.825146	9	1.592944	1.685824	1.778705	24	40.55267	68.66791	1.689694	0.078075	1.670344	-2.44545	-1.14519	1.6858344	1.7320508
4	2	S T U	11	1.825146	1.918026	7	12	1.918026	2.010907	15	13	2.010907	2.103788	14	1.871586	1.964467	2.057348	36	71.37097	141.6643	1.982527	0.068601	2.080587	-0.87365	0.910966	1.964468	2
5	2.236068	V W X	14	2.103788	2.196669	10	15	2.196669	2.28955	3	16	2.28955	2.38243	8	2.150228	2.243109	2.33599	21	46.91953	104.9842	2.234263	0.085535	2.250849	-0.0807	0.742341	2.2431992	2.236068
6	2.44949	X Y Z	16	2.28955	2.38243	8	17	2.38243	2.475311	1	18	2.475311	2.568192	1	2.33599	2.428871	2.521752	10	23.63854	55.91344	2.363854	0.059473	2.475311	-3.49605	4.715052	2.4288708	2.4494897
7	2.645751	AA AB	18	2.475311	2.568192	1	19	2.568192	2.661073	0	20	2.661073	2.753954	0	2.521752	2.616632	2.707513	1	2.521752	6.359231	2.521752	0	2.661073	-4.68675	5.524779	2.6146424	2.6457513
8	2.828427	AB AC AD	20	2.661073	2.753954	0	21	2.753954	2.846834	1	22	2.846834	2.939715	0	2.707513	2.800394	2.893275	1	2.800394	7.842207	2.800394	0	2.800394	-0.99112	0	2.800894	2.8284271
9	3	AC AD AE	21	2.753954	2.846834	0	22	2.846834	2.939715	0	23	2.939715	3.032596	1	2.800394	2.893275	2.986156	1	2.800394	7.842207	2.800394	0	2.939715	-6.65353	4.975057	2.8932948	3
9			9			9	9			9	9			9	9	9	9	9	9	9	9	9	9	9	9	9	9 Count
45	19.306	0 0 0	114	17.81952	18.65545	42	123	18.65545	19.49138	50	132	19.49138	20.3273	39	18.23749	19.07341	19.90934	131	231.7101	439.9559	18.82931	0.411525	16.2378	-16.7991	16.02307	19.073813	19.306001 Sum
5	2.145111	## ## ##	12.67	1.979947	2.072828	4.8	13.67	2.072828	2.165709	5.6	14.67	2.165709	2.258589	4.2	2.026387	2.119268	2.212149	15	25.74556	48.88399	2.092146	0.045725	2.137533	-1.86657	1.780341	2.1195581	2.1451112 Mean
2.738613	0.66956	## ## ##	6.928	0.643497	0.643497	3.8	6.928	0.643497	0.643497	6.9	6.928	0.643497	0.643497	5	0.643497	0.643497	0.643497	13	23.90803	48.46363	0.617931	0.035506	0.569146	2.820412	2.543668	0.643971	0.6695598 Sample SD

Table E2
Single Ω-Inclusive Triads Table
For the Nine Triads with the
Square Root of a Natural Number included in the Modal Bin

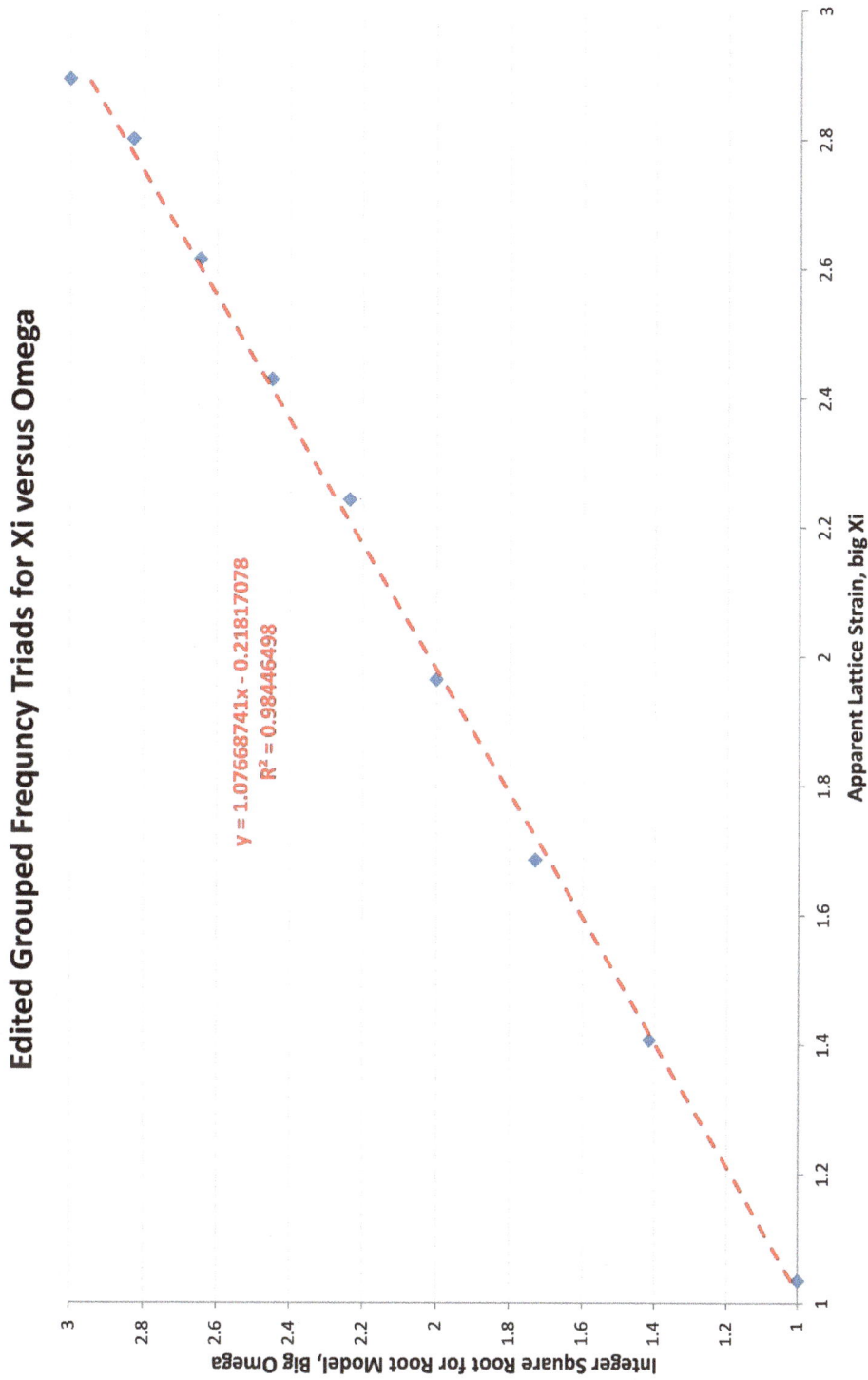

Edited Grouped Frequncy Triads for Xi versus Omega

y = 1.07668741x - 0.21817078
R² = 0.9846498

Apparent Lattice Strain, big Xi

Integer Square Root for Root Model, Big Omega

Figure E2
Plot of the Single Ω-Inclusive Triads
For the Nine Triads with the
Square Root of a Natural Number included in the Modal Bin

APPENDIX F
CHI-SQUARED TEST OUTCOMES FOR
VARIOUS PARAMETERS OF
ONE-HUNDRED AND TWENTY-FIVE MINERALS
AGAINST UNIFORM DISTRIBUTION

TEST DESCRIPTION: RAND() is a Uniform Distribution

Total

LB	0.872599724	1.455725106	2.038850488	2.62197587	
UB	1.455725106	2.038850488	2.62197587	3.205101252	
Bin Width	0.583125382	0.583125382	0.583125382	0.583125382	
Fraction	0.25	0.25	0.25	0.25	1
Observed	31	32	31	31	125
Expected	31.25	31.25	31.25	31.25	125
Chi-Square Contribution	0.002	0.018	0.002	0.002	0.024
Chi-Square	0.024				
Degrees of Freedom	3				
α	0.001				
p	9.990182E-01				
CHISQ.INV.RT	16.2662362				

DATA CONFORMS TO MODEL

Table F1
Chi-Squared
Trail of Uniform Distribution to
Prove Uniform Distribution

TEST DESCRIPTION: Xi is a Uniform Distribution

Total

LB	0.896337622	1.476842618	2.057347615	2.637852611	
UB	1.476842618	2.057347615	2.637852611	3.218357608	
Bin Width	0.580504996	0.580504996	0.580504996	0.580504996	
Fraction	0.25	0.25	0.25	0.25	1
Observed	38	55	30	2	125
Expected	31.25	31.25	31.25	31.25	125
Chi-Square Contribution	1.458	18.05	0.05	27.378	46.936
Chi-Square	46.936				
Degrees of Freedom	3				
α	0.001				
p	3.586279E-10				
CHISQ.INV.RT	16.2662362				

DATA DOES NOT CONFORM TO MODEL

Table F2
Chi-Squared
Trial of a Three-Cycle sine Curve to
Prove Uniform Distribution

TEST SERIAL: 02

TEST DESCRIPTION: Psi is a Uniform Distribution

					Total
LB	0.813506961	0.860130221	0.906753481	0.95337674	
UB	0.860130221	0.906753481	0.95337674	1.0000001	
Bin Width	0.04662326	0.04662326	0.04662326	0.04662336	
Fraction	0.249999866	0.249999866	0.249999866	0.250000402	1
Observed	4	32	9	80	125
Expected	31.24998324	31.24998324	31.24998324	31.25005027	125
Chi-Square Contribution	23.76198352	0.018000814	15.84198463	76.04972082	115.6717
Chi-Square	115.6716898				
Degrees of Freedom	3				
α	0.001				
p	6.598912E-25				
CHISQ.INV.RT	16.2662362				

DATA DOES NOT CONFORM TO MODEL

Table F3
Chi-Squared
Trial of Apparent Lattice Strain Ξ to
Prove Uniform Distribution

TEST SERIAL: 03

TEST DESCRIPTION: Product Xi.Psi is a Uniform Distribution

					Total
LB	0.850279487	1.442299017	2.034318547	2.626338078	
UB	1.442299017	2.034318547	2.626338078	3.218357608	
Bin Width	0.59201953	0.59201953	0.59201953	0.59201953	
Fraction	0.25	0.25	0.25	0.25	1
Observed	39	64	20	2	125
Expected	31.25	31.25	31.25	31.25	125
Chi-Square Contribution	1.922	34.322	4.05	27.378	67.672
Chi-Square	67.672				
Degrees of Freedom	3				
α	0.001				
p	1.344738E-14				
CHISQ.INV.RT	16.2662362				

DATA DOES NOT CONFORM TO MODEL

Table F4
Chi-Squared
Trial of Lattice Angle Multiplier Ψ to
Prove Uniform Distribution

CHAPTER FIVE

Name Enunciability

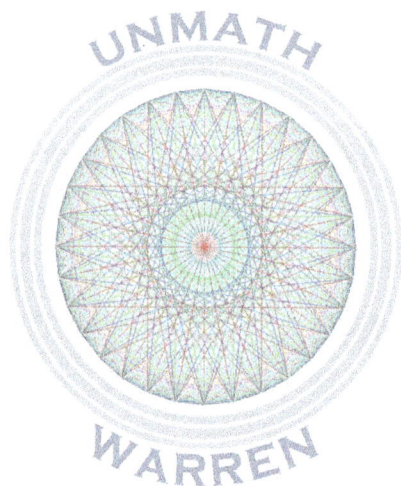

On the Enunciability of Some City Names

by
James R Warren BSc MSc PhD PGCE

PART ONE
MATHEMATICAL PRINCIPLES

Consider a collection of some symbols:-

444444   **Series 1**

It is not clear (at least to me) whether these six fours are *distinct individuals* (notwithstanding their apparent identity) or whether they are *replicas* of the same object. Secondly, I am not sure whether these are stand-alone objects that happen to be written together for convenience, or whether they are somehow members of corrals or sub-groups impossible to discern.

Some light may or may not be thrown upon such perplexities by periodising the list of six objects in one way or another. For example:-

44.444.4   **Series 2**

I should say immediately that the period marks do not really exist: They are only there to clarify the argument. (If you want the periods to actually exist and participate as symbols or surrogates, then special arrangements will have to be made, and there will be a cost mathematically-speaking).

For the sake of argument, Series One consists only of members drawn repeatedly from the single-member set {4}, as does Series Two.

We could of course extend our repertoire of symbols to include {4Hc}. A possible series of six symbols would then be:-

H4HccH   **Series 3**

The English Language comprises a repertoire or set of twenty-six written symbols:-

L = {abcdefghijklmnopqrstuvwxyz}   **Set 3**

The order is conventional, but implies conditions which we may study later.
Of these twenty-six symbols, confusingly often called letters, only five are vowels:-

V = {aeiou}   **Set 4**

and the rest are consonants:-

$$C = \{\text{bcdfghjklmnpqrstvwxyz}\} \qquad \textbf{Set 5}$$

Additionally, some of the voiced sounds in the language are not specified by unique symbols, but are represented by diphthongs or digraphs. There are no superscript or subscript accent marks in written English. The dual-symbol vocalisations are usually post-dental hisses of various sorts including "ch" and "th", the latter cognate with the Greek θ (theta) form or the Anglo-Saxon "thorn" which was written "Y", as in the facetious phrase "Ye Olde English Tea-shop". The hyphen fulfils some of the unvoiced periodisation displayed in Series Two, but is increasingly obsolete and is absent in American English.

Notwithstanding, English, unlike Welsh or (say) Polish is incapable of vocalising protracted series of consonants, and in the case of Welsh at least several digraphs, notably "dd", are voiced as theta.

With many exceptions, English words are written of vowels alternated with consonants in a manner that makes them readily enunciable. Take for example the members of Set Three composed "cow", "cat", "dog" and "shag", or more complex multi-syllable forms like "London" (the upper-case letter, helpfully imposed by MicroSoft®, is not mathematically-distinct from a lower-case l).

These considerations, infantile as they seem, bear profoundly upon the likelihood of English words, and for the purposes of this disquisition, the names of capital cities as rendered in English rather than their native languages.

And for the purposes of our researches a capital city is not "of a sovereign state that is a member of The United Nations in its own right" as so inimitably qualified by Mr Richard Osman, but any administrative chief town of any distinct territory as listed by Wikipedia[1], though this excludes American and other federal states, but includes such anomalies as Scotland and Tristan da Cunha. Remarkably, both those territories have capital towns of the same name.

Vocability

It is clearly not the case that all combinations of English letters, even if they include vowels, are vocable. The names of major cities are usually modified in such a manner that however spelt and vocalised in the native language they are readily enunciable by the average American or Briton. On the whole, the more frequently-referenced cities like PARIS and MOSCOW have easy BABAB or BABBAB vowel-consonant strictures where A is any vowel and B any consonant. On the other hand, MAMOUDZOU and especially MBABANE are less referenceable though their structures are BABAABBAA and BBABABA. Typically, easy English renditions are short and alternate vowels and consonants restricting the numbers of character runs within the word.

Some theories focus upon the vocable core (loosely we may say a syllable) and its combinations. The structure of many or most vocable cores being BAB. Accordingly, LONDON, MADRID and MOSCOW are all abutments of a pair of vocable cores   and are in some sense ideal compositions. If we consider six-letter combinations, sixty-four of which constitute a certain name genus, then at the extremes AAAAAA and ZZZZZZ are not enunciable whereas better-alternated structures such as ABABAB and BABABA are.

There are approximately one million English words of which some three hundred are the names of governing towns of self-governing territories. As we shall compute, there are many more *possible* and *pronounceable* letter-combinations, and whilst they are listably finite, they run into the many millions.

A Taxonomy of Words

Set Three L clearly has a repertoire of 26 symbols: It is of course the common English alphabet. Some European languages have slightly more or fewer letters.

Correspondingly, the subsets V and C have respectively 5 and 21 symbols. L is of course the set addition of V and C.

Accordingly, the Repertoires of Set X are defined as $r_L = 26$, $r_V = 5$ and $r_C = 21$. The population (or potential population) of distinct words is controlled by such symbol repertoires together with the number of component symbols in that repertoire. Let us call a word n characters long a Species. Then the Species Population, S, is given by:-

$$S = \prod_{i=1}^{m} r_i^{n_i}$$

Equation 1

where S is the Limiting Number of Individuals in the Species; r_i is the ith Symbol Repertoire and n_i is the Number of Symbols of Repertoire r_i in the Species term.

Note that this model considers all words of length n to be members of Genus G.

For example LONDON is an individual of the Species BABBAB, where A is some Vowel and B is some Consonant. Vowels A have Repertoire $r_V=5$ and Consonants B have the Repertoire $r_C=21$. ZAGREB is also an Individual of this Species. The Limiting Population of BABBAB is given by:-

$$S = r_B.r_A.r_B.r_B.r_A.r_B = r_A^{n_A}.r_B^{n_a} = 5^2.21^4 = 25.194481 = 4862025$$

Equation 2

Therefore, there are 4,862,025 potential enunciable names of the structure BABBAB.

The converse of Population is Probability such that:-

$$p(I) = \frac{1}{S}$$

Equation 3

where p(I) is the Probability of a given Individual in Species S.
Therefore:-

$$p(LONDON) = \frac{1}{S} = \frac{1}{4862025} = 0.000000205675618698$$

Equation 4

Alternatively:-

$$p(LONDON) = \frac{1}{\prod_{i=1}^{6} r_i} = \frac{1}{21 \times 5 \times 21 \times 21 \times 5 \times 21} = \frac{1}{4862025} = 0.000000205675618698$$

Equation 5

With regard to Symbol Repertoires we may observe that the Probability of a randomly-found character being a Vowel, p(V), is given by:-

$$p(V) = \frac{r_A}{r_A + r_B} = \frac{5}{5 + 21} = \frac{5}{26} = 0.1923076923$$

Equation 6

and the Probability of a Consonant, p(C), is:-

$$p(C) = \frac{r_B}{r_A + r_B} = \frac{21}{5 + 21} = \frac{21}{26} = 0.80769230769$$

Equation 7

Note that:-

$$p(V) + p(C) = 1$$

Equation 8

Because a character in any word can only either be a vowel or a consonant.

We are now in a position to treat a probabilistic interpretation of the likelihood the species of six-letter word of pattern BABBAB inserting the probabilities of Equations Six and Seven to achieve:-

$$p(BABBAB) = p(C).p(V).p(C).p(C).p(V).p(C)$$

$$= \frac{21}{26} \cdot \frac{5}{26} \cdot \frac{21}{26} \cdot \frac{21}{26} \cdot \frac{5}{26} \cdot \frac{21}{26} = \left(\frac{5}{26}\right)^2 \left(\frac{21}{26}\right)^4 = 0.01573899871$$

Equation 8

The value of G (i.e. the Genus Population) is given by:-

$$G = r^n$$

Equation 9

Table One shows the elaboration of the 64 members of G(6), the six-letter word Genus. Note that the Population considerably exceeds both the Combinations ^nC_r and Permutations ^nP_r resolvable with this set of objects.

It is also the case that:-

$$\sum_{i=1}^{N}\prod_{j=1}^{r} n_j^{p(r_j)} = 1$$

Equation 10

where:-

$$N = r^n$$

Equation 11

in which r is the Number of Repertoires (i.e. 2) and n is the Number of Objects (i.e. 6) whilst r_j is the repertoire of the jth distinct component (that is, 5 for Vowels r_1 and 21 for Consonants r_2); and n_j is the number of the relevant components in the name. N is clearly equivalent to the Genus Population G.

For example, in the tabulated case of G(6) we may specify:-

$$\sum n_A^{p(A)} n_B^{p(B)} n_C^{p(C)} n_D^{p(D)} = 1$$

Equation 12

Now the repertoires C and D could be punctuation symbols or Cyrillic characters or anything but in our study they do not exist because by definition we are only considering English vowels and consonants, the repertoires A and B.

Accordingly, Equation Twelve reduces to:-

$$\sum n_A^{p(A)} n_B^{p(B)} = 1$$

Equation 13

which by substitution we may elaborate as:-

$$1 = 1.p(A)^6 p(B)^0 + 6.p(A)^5 p(B)^1 + 15.p(A)^4 p(B)^2 + 20.p(A)^3 p(B)^3$$
$$+ 15.p(A)^4 p(B)^2 + 6.p(A)^5 p(B)^1 + 1.p(A)^6 p(B)^0$$

Equation 14

or alternatively:-

$$\sum_{l=0}^{n} \frac{n!}{l!(n-l)!}.p(A)^l.p(b)^{n-l} = (p(A)+p(B))^n = 1$$

Equation 15

Set   A,B

Serial	Binary Serial	1 2 3 4 5 6	Denary	Letter Text Values	Codes	Conjoint p(XXX) Probability	Count A 0.19	B 0.81	Check
0	000000	0 0 0 0 0 0	0 0 0 0 0 0	A A A A A A	AAAAAA	5.058E-05	6	0	6
1	000001	0 0 0 0 0 1	0 0 0 0 0 1	A A A A A B	AAAAAB	0.00021244	5	1	6
2	000010	0 0 0 0 1 0	0 0 0 0 1 0	A A A A B A	AAAABA	0.00021244	5	1	6
3	000011	0 0 0 0 1 1	0 0 0 0 1 1	A A A A B B	AAAABB	0.00089223	4	2	6
4	000100	0 0 0 1 0 0	0 0 0 1 0 0	A A A B A A	AAABAA	0.00021244	5	1	6
5	000101	0 0 0 1 0 1	0 0 0 1 0 1	A A A B A B	AAABAB	0.00089223	4	2	6
6	000110	0 0 0 1 1 0	0 0 0 1 1 0	A A A B B A	AAABBA	0.00089223	4	2	6
7	000111	0 0 0 1 1 1	0 0 0 1 1 1	A A A B B B	AAABBB	0.00374738	3	3	6
8	001000	0 0 1 0 0 0	0 0 1 0 0 0	A A B A A A	AABAAA	0.00021244	5	1	6
9	001001	0 0 1 0 0 1	0 0 1 0 0 1	A A B A A B	AABAAB	0.00089223	4	2	6
10	001010	0 0 1 0 1 0	0 0 1 0 1 0	A A B A B A	AABABA	0.00089223	4	2	6
11	001011	0 0 1 0 1 1	0 0 1 0 1 1	A A B A B B	AABABB	0.00374738	3	3	6
12	001100	0 0 1 1 0 0	0 0 1 1 0 0	A A B B A A	AABBAA	0.00089223	4	2	6
13	001101	0 0 1 1 0 1	0 0 1 1 0 1	A A B B A B	AABBAB	0.00374738	3	3	6
14	001110	0 0 1 1 1 0	0 0 1 1 1 0	A A B B B A	AABBBA	0.00374738	3	3	6
15	001111	0 0 1 1 1 1	0 0 1 1 1 1	A A B B B B	AABBBB	0.015739	2	4	6
16	010000	0 1 0 0 0 0	0 1 0 0 0 0	A B A A A A	ABAAAA	0.00021244	5	1	6
17	010001	0 1 0 0 0 1	0 1 0 0 0 1	A B A A A B	ABAAAB	0.00089223	4	2	6
18	010010	0 1 0 0 1 0	0 1 0 0 1 0	A B A A B A	ABAABA	0.00089223	4	2	6
19	010011	0 1 0 0 1 1	0 1 0 0 1 1	A B A A B B	ABAABB	0.00374738	3	3	6
20	010100	0 1 0 1 0 0	0 1 0 1 0 0	A B A B A A	ABABAA	0.00089223	4	2	6
21	010101	0 1 0 1 0 1	0 1 0 1 0 1	A B A B A B	ABABAB	0.00374738	3	3	6
22	010110	0 1 0 1 1 0	0 1 0 1 1 0	A B A B B A	ABABBA	0.00374738	3	3	6
23	010111	0 1 0 1 1 1	0 1 0 1 1 1	A B A B B B	ABABBB	0.015739	2	4	6
24	011000	0 1 1 0 0 0	0 1 1 0 0 0	A B B A A A	ABBAAA	0.00089223	4	2	6
25	011001	0 1 1 0 0 1	0 1 1 0 0 1	A B B A A B	ABBAAB	0.00374738	3	3	6
26	011010	0 1 1 0 1 0	0 1 1 0 1 0	A B B A B A	ABBABA	0.00374738	3	3	6
27	011011	0 1 1 0 1 1	0 1 1 0 1 1	A B B A B B	ABBABB	0.015739	2	4	6
28	011100	0 1 1 1 0 0	0 1 1 1 0 0	A B B B A A	ABBBAA	0.00374738	3	3	6
29	011101	0 1 1 1 0 1	0 1 1 1 0 1	A B B B A B	ABBBAB	0.015739	2	4	6
30	011110	0 1 1 1 1 0	0 1 1 1 1 0	A B B B B A	ABBBBA	0.015739	2	4	6
31	011111	0 1 1 1 1 1	0 1 1 1 1 1	A B B B B B	ABBBBB	0.06610379	1	5	6
32	100000	1 0 0 0 0 0	1 0 0 0 0 0	B A A A A A	BAAAAA	0.00021244	5	1	6
33	100001	1 0 0 0 0 1	1 0 0 0 0 1	B A A A A B	BAAAAB	0.00089223	4	2	6
34	100010	1 0 0 0 1 0	1 0 0 0 1 0	B A A A B A	BAAABA	0.00089223	4	2	6
35	100011	1 0 0 0 1 1	1 0 0 0 1 1	B A A A B B	BAAABB	0.00374738	3	3	6
36	100100	1 0 0 1 0 0	1 0 0 1 0 0	B A A B A A	BAABAA	0.00089223	4	2	6
37	100101	1 0 0 1 0 1	1 0 0 1 0 1	B A A B A B	BAABAB	0.00374738	3	3	6
38	100110	1 0 0 1 1 0	1 0 0 1 1 0	B A A B B A	BAABBA	0.00374738	3	3	6
39	100111	1 0 0 1 1 1	1 0 0 1 1 1	B A A B B B	BAABBB	0.015739	2	4	6
40	101000	1 0 1 0 0 0	1 0 1 0 0 0	B A B A A A	BABAAA	0.00089223	4	2	6
41	101001	1 0 1 0 0 1	1 0 1 0 0 1	B A B A A B	BABAAB	0.00374738	3	3	6
42	101010	1 0 1 0 1 0	1 0 1 0 1 0	B A B A B A	BABABA	0.00374738	3	3	6
43	101011	1 0 1 0 1 1	1 0 1 0 1 1	B A B A B B	BABABB	0.015739	2	4	6
44	101100	1 0 1 1 0 0	1 0 1 1 0 0	B A B B A A	BABBAA	0.00374738	3	3	6
45	101101	1 0 1 1 0 1	1 0 1 1 0 1	B A B B A B	BABBAB	0.015739	2	4	6
46	101110	1 0 1 1 1 0	1 0 1 1 1 0	B A B B B A	BABBBA	0.015739	2	4	6
47	101111	1 0 1 1 1 1	1 0 1 1 1 1	B A B B B B	BABBBB	0.06610379	1	5	6
48	110000	1 1 0 0 0 0	1 1 0 0 0 0	B B A A A A	BBAAAA	0.00089223	4	2	6
49	110001	1 1 0 0 0 1	1 1 0 0 0 1	B B A A A B	BBAAAB	0.00374738	3	3	6
50	110010	1 1 0 0 1 0	1 1 0 0 1 0	B B A A B A	BBAABA	0.00374738	3	3	6
51	110011	1 1 0 0 1 1	1 1 0 0 1 1	B B A A B B	BBAABB	0.015739	2	4	6
52	110100	1 1 0 1 0 0	1 1 0 1 0 0	B B A B A A	BBABAA	0.00374738	3	3	6
53	110101	1 1 0 1 0 1	1 1 0 1 0 1	B B A B A B	BBABAB	0.015739	2	4	6
54	110110	1 1 0 1 1 0	1 1 0 1 1 0	B B A B B A	BBABBA	0.015739	2	4	6
55	110111	1 1 0 1 1 1	1 1 0 1 1 1	B B A B B B	BBABBB	0.06610379	1	5	6
56	111000	1 1 1 0 0 0	1 1 1 0 0 0	B B B A A A	BBBAAA	0.00374738	3	3	6
57	111001	1 1 1 0 0 1	1 1 1 0 0 1	B B B A A B	BBBAAB	0.015739	2	4	6
58	111010	1 1 1 0 1 0	1 1 1 0 1 0	B B B A B A	BBBABA	0.015739	2	4	6
59	111011	1 1 1 0 1 1	1 1 1 0 1 1	B B B A B B	BBBABB	0.06610379	1	5	6
60	111100	1 1 1 1 0 0	1 1 1 1 0 0	B B B B A A	BBBBAA	0.015739	2	4	6
61	111101	1 1 1 1 0 1	1 1 1 1 0 1	B B B B A B	BBBBAB	0.06610379	1	5	6
62	111110	1 1 1 1 1 0	1 1 1 1 1 0	B B B B B A	BBBBBA	0.06610379	1	5	6
63	111111	1 1 1 1 1 1	1 1 1 1 1 1	B B B B B B	BBBBBB	0.27763594	0	6	6

$\sum n_A{}^{p(A)} n_B{}^{p(B)} n_C{}^{p(C)} n_D{}^{p(D)}$ 1.0000   1   192   192   384

Gather 6 Objects from 2

Entities, r	2
Objects, n	6
Combinations, nCr	15
Permutations, nPr	30
n/r	3
r^n	64
$n.r^n$	384
r^{n+1}	128
$(n/r).r^{n+1}$	384
p	0.1923
q	0.8077
p^r	0.0370
q^{n-r}	0.4256
$p^r q^{n-r}$	0.0157 prob(BABBAB)
$1/p^r q^{n-r}$	63.5364
p+q	1.0000
$q^{n-r}+1/p^r q^{n-r}$	63.9620
$-2.468*p^r q^{n-r}+q^{n-r}+1/p^r q^{n-r}$	64.0009
$(p+q)/2$	0.5000
$sqrt(pq)$	0.3941

Table One
The Members of Genus G(6) and Some of their Statistics

Essentially, we have defined a Genus as being the family of names of a set length n. Above the Genus we may define a Class of words that group length-families. Given that capital city names of less than three letters are unsatisfactorily-enunciable they are indeed absent and as the longest actual capital city name in English is nineteen letters we may define the total Population of the Class of capital town names as:-

$$C = \sum_{i=k_1}^{k_2} r_i$$

Equation 16

where $k_1 = 3$ and $k_2 = 19$.

(Because spaces between words, punctuation and elision marks, and accentuations are all ineffable in spoken English we strip them out computationally so that only vowels and consonants are conserved for analysis).

Table Two presents the Genus Populations G_k for the word Genera $G(3)$ through $G(19)$.

In the range of possible names, which total 1048568, a mere 251 are actually logged as capital names by Wikipedia, which is only 0.023937408 percent of the available letter combinations.

Because Equation Sixteen constitutes an integer geometric series we are able to write for any Genus Population G_k:-

$$G_k = a.\frac{\left(1 - r^k\right)}{\left(1 - r\right)}$$

Equation 17

where a is the Starting Value of the Series which identifies with r in our context: And r is of course 2.

Therefore, referent to the range $3 \leq r \leq 19$, we may write:-

$$G_{k_1,k_2} = a\frac{\left(1 - r^{k_2}\right)}{\left(1 - r\right)} - a\frac{\left(1 - r^{k_1 - 1}\right)}{\left(1 - r\right)} = \frac{a}{r - 1}\left(r^{k_2} - r^{k_1 - 1}\right)$$

Equation 18

or:-

$$G_{k_1,k_2} = \frac{a}{r - 1}\left(r^{k_2} - r^{k_1 - 1}\right) = 2\left(2^{19} - 2^2\right) = 1048568$$

Equation 19

	r	2
	Start n	3
	Finish n	19
	Tabulated Sum of r^n	1048568
	Formula Sum of r^n	1048568
	Sum(3:19)	1048568
	Mean(r^n)	2581110
	Mean(Serial)	18.00013
	p	0.192308
	q	0.807692

Class

Serial	Sum Criterion	r^n	x	f	fx	Cumulate of r^n	$1/p^r q^{n-r}$
0	0	1	0	0	0	0	17.64
1	0	2	0	0	0	0	21.84
2	0	4	0	0	0	0	27.04
3	1	8	3	8	24	8	33.4781
4	1	16	4	16	64	24	41.44907
5	1	32	5	32	160	56	51.3179
6	1	64	6	64	384	120	63.53644
7	1	128	7	128	896	248	78.66417
8	1	256	8	256	2048	504	97.39373
9	1	512	9	512	4608	1016	120.5827
10	1	1024	10	1024	10240	2040	149.2929
11	1	2048	11	2048	22528	4088	184.8388
12	1	4096	12	4096	49152	8184	228.8481
13	1	8192	13	8192	106496	16376	283.3357
14	1	16384	14	16384	229376	32760	350.7966
15	1	32768	15	32768	491520	65528	434.3196
16	1	65536	16	65536	1048576	131064	537.729
17	1	131072	17	131072	2228224	262136	665.7597
18	1	262144	18	262144	4718592	524280	824.2739
19	1	524288	19	524288	9961472	1048568	1020.53
20	0	1048576	0	0	0	0	1263.513
21	0	2097152	0	0	0	0	1564.349
22	0	4194304	0	0	0	0	1936.813
23	0	8388608	0	0	0	0	2397.959
24	0	16777216	0	0	0	0	2968.902
25	0	33554432	0	0	0	0	3675.783

Table Two
Genus Populations and their Cumulates

PART TWO
LETTER-SYLLABLE ADDRESS TRANSFORMATIONS

A (three-dimensional) Address Transformation Matrix MT[] arises from a specific treatment of a list of paired integers. We can of course assume and test some relationship between the integers, or conversely generate the paired-list and study the transformed results. We shall initially review simple examples of both procedures.

In general terms we may define:-

$$MA\big[ix,iy,\sum(ix,iy)\big]= Trans\{IA\big[ent(x),ent(f(x))\big]\}= Trans\{IA\big[ix,iy\big]\}$$

Equation 20

By way of example firstly consider a list of sixteen paired integers (representing array addresses) denoted as x and y or more properly ix and iy:-

x	y
4	0
0	1
2	1
2	1
1	2
1	2
1	2
1	2
1	2
3	3
3	3
3	3
3	3
4	4
4	4
4	4

Table Three
Sixteen Paired Integer Data for Address Transformation

Interpret each integer pair as an (x,y) co-ordinate. Note that the pairs may be repeated though not necessarily sorted as they happen to be in the above list. Where repetitions occur cumulate them to the relevant element of the two-dimensional projection of MA[]:

Singletons are awarded to the relevant element on their own. Where a given pairing is absent, the addressed element is kept null.

The transformed address matrix MA[] resulting from this list is given below. The Total Number of Instances is of course sixteen:-

Instances:	16						
		x					
			0	1	2	3	4
y	0						1
	1		1		2		
	2			5			
	3					4	
	4						3

Table Four
Address Transformation Matrix for the Sixteen Paired Integers

For illustrative purposes the Address Transformation Matrix MA[] is plotted in the three-dimensional bar chart below:-

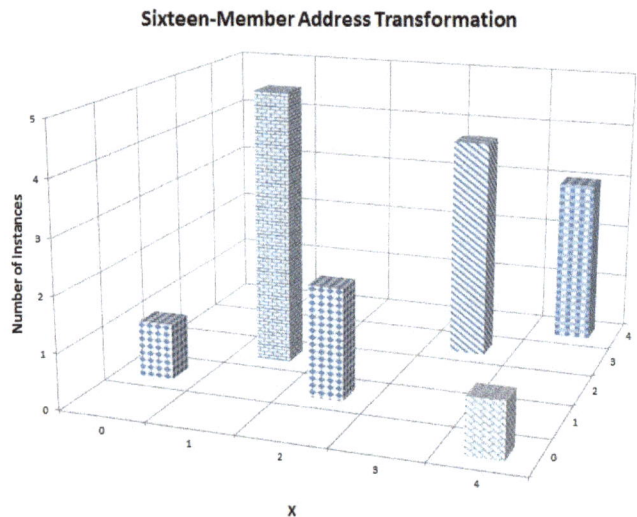

Sixteen-Member Address Transformation

Plot One
The Cumulative Plot of the Sixteen Address Transformed Data

The topography of such charts reflects the generative mathematics, including any error inherent in that.

As a second, slightly more elaborate, example consider the 1024 paired (ix,iy) data that issue from the elaboration of sufficient instances of the polynomial equation:-

$$y = c_0 + c_1 x + c_2 x^2 + \kappa\varepsilon$$

Equation 21

where c_0, c_1 and c_2 or all Constants, as is the Scaling Constant κ. ε is a Uniformly-Distributed Random Variable such that $0 \le \varepsilon \le 1$.

For our set of paired-integers we shall assume $c_0 = 0.2$, $c_1 = 0.6$, $c_2 = 0$ and $\kappa = 0.1$. Accordingly we have to hand a First-Degree Algebraic Polynomial, the equation of a straight line.

The outputs of Equation Twenty-One are of course real but we may transform them to suitable address integers using any suitable convention. I used:-

$$ix = ent\left\{ \left(ix_{min} + x.ix_{max}\right) + \frac{1}{2} \right\} \qquad \textbf{Equation 22a}$$

$$iy = ent\left\{ \left(iy_{min} + y.iy_{max}\right) + \frac{1}{2} \right\} \qquad \textbf{Equation 22b}$$

where x_{min} and x_{max} are integral Scaling Constants respectively the Lower and Upper Bounds of the Output Range of ix; and y_{min} and y_{max} are the corresponding values applicable to iy. In this context "ent" should be read as the Integer Part of the Value (i.e. a "floor" function), so that ent($x+1/2$) rounds to the nearest integer.

A list of all 1024 pairs would be wasteful but for checking purposes I present the first eight and last eight members of the list below:-

ATDEMO	
1024	
9	12
21	20
20	21
12	14
13	15
10	14
23	21
5	12
>	>
>	>
32	26
12	15
32	26
5	13
1	8
29	26
27	23
25	22

Table Five
The First Eight and Last Eight Paired Integers Data
for the
Polynomial Address Transformation

Table Six shows the generated Address Transformation Matrix MA[].

Table Six
Address Transformation Matrix for the
1024 Paired Integers Data
Generated by the Entier Expression of a Defined First-Degree Polynomial

Matrix	Total	1024
	Mean	7.262411348
Column	Total	1024
	Mean	31.03030303
Row	Total	1024
	Mean	34.13333333

For illustrative purposes the Address Transformation Matrix MA[] is plotted in the three-dimensional bar chart below:-

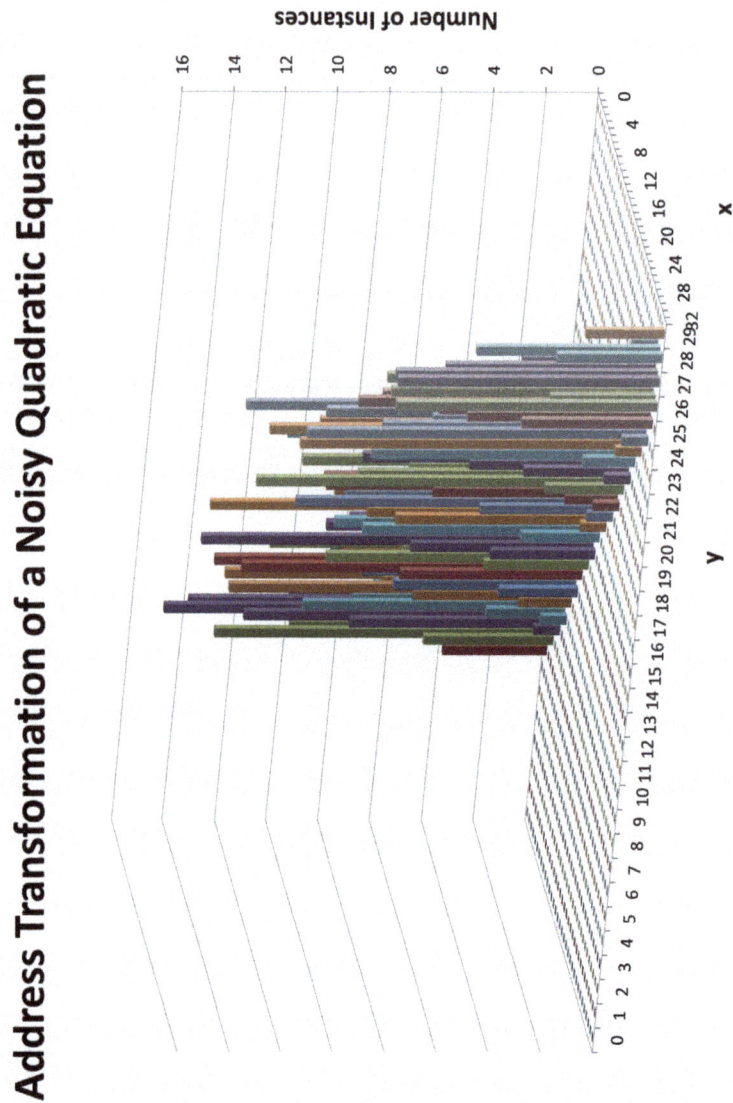

Plot Two
The Cumulative Plot of the Address Transformation Matrix
for the
1024 Paired Integers Data
Generated by the Entier Expression of a Defined First-Degree Polynomial

Letter-Syllable Address Transformations

When the technique of Address Transformation is applied to the Letter and Syllable Counts of the Capital Names as listed in Appendix One some enigmatic tendencies appear.

As one might suspect a vaguely linear relationship seems to associate letter and syllable numbers, whilst some incondite Gaussian distribution seems to characterise the spread of syllables at any letter count, or the converse spread of letters at any syllable count. But the number of names involved, a maximum of 251, is too scarce to bear any definite rule: At least one arising from this technique.

As with the above 1024-point exercise my Program TRANSUM.BAS was applied to the mechanics of address matrix formation and this routine is reproduced in Appendix Two.

The resulting address matrices, for European Capitals Only and then for World Capitals including Europe are respectively presented in Tables Seven and Eight.

I can detect no obvious difference between European and World component frequencies.

Plots Five and Six present the actual three-dimensional plots of the address transformation matrices.

Matrix		
	Total	55
	Mean	2.391304348
Column		
	Total	55
	Mean	7.857142857
Row		
	Total	55
	Mean	2.894736842
CORREL Columns	-0.69398	
CORREL Rows	-0.11631	
R^2 Columns	0.481606	
R^2 Rows	0.013527	

Letters	Syllables 1	2	3	4	5	6	7	Row Totals	Normalised Poisson PDF	Poisson PDF
1								0	8.807	0.160
2								0	12.746	0.232
3								0	12.299	0.224
4	2	3						5	8.901	0.162
5	1	2	1					4	5.153	0.094
6	1	11	3					15	2.486	0.045
7		2	2	1				5	1.028	0.019
8		3	5	1				9	0.372	0.007
9		1	4	2				7	0.120	0.002
10			3	3				6	0.035	0.001
11								0	0.009	0.000
12				1				1	0.002	0.000
13								0	0.000	0.000
14				1				1	0.000	0.000
15								0	0.000	0.000
16						1		1	0.000	0.000
17								0	0.000	0.000
18								0	0.000	0.000
19							1	1	0.000	0.000
Column Totals	4	22	18	8	1	1	1	55	51.958	0.945
Normalised Poisson PDF	0.17	0.66	1.72	3.38	5.31	6.96	7.81			25.998
Poisson PDF	0	0.01	0.03	0.06	0.1	0.13	0.14			0.473

Table Seven

Address Transformation Matrix for the
Fifty-Five European Capital City Names

Matrix

	Total	251
	Mean	5.976190476

Column

	Total	251
	Mean	35.85714286

Row

	Total	251
	Mean	13.21052632

CORREL Columns	-0.45663
CORREL Rows	-0.15282
R^2 **Columns**	0.208507
R^2 **Rows**	0.023353

Letters	Syllables 1	2	3	4	5	6	7	Row Totals	Normalised Poisson PDF	Poisson PDF
1								0	0.006	0.000
2								0	0.040	0.000
3								0	0.177	0.001
4	3	11	1					15	0.583	0.002
5	2	16	3					21	1.541	0.006
6	1	23	26					50	3.393	0.014
7		10	14	6		1		31	6.404	0.026
8		13	17	12				42	10.574	0.042
9		6	11	13				30	15.521	0.062
10		5	12	6	3			26	20.505	0.082
11		2	4	4	4			14	24.625	0.098
12			2	4	1	1		8	27.109	0.108
13			2		1			3	27.548	0.110
14			2	1	1	1		5	25.995	0.104
15								0	22.894	0.091
16				1	1	1		3	18.902	0.075
17				1				1	14.689	0.059
18								0	10.780	0.043
19							2	2	7.496	0.030
Column Totals	6	86	94	48	11	4	2	251	238.783	0.951
Normalised Poisson PDF	0	0	0	0	0	0	0			0.000
Poisson PDF	0	0	0	0	0	0	0			0.000

Table Eight
Address Transformation Matrix for the
Two Hundred and Fifty-One World Capital City Names

The resulting Letter Frequency and Capital Frequency Curves for Europe and the World are respectively Plot Three and Plot Four.

Frequencies of Capital Name Letter Counts

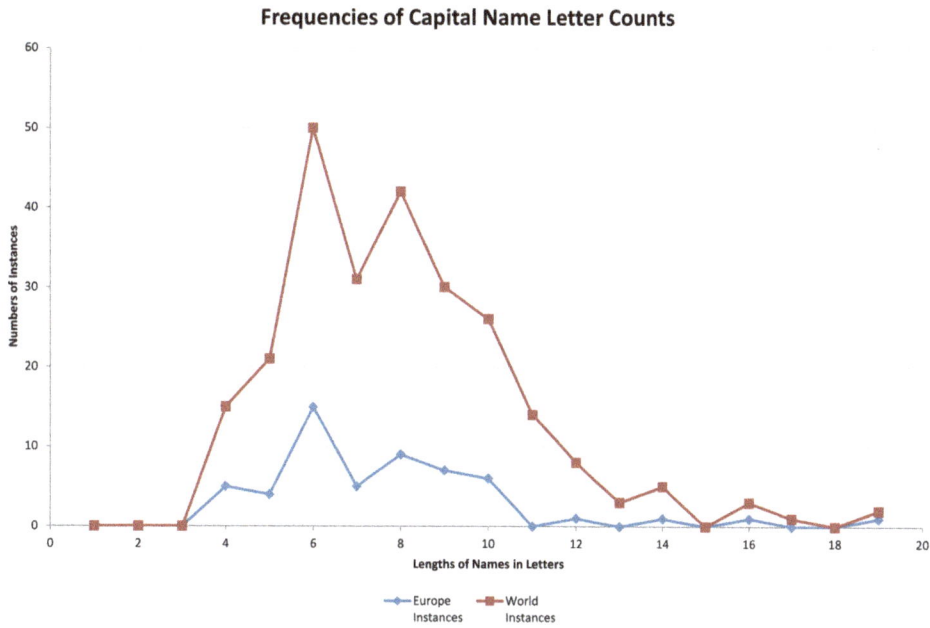

Plot Three
The Letter Frequency Plots for the Capital City Names

Frequencies of Capital Name Syllable Counts

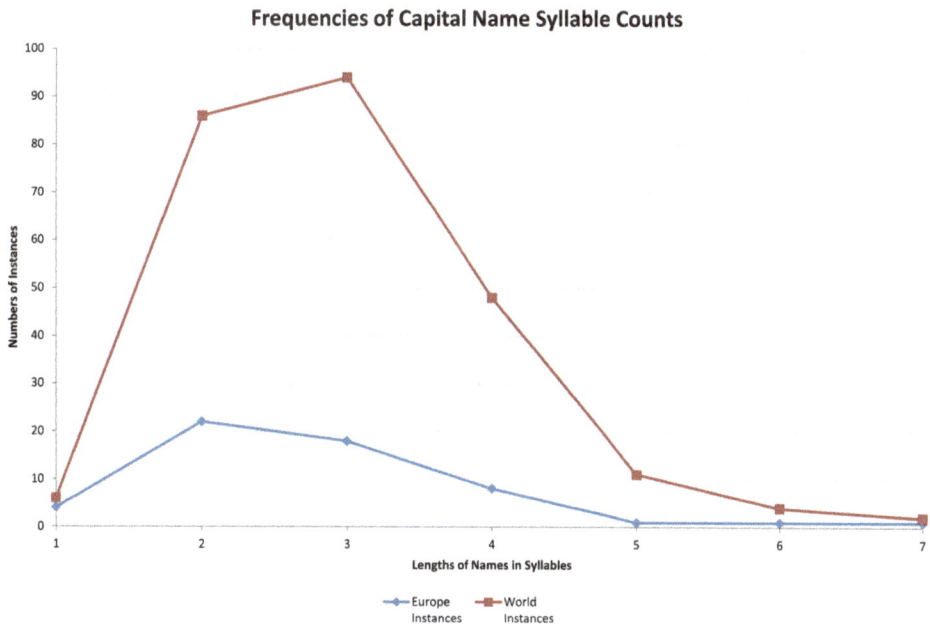

Plot Four
The Syllable Frequency Plots for the Capital City Names

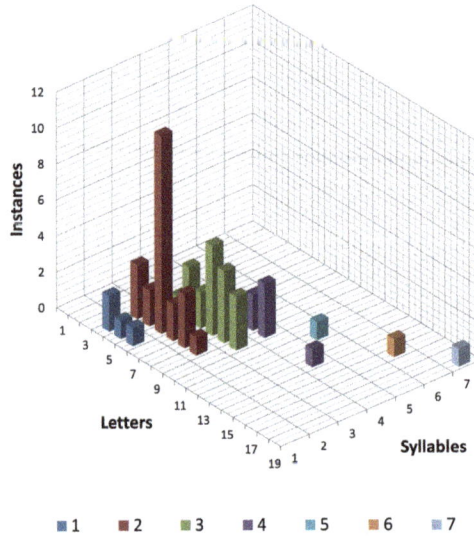

Plot Five
The Cumulative Plot of the Address Transformation Matrix
for the Letter-Syllable Relations of the European Capital City Names

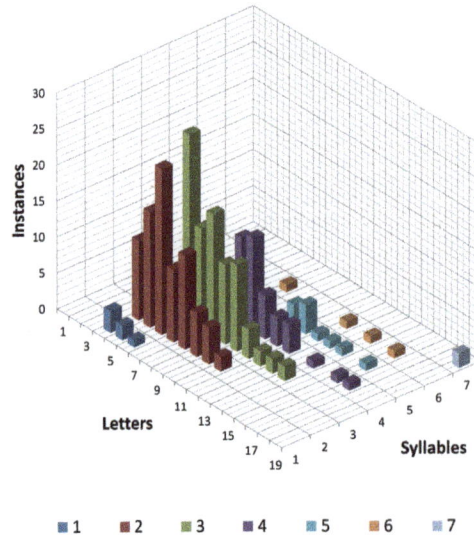

Plot Six
The Cumulative Plot of the Address Transformation Matrix
for the Letter-Syllable Relations of the World Capital City Names

APPENDIX ONE

TERRITORY NAMES AND CAPITAL NAMES
WITH LETTER-SYLLABLE COUNTS
AND REGION CODES

Region	Name Length in Syllables	Name Length in Characters	Country	Capital City
AF	2	4	South Sudan	Juba
AF	2	4	Togo	Lomé
AF	2	5	Ghana	Accra
AF	2	5	Egypt	Cairo
AF	2	5	Senegal	Dakar
AF	2	5	Cape Verde	Praia
AF	2	5	Morocco	Rabat
AF	2	5	Tunisia	Tunis
AF	2	6	Gambia	Banjul
AF	2	6	Guinea-Bissau	Bissau
AF	2	6	Niger	Niamey
AF	2	7	Cameroon	Yaoundé
AF	2	8	Sierra Leone	Freetown
AF	2	8	Sudan	Khartoum
AF	2	8	Namibia	Windhoek
AF	3	5	Nigeria	Abuja
AF	3	6	Eritrea	Asmara
AF	3	6	Mali	Bamako
AF	3	6	Central African Republic	Bangui
AF	3	6	Tanzania	Dodoma
AF	3	6	Zimbabwe	Harare
AF	3	6	Rwanda	Kigali
AF	3	6	Angola	Luanda
AF	3	6	Zambia	Lusaka
AF	3	6	Equatorial Guinea	Malabo
AF	3	6	Mozambique	Maputo
AF	3	6	Lesotho	Maseru
AF	3	6	Comoros	Moroni
AF	3	7	Guinea	Conakry
AF	3	7	Uganda	Kampala
AF	3	7	Kenya	Nairobi
AF	3	7	Libya	Tripoli
AF	3	8	Djibouti	Djibouti
AF	3	8	Democratic Republic of the Congo	Kinshasa
AF	3	8	Malawi	Lilongwe
AF	3	9	Mayotte	Mamoudzou
AF	3	10	Gabon	Libreville
AF	3	10	Mauritania	Nouakchott
AF	3	10	Mauritius	Port Louis
AF	3	11	Republic of the Congo	Brazzaville
AF	3	11	Réunion	Saint-Denis
AF	4	7	Swaziland	Mbabane
AF	4	8	Botswana	Gaborone
AF	4	8	Somaliland	Hargeisa
AF	4	8	Liberia	Monrovia
AF	4	8	South Africa	Pretoria
AF	4	8	São Tomé and Príncipe	São Tomé
AF	4	8	Seychelles	Victoria
AF	4	9	Burundi	Bujumbura
AF	4	9	Somalia	Mogadishu
AF	4	9	Chad	N'Djamena
AF	4	10	Benin	Porto-Novo
AF	4	11	Burkina Faso	Ouagadougou
AF	4	12	Côte d'Ivoire	Yamoussoukro
AF	5	11	Ethiopia	Addis Ababa
AF	6	7	Algeria	Algiers
AF	6	12	Madagascar	Antananarivo
Maxima	6	12		

Table A1
Region Africa (AF)
Letter-Syllable Data

Region	Name Length in Syllables	Name Length in Characters	Country	Capital City
AS	1	5	South Korea	Seoul
AS	2	4	East Timor (Timor-Leste)	Dili
AS	2	4	Maldives	Malé
AS	2	5	Bangladesh	Dhaka
AS	2	5	Vietnam	Hanoi
AS	2	5	Afghanistan	Kabul
AS	2	5	Japan	Tokyo
AS	2	6	Taiwan	Taipei
AS	2	7	Thailand	Bangkok
AS	2	7	China	Beijing
AS	2	7	Kyrgyzstan	Bishkek
AS	2	7	Bhutan	Thimphu
AS	2	8	Uzbekistan	Tashkent
AS	2	9	Hong Kong	Hong Kong
AS	2	9	North Korea	Pyongyang
AS	2	10	Cambodia	Phnom Penh
AS	3	6	Kazakhstan	Astana
AS	3	6	Philippines	Manila
AS	3	7	Indonesia	Jakarta
AS	3	7	Ceylon	Colombo
AS	3	8	Turkmenistan	Ashgabat
AS	3	8	Tajikistan	Dushanbe
AS	3	9	Nepal	Kathmandu
AS	3	9	Myanmar	Naypyidaw
AS	3	9	India	New Delhi
AS	3	9	Singapore	Singapore
AS	3	12	Papua New Guinea	Port Moresby
AS	4	9	Pakistan	Islamabad
AS	4	9	Laos	Vientiane
AS	4	11	Mongolia	Ulaanbaatar
AS	4	12	Malaysia	Kuala Lumpur
AS	7	19	Brunei	Bandar Seri Begawan
Maxima	7	19		

Table A2
Region Asia (AS)
Letter-Syllable Data

Region	Name Length in Syllables	Name Length in Characters	Country	Capital City
AU	3	8	Australia	Canberra
AU	3	10	New Zealand	Wellington
AU	4	16	Christmas Island	Flying Fish Cove
Maxima	4	16		

Table A3
Region Australasia (AU)
Letter-Syllable Data

Region	Name Length in Syllables	Name Length in Characters	Country	Capital City
CA	2	6	Bahamas	Nassau
CA	2	6	Dominica	Roseau
CA	2	8	Saint Lucia	Castries
CA	2	8	Jamaica	Kingston
CA	2	8	Montserrat	Plymouth
CA	2	8	Puerto Rico	San Juan
CA	2	9	Saint Vincent and the Grenadines	Kingstown
CA	2	9	British Virgin Islands	Road Town
CA	2	10	Barbados	Bridgetown
CA	2	10	Antigua and Barbuda	St. John's
CA	2	11	Guadeloupe	Basse-Terre
CA	2	11	Cayman Islands	George Town
CA	3	6	Cuba	Havana
CA	3	7	Saint Martin	Marigot
CA	3	8	Belize	Belmopan
CA	3	8	Saint Barthélemy	Gustavia
CA	3	8	Costa Rica	San José
CA	3	10	Anguilla	The Valley
CA	3	10	Curaçao	Willemstad
CA	3	11	Sint Maarten	Philipsburg
CA	3	12	Grenada	St. George's
CA	3	13	Turks and Caicos Islands	Cockburn Town
CA	3	13	Trinidad and Tobago	Port of Spain
CA	3	14	Martinique	Fort-de-France
CA	3	14	Haiti	Port-au-Prince
CA	4	7	Nicaragua	Managua
CA	4	10	Saint Kitts and Nevis	Basseterre
CA	4	10	Aruba	Oranjestad
CA	4	12	El Salvador	San Salvador
CA	5	11	Panama	Panama City
CA	5	11	Honduras	Tegucigalpa
CA	5	13	Dominican Republic	Santo Domingo
CA	5	16	United States Virgin Islands	Charlotte Amalie
CA	6	14	Guatemala	Guatemala City
Maxima	6	16		

Table A4
Region Central America (CA)
Letter-Syllable Data

Region	Name Length in Syllables	Name Length in Characters	Country	Capital City
E	1	4	Switzerland	Bern
E	1	4	Italy	Rome
E	1	5	Belarus	Minsk
E	1	6	Czech Republic	Prague
E	2	4	Ukraine	Kiev
E	2	4	Norway	Oslo
E	2	4	Latvia	Riga
E	2	5	France	Paris
E	2	5	Liechtenstein	Vaduz
E	2	6	Greece	Athens
E	2	6	Germany	Berlin
E	2	6	Ireland	Dublin
E	2	6	Chechnya	Grozny
E	2	6	Portugal	Lisbon
E	2	6	United Kingdom	London
E	2	6	Spain	Madrid
E	2	6	Russia	Moscow
E	2	6	Republic of Macedonia	Skopje
E	2	6	Poland	Warsaw
E	2	6	Croatia	Zagreb
E	2	7	Isle of Man	Douglas
E	2	7	Estonia	Tallinn
E	2	8	Serbia	Belgrade
E	2	8	Belgium	Brussels
E	2	8	Faroe Islands	Tórshavn
E	2	9	Sweden	Stockholm
E	3	5	Bulgaria	Sofia
E	3	6	Monaco	Monaco
E	3	6	Albania	Tirana
E	3	6	Austria	Vienna
E	3	7	Lithuania	Vilnius
E	3	7	Armenia	Yerevan
E	3	8	Hungary	Budapest
E	3	8	Moldova	Chișinău
E	3	8	Finland	Helsinki
E	3	8	Kosovo[g]	Pristina
E	3	8	Malta	Valletta
E	3	9	Netherlands	Amsterdam
E	3	9	Romania	Bucharest
E	3	9	Gibraltar	Gibraltar
E	3	9	Iceland	Reykjavík
E	3	10	Luxembourg	Luxembourg
E	3	10	San Marino	San Marino
E	3	10	Canary Islands	Santa Cruz
E	4	7	Georgia	Tbilisi
E	4	8	Bosnia and Herzegovina	Sarajevo
E	4	9	Slovenia	Ljubljana
E	4	9	Montenegro	Podgorica
E	4	10	Slovakia	Bratislava
E	4	10	Denmark	Copenhagen
E	4	10	Jersey	St. Helier
E	4	14	Guernsey	St. Peter Port
E	5	12	Vatican City	Vatican City
E	6	16	Andorra	Andorra la Vella
E	7	19	Akrotiri and Dhekelia	Episkopi Cantonment
Maxima	7	19		

Table A5
Region Europe (E)
Letter-Syllable Data

Region	Name Length in Syllables	Name Length in Characters	Country	Capital City
ME	2	4	Azerbaijan	Baku
ME	2	4	Qatar	Doha
ME	2	5	Jordan	Amman
ME	2	6	Lebanon	Beirut
ME	2	6	Oman	Muscat
ME	2	6	Saudi Arabia	Riyadh
ME	2	6	Yemen	Sana'a
ME	2	6	Iran	Tehran
ME	2	7	Iraq	Baghdad
ME	2	8	Sahrawi Arab Democratic Republic [c]	Laayoune
ME	3	6	Turkey	Ankara
ME	3	6	Bahrain	Manama
ME	3	7	Abkhazia	Sukhumi
ME	3	8	Syria	Damascus
ME	3	8	Transnistria	Tiraspol
ME	3	10	South Ossetia	Tskhinvali
ME	4	7	Cyprus	Nicosia
ME	4	7	Northern Cyprus	Nicosia
ME	4	9	United Arab Emirates	Abu Dhabi
ME	4	9	Israel	Jerusalem
ME	4	11	Kuwait	Kuwait City
ME	4	11	Nagorno-Karabakh Republic	Stepanakert
ME	5	14	State of Palestine	East Jerusalem
Maxima	5	14		

Table A6
Region Middle East (ME)
Letter-Syllable Data

Region	Name Length in Syllables	Name Length in Characters	Country	Capital City
NA	1	4	Greenland	Nuuk
NA	3	6	Canada	Ottawa
NA	3	8	Bermuda	Hamilton
NA	3	10	Saint Pierre and Miquelon	St. Pierre
NA	3	10	United States	Washington
NA	5	11	Mexico	Mexico City
Maxima	5	11		

Table A7
Region North America (NA)
Letter-Syllable Data

Region	Name Length in Syllables	Name Length in Characters	Country	Capital City
SA	2	4	Peru	Lima
SA	2	5	Ecuador	Quito
SA	2	5	Bolivia	Sucre
SA	2	7	French Guiana	Cayenne
SA	2	7	Falkland Islands	Stanley
SA	2	9	Saint Helena	Jamestown
SA	2	10	Ascension Island	Georgetown
SA	2	10	Guyana	Georgetown
SA	3	6	Colombia	Bogotá
SA	3	7	Venezuela	Caracas
SA	4	8	Paraguay	Asunción
SA	4	8	Brazil	Brasília
SA	4	8	Chile	Santiago
SA	4	9	Tristan da Cunha	Edinburgh
SA	4	9	Easter Island	Hanga Roa
SA	4	12	Argentina	Buenos Aires
SA	4	17	South Georgia and the South Sandwich Islands	King Edward Point
SA	5	10	Uruguay	Montevideo
SA	5	10	Suriname	Paramaribo
Maxima	5	17		

Table A8
Region South America (SA)
Letter-Syllable Data

Region	Name Length in Syllables	Name Length in Characters	Country	Capital City
SP	2	4	Fiji	Suva
SP	2	5	Nauru	Yaren
SP	2	6	Northern Mariana Islands	Saipan
SP	2	8	Norfolk Island	Kingston
SP	3	4	Samoa	Apia
SP	3	5	Niue	Alofi
SP	3	6	Cook Islands	Avarua
SP	3	6	Marshall Islands	Majuro
SP	3	6	New Caledonia	Nouméa
SP	3	6	Kiribati	Tarawa
SP	3	7	Guam	Hagåtña
SP	3	7	Federated States of Micronesia	Palikir
SP	3	7	French Polynesia	Papeete
SP	3	9	Pitcairn	Adamstown
SP	3	9	Vanuatu	Port Vila
SP	3	11	Cocos (Keeling) Islands	West Island
SP	4	7	Solomon Islands	Honiara
SP	4	8	Tuvalu	Funafuti
SP	4	8	Wallis and Futuna	Mata-Utu
SP	4	9	Palau	Ngerulmud
SP	4	9	American Samoa	Pago Pago
SP	5	10	Tonga	Nuku'alofa
Maxima	5	11		

Table A9
Region South Pacific (SP)
Letter-Syllable Data

APPENDIX TWO

THE PROGRAM TRANSUM.BAS

```
'          PROGRAM TRANSUM.BAS
'          A PROGRAM TO SUM TWO MATCHED SERIES INTO A TWO-DIMENSIONAL ARRAY
'
'          WRITTEN BY:-
'
'          JAMES R WARREN BSc MSc PhD PGCE
'          "SOUTHGATE"
'          31 VICTORIA AVENUE
'          BLOXWICH
'          WS3 3HS
'          UNITED KINGDOM
'
'          20 MARCH 2017
'
'          THIS PROGRAM IS WRITTEN IN MICROSOFT QBASIC
'
' VARIABLE TYPE DEFAULTS
      DEFDBL A-H, O-R, T-Z
      DEFSTR S
      DEFINT I-K, M-N
      DEFLNG L
' SEGMENT DECLARATIONS
      DECLARE SUB DATAIN(SF,N,IX(),IY())
      DECLARE SUB DATAOUT(N,IX(),IY(),IXMAX,IYMAX)
      DECLARE SUB MAXI(N,IX(),IY(),IXMAX,IYMAX)
      DECLARE SUB META(NMET,SCA)
      DECLARE SUB TRANSMO(N,IXMAX,IYMAX,IX(),IY(),MO())
' COMMON VARIABLES
      COMMON SHARED IU, IV, SXU, SXV, SP, SM, SCR
' STATIC ARRAY DEFINITIONS
'      ( none )
' DYNAMIC ARRAY DEFINITIONS
      M1 = 256: M2 = 32
      DIM IX(M1), IY(M1), MO(M2, M2)
' DYNAMIC ARRAY LIMITS
'      ( none )
' DEVICE ATTRIBUTIONS
'      ( none )
' LOGICAL UNIT, EXTENSION AND PATHNAME SETTINGS
      IU = 1: IV = 2
      SXU = ".CSV": SXV = ".CSV"
      SP = "C:\QBASIC\QBFILES\"
' FORMAT DEFINITIONS
      SI6 = "######"
' NUMERICAL CONSTANT DEFINITIONS
'      ( none )
' STRING CONSTANT DEFINITIONS
      SC = ":": SM = ",": SCR = CHR$(13) + CHR$(10)
' TEXT VARIABLE DEFINITIONS
'      ( none )
'
' *** THE ALGORITHM ***
'
'      CLS
      SINP="WORLD"
      DATAIN SINP, N, IX(), IY()
      MAXIXY N, IX(), IY(), IXMAX, IYMAX
      TRANSMO N, IXMAX, IYMAX, IX(), IY(), MO()
      DATAOUT SINP, IXMAX, IYMAX, MO()
      END
```

Program TRANSUM.BAS
Master Segment

```
    SUB DATAIN (SF, N, IX(), IY())
' A SUBROUTINE TO INPUT THE MATCHED DATA SERIES
' ARGUMENTS:
'    SF    THE INPUT FILE NAME
'    N     THE NUMBER OF MATCHED DATA PAIRS
'    IX()  THE FIRST  SERIES INTEGER ARRAY
'    IY()  THE SECOND SERIES INTEGER ARRAY
' (IU, IV, SXU, SXV AND SP ARE COMMON SHARED)
'
    SINF="TIN-"+SF
    OPEN "I", IU, SP + SINF + SXU
    INPUT #IU, SF
    INPUT #IU, N
    FOR I = 1 TO N
      INPUT #IU, IX(I), IY(I)
    NEXT I
    CLOSE IU
    END SUB

    SUB DATAOUT (SF, IXMAX, IYMAX, MO())
' A SUBROUTINE TO OUTPUT THE ADDRESSED TOTALS
' ARGUMENTS:
'    SF    THE OUTPUT FILE NAME
'    IXMAX  THE LARGEST ELEMENT IN IX()
'    IYMAX  THE LARGEST ELEMENT IN IY()
'    MO()   THE ( TWO-DIMENSIONAL ) ARRAY OF
'           PAIR-WISE INPUT ARRAY ADDRESS TOTALS
' ( IU, IV, SXU, SXV, SM, SCR AND SP ARE COMMON SHARED )
'
    SOUF="TSUM-"+SF
    OPEN "O", IV, SP + SOUF + SXV
' PRINT X-HEADER
    PRINT #IV, SF
    PRINT #IV, IXMAX;SM;IYMAX
    PRINT #IV, SM; 0;
    FOR I = 1 TO IXMAX
      PRINT #IV, SM; I;
    NEXT I
    PRINT #IV, SCR
' PRINT MATRIX MO() WITH Y-LABELS
    FOR J = 0 TO IYMAX
      PRINT #IV, J;
      FOR I = 0 TO IXMAX-1
        PRINT #IV, SM; MO(I, J);
      NEXT I
      PRINT #IV, SM;MO(IXMAX,J)
'      PRINT #IV, SCR
    NEXT J
' TERMINATE OUTPUT
    CLOSE IV
    END SUB

    SUB MAXIXY (N, IX(), IY(), IXMAX, IYMAX)
' A SUBROUTINE TO ESTABLISH THE MAXIMA OF THE ELEMENTS IN ARRAYS
' IX() AND IY()
' ARGUMENTS:
'    N     THE NUMBER OF MATCHED DATA PAIRS
'    IX()  THE FIRST  SERIES INTEGER ARRAY
'    IY()  THE SECOND SERIES INTEGER ARRAY
'    IXMAX  THE LARGEST ELEMENT IN IX()
'    IYMAX  THE LARGEST ELEMENT IN IY()
'
    IXMAX = -(2 ^ 15 - 1): IYMAX = -(2 ^ 15 - 1)
    FOR I = 1 TO N
      IF IX(I) > IXMAX THEN IXMAX = IX(I)
      IF IY(I) > IYMAX THEN IYMAX = IY(I)
    NEXT I
    END SUB

    SUB META(NMET,SCA)
' A SUBROUTINE TO RE[PORT THE STAGE OF PROGRESS TO THE SCREEN
' ARGUMENTS:
'    NMET   THE META NUMBER
'    SCA    THE COMMENT FOR THE STAGE
'
    PRINT NMET,SCA
    END SUB

    SUB TRANSMO (N, IXMAX, IYMAX, IX(), IY(), MO())
' A SUBROUTINE TO ASSIGN SUMMATIONS OF (IX(),IY()) PAIRED VALUES
' TO THE ( TWO-DIMENSIONAL ) ARRAY MO()
' ARGUMENTS:
'    N     THE NUMBER OF MATCHED DATA INPUT PAIRS
'    IXMAX  THE LARGEST ELEMENT IN IX()
'    IYMAX  THE LARGEST ELEMENT IN IY()
'    IX()  THE FIRST  SERIES INTEGER ARRAY
'    IY()  THE SECOND SERIES INTEGER ARRAY
'    MO()   THE ( TWO-DIMENSIONAL ) ARRAY OF
'           PAIR-WISE INPUT ARRAY ADDRESS TOTALS
'
' EXPLICIT INITIAL CLEARDOWN OF MO()
    FOR I = 0 TO IXMAX
      FOR J = 0 TO IYMAX
        MO(I, J) = 0
      NEXT J
    NEXT I
' ASSIGN MO() TOTALS
    FOR I = 1 TO N
      MO(IX(I), IY(I)) = MO(IX(I), IY(I)) + 1
    NEXT I
' TERMINATE ASSIGNMENTS
    END SUB
```

Program TRANSUM.BAS
Subroutines

CHAPTER SIX

Polygon Mensuration

On the Mensuration and Quadrature of
Certain Concave Polygons

by
James R Warren BSc MSc PhD PGCE

PART ONE
SOME PROPERTIES OF {n,2} REGULAR POLYGRAMS

A {n,2} regular polygram is a star-shaped polygon of n extreme vertices whose radii alternate in length such that intermediate radial lines of common length Intersectoral Radius P stand between the n dominant Apical Radii which are all of length r. The Radial Interangle, β, between each r and P radius is constant with a value π/n.

Sometimes a polygram is called a concave polygon, a sideroid, a star polygon or whatever.

The "2" in the specifier {n,2} denotes that the perimeter if the shape is generated by connecting each second extreme vertex with a chord.

Figure One shows the geometrical relations for the pentagram, a regular five-pointed {5,2} sideroid.

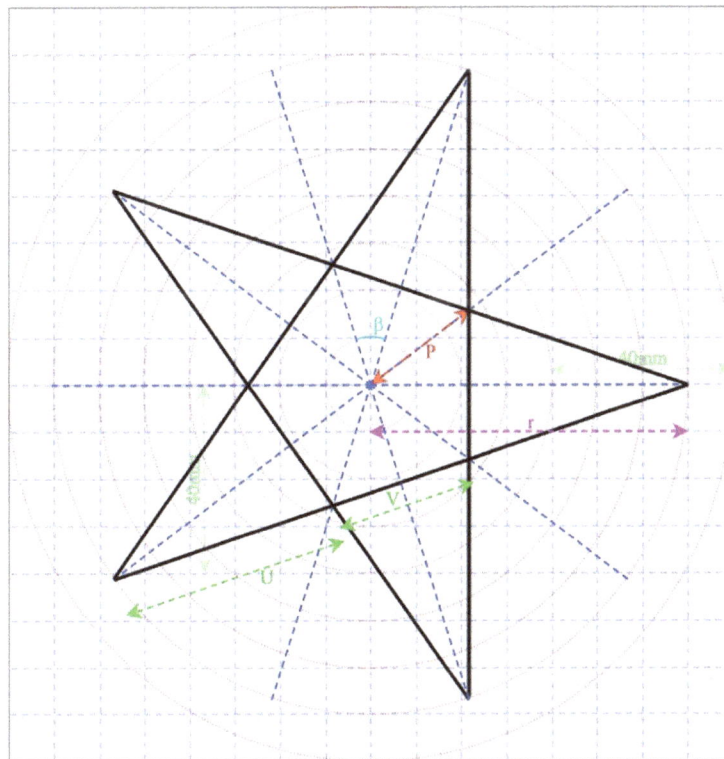

Figure 1
A Regular Pentagram

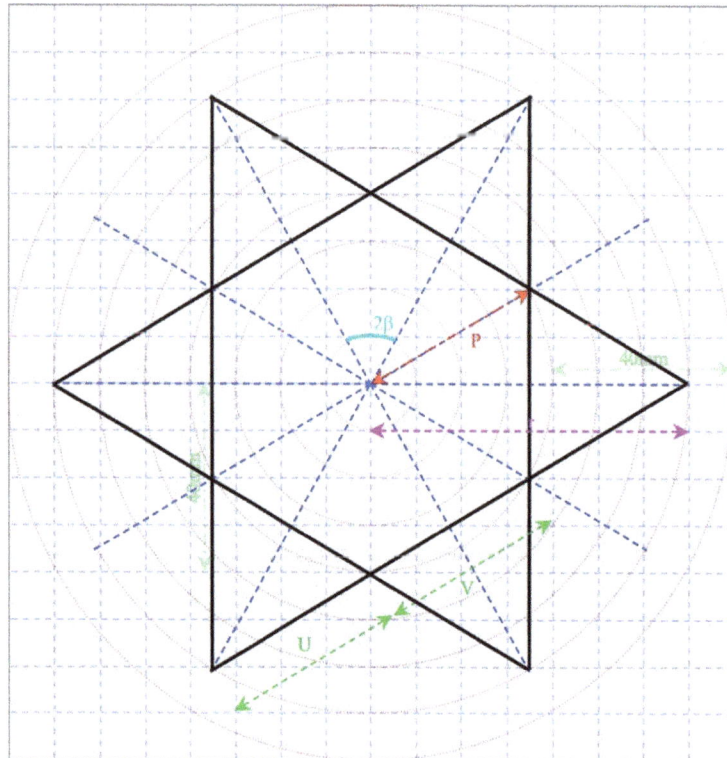

Figure 2
A Regular Hexagram

For the Regular Pentagram of Figure One, n, the Number of (Outer or Apical) Vertices is 5 and the Apical Radius is 70 units. The shape may be generalised by defining a Unit Radius, $r = 1$. But in our example $r = 70$.

A third definitional parameter is the Radial Interangle (or Sectoral Half-Angle) β noted above:

$$\beta = \frac{\pi}{n}$$
Equation 1

Further, derivative, angular values of use include:-

$$\alpha = \frac{2.\pi}{n}$$
Equation 2

which is the Sectoral Angle, and is not to be confused with $\alpha_{lapse(i)}$ or ψ_i which is the Lapsed Angle, that is the Sum of the Angular Displacements swept from the easternmost point of the circle.

And also:-

$$\gamma = \frac{4.\pi}{n}$$
Equation 3

$$\delta = \frac{\pi}{2}\left(1 - \frac{4}{n}\right)$$
Equation 4

$$\eta = \pi\left(\frac{1}{2} + \frac{1}{n}\right)$$
Equation 5

We may note in passing that at least for n = 5 the Chord Length, B, is given by The Cosine Rule as:-

$$B = \sqrt{2r^2(1 - \cos\gamma)}$$
Equation 6

The important Intersectoral Radius P (or P_{an}) may for the *regular* polygram be analytically-determined as:-

$$P_{an} = r.\frac{\sin\delta}{\sin\eta} = r.\frac{\cos 2\beta}{\cos\beta}$$
Equation 7

It follows that the polygram's Sectoral Part-Area, a, is:-

$$a = \frac{1}{2}P_{an}r\sin\beta$$
Equation 8

Accordingly the full Area of the Polygram, A, is yielded by:-

$$A = 2na = nrP_{an}\sin\beta$$
Equation 9

The Sides Ratio, ϕ_n

The Sides Ratio ϕ_n is the ratio of the polygram linear elements U and V as illustrated in Figure One and defined as a ratio of Cosine-Rule lineaments below:-

$$\phi_n = \frac{U}{V} = \frac{\sqrt{2P^2(1 - \cos 2\beta)}}{\sqrt{r^2 + P^2 - 2rP\cos\beta}}$$
Equation 10

We may expand and simplify Equation Ten in terms of:-

$$\phi_n = \sqrt{\frac{2P^2(1 - \cos 2\beta)}{r^2 + P^2 - 2rP\cos\beta}}$$

$$= \sqrt{\frac{2\left(r.\frac{\sin\delta}{\sin\eta}\right)^2(1 - \cos 2\beta)}{r^2 + \left(r.\frac{\sin\delta}{\sin\eta}\right)^2 - 2r\left(r.\frac{\sin\delta}{\sin\eta}\right)\cos\beta}}$$

$$= \sqrt{\frac{r^2\left[2\left(r.\frac{\sin\delta}{\sin\eta}\right)^2(1 - \cos 2\beta)\right]}{r^2\left[1 + \left(\frac{\sin\delta}{\sin\eta}\right)^2 - 2\left(\frac{\sin\delta}{\sin\eta}\right)\cos\beta\right]}}$$

$$= \sqrt{\frac{\left[2\left(r.\frac{\sin\delta}{\sin\eta}\right)^2(1 - \cos 2\beta)\right]}{\left[1 + \left(\frac{\sin\delta}{\sin\eta}\right)^2 - 2\left(\frac{\sin\delta}{\sin\eta}\right)\cos\beta\right]}}$$

Equation 11

Continued manipulation allows us to formulate:-

$$\phi_n = \frac{\sqrt{2}.\left(\frac{\cos\frac{2\pi}{n}}{\cos\frac{\pi}{n}}\right)\sqrt{1 - \cos\frac{2\pi}{n}}}{\sqrt{\left(\frac{\cos\frac{2\pi}{n}}{\cos\frac{\pi}{n}} - \cos\frac{\pi}{n} - \sqrt{\left(\cos\frac{\pi}{n}\right)^2 - 1}\right)\left(\frac{\cos\frac{2\pi}{n}}{\cos\frac{\pi}{n}} - \cos\frac{\pi}{n} + \sqrt{\left(\cos\frac{\pi}{n}\right)^2 - 1}\right)}}$$

Equation 12

Or alternatively:-

$$\phi_n = \frac{\sqrt{2}.\left(\frac{P}{r}\right)\sqrt{1 - \cos\frac{2\pi}{n}}}{\sqrt{\left(\frac{P}{r} - \cos\beta - \sqrt{(\cos\beta)^2 - 1}\right)\left(\frac{P}{r} - \cos\beta + \sqrt{(\cos\beta)^2 - 1}\right)}}$$

Equation 13

For the sake of convenience and computational clarity, allow that:-

$$s = \cos\frac{2\pi}{n} = \cos\alpha$$

Equation 14a

$$t = \cos\frac{\pi}{n} = \cos\beta$$

Equation 14b

$$u = \frac{s}{t} = \frac{\cos\alpha}{\cos\beta} = \frac{P}{r}$$

Equation 14c

Which permits the following substitutional arrangement:-

$$\phi_n = \frac{u\sqrt{2(1-s)}}{\sqrt{\left(u - t - \sqrt{t^2 - 1}\right)\left(u - t + \sqrt{t^2 - 1}\right)}}$$

Equation 15

The Ratio of Phidias, R_ϕ, is given by:-

$$R_\phi = \frac{1 + \sqrt{5}}{2} = 1 + \phi_n$$

Equation 16

and analogous geometric ratios may be developed for higher {n,2} polygrams by employment of Equation Fifteen in its various forms.

<u>Trigonometrical Developments of the Areal Formula</u>

By the substitution of Equation Seven into Equation Nine we may establish:-

$$A = nr^2 \frac{\sin\delta}{\sin\eta}\sin\beta$$

Equation 17

from which we have:-

$$A = nr^2 \frac{\sin\left[\frac{\pi}{2}\left(1 - \frac{4}{n}\right)\right]}{\sin\left[\pi\left(\frac{1}{2} + \frac{1}{n}\right)\right]}\sin\frac{\pi}{n}$$

Equation 18

By applying the Sine Identities for the Sums and Differences of Angles to the fractional term we may re-express Equation Eighteen as:-

$$A = nr^2 \frac{\cos\dfrac{2\pi}{n}}{\cos\dfrac{\pi}{n}} \sin\frac{\pi}{n}$$

Equation 19

or:-

$$A = nr^2 \cos\frac{2\pi}{n} \tan\frac{\pi}{n}$$

Equation 20

Questions of computational efficiency and numerical stability will of course color your choice of methods, especially for large n or its statistical surrogates. Therefore, a number (perhaps an infinitude?) of algorithmic avenues may be explored.

Accordingly, a third approach is to exploit the pair of identities:-

$$\cos 2\beta = (\cos\beta)^2 - (\sin\beta)^2 \qquad\qquad \textbf{Identity A}$$

$$\cos\beta - \frac{(\sin\beta)^2}{\cos\beta} = 2\cos\beta - \frac{1}{\cos\beta} \qquad\qquad \textbf{Identity B}$$

By reference to Equation Nineteen we may now write:-

$$
\begin{aligned}
A &= nr^2 \frac{\cos\dfrac{2\pi}{n}}{\cos\dfrac{\pi}{n}} \sin\frac{\pi}{n} \\
&= nr^2 \cos 2\beta \frac{\sin\beta}{\cos\beta} \\
&= nr^2 \left[(\cos\beta)^2 - (\sin\beta)^2\right]\frac{\sin\beta}{\cos\beta} \\
&= nr^2 \sin\beta \left(2\cos\beta - \frac{1}{\cos\beta}\right)
\end{aligned}
$$

Equation 21

From these facts it is immediately apparent that a third trigonometric identity may be specified as:-

$$\cos 2\beta \tan\beta = \sin\beta \left(2\cos\beta - \frac{1}{\cos\beta}\right) \qquad\qquad \textbf{Identity C}$$

Corroboration of Polygram Area using Meister's Rule

Meister's Rule is an exact method for the area of an arbitrary closed planar polygon. Meister's Rule is sometimes called the Surveyor's Rule, the Shoelace Formula, Gauss's Formula for Area, or something else. (Johann) Carl Friedrich Gauss (1777-1855) was a prolific German genius. There are several methods for computing area due to Gauss, and therefore, to avoid confusion it is best to call the method Meister's Rule after one of the method's early advocates.

Meister's Rule operates upon the figure's m Cartesian co-ordinates (x_i, y_i) for the series $i = 0 \ldots m$. In this system the zeroth point should identify with the mth point, i.e. the terminal points should coincide and the figure be closed.

The formula for the Meister's Rule Area Estimate, $A_{meister}$, is then given by:-

$$A_{meister} = \frac{1}{2} \cdot \left| \sum_{i=0}^{m-1} x_i \cdot y_{i+1} + x_m \cdot y_0 - \sum_{i=0}^{m-1} x_{i+1} \cdot y_i - x_0 \cdot y_m \right|$$
Equation 22

Our application involves the mensuration of a regular multilateral whose two defining radii, r and P, alternate at a constant angular interval. This simplifies our computational scheme so that the following formulae apply:-

$$m = 2n$$
Equation 23

$$i = 0 \ldots m$$
Equation 24

$$\theta_i = 2\pi . \frac{i}{m}$$
Equation 25

where θ_i is the Lapsed Sweep Angle tantamount to $\alpha_{lapse(i)}$ and in this regular polygram context is incremented by a constant integral dividend of 2π. (i.e. the increment is β).

It is convenient to assign the alternating radial values of r and P using some simple algebraic alternator such as:-

$$j_i = -1^i = \cos i\pi$$
Equation 26

Equation Twenty-Six gives minus unity for odd i and positive unity for even i including $i = 0$.

In regard to PTC Mathcad Express Prime 3.1® syntax we can then specify the Step Radial, D_i, using:-

$$D_i = if(j_i > 0, r, P)$$
Equation 27

Which then permits co-ordinate definitions in terms of:-

$$x_i = D_i \cdot \cos \theta_i \qquad \textbf{Equation 28a}$$
$$y_i = D_i \cdot \sin \theta_i \qquad \textbf{Equation 28b}$$

As per our requirement for closure of the n = 5 and r = 70 polygram we can also note the terminal conditions:-

$$x_m = x_0 = 70 \qquad \textbf{Equation 29a}$$
$$y_m = y_0 = 0 \qquad \textbf{Equation 29b}$$

In this instance P = 26.737621, and Mathcad quotes the analytic Equation Nine Area as 5500.59271309992 units. The same software yields up the Meister Estimate as 5500.59271309992 units.

<u>Aspects of Subscript Control applied to Intercalation</u>

Allow that kd is the numerical output of a sufficiently-long algebraic polynomial defined as:-

$$kd = \sum_{i=0}^{lt} c_i \cdot n^i$$
Equation 30

kd is a real number.

Limit lt is the Number of Terms in the polynomial. Experiment determined that for Polygram Vertex Count n, lt was six for reliable subscript identification when n was up to sixteen. That is, Equation Thirty should be a sextic equation.

As further predicate we shall adopt the PTC Mathcad Express® syntax for the expression of the entier which is the floor() function:-

$$j = floor\left(\frac{i}{2} \cdot kd + 1\right)$$
Equation 31

In the pentagram's n = 5 context, kd was found to be adequately-near 1.77778, for which you may compute j at point i = 1 to be the entier of 1.88889 which is of course 1.

For various n-vertex regular polygrams (n = 5...16) the useful values of kd were found by trial-and-error.

If trial kd was too low corroborative application of Meister's Rule failed: It also failed if kd was too high.

For discriminatory convenience, allow the Cartesian co-ordinates (u_i, v_i) to surrogate (x_i, y_i).

$alt_{k,f(i)}$ is an Alternator Function such that:-

$$alt_{0,f(k)} = \frac{1+\cos(i\pi)}{2} \cdot x_{floor(i)} \qquad \textbf{Equation 32a}$$

$$alt_{1,f(k)} = \frac{1-\cos(i\pi)}{2} \cdot Px_{floor(j)} \qquad \textbf{Equation 32b}$$

and:-

$$alt_{0,f(k)} = \frac{1+\cos(i\pi)}{2} \cdot y_{floor(i)} \qquad \textbf{Equation 33a}$$

$$alt_{1,f(k)} = \frac{1-\cos(i\pi)}{2} \cdot Py_{floor(j)} \qquad \textbf{Equation 33b}$$

The expression floor(i) is a concession to the syntax of Mathcad Express®: i is of course an integer.

(x_i, y_i) are the Cartesian co-ordinates of successive polygram radial vertices and (Px_i, Py_i) the co-ordinates of successive indent points at the (constant) radial distance P_{an}.

Taken together, these Alternator Functions change between zero and unity with changing values of integer subscript i in the range i = 0,1...jn.

So we may now define u_i and v_i in the following way:-

$$u_i = \frac{1 + \cos i\pi}{2} \cdot x_{floor(i)} + \frac{1 - \cos i\pi}{2} \cdot Px_{floor\left(\frac{i}{2} \cdot kd+1\right)}$$

$$\textbf{Equation 34a}$$

$$v_i = \frac{1 + \cos i\pi}{2} \cdot y_{floor(i)} + \frac{1 - \cos i\pi}{2} \cdot Py_{floor\left(\frac{i}{2} \cdot kd+1\right)}$$

$$\textbf{Equation 34b}$$

These devices enable us to move directly to the Meister's Rule corroboration of the polygram area in terms of:-

$$A2_{meister} = \frac{1}{2} \cdot \left| \sum_{i=0}^{2n-1} u_i \cdot v_{i+1} + u_{2n} \cdot v_0 - \sum_{i=0}^{2n-1} u_{i+1} \cdot v_i - u_0 \cdot v_{2n} \right|$$

$$\textbf{Equation 35}$$

For the case of the n = 5 pentagram of apical radius 50 units the analytical P_{an} is 19.098301 units and the analytical Area A is 2806.424854 square units. The Meister area $A2_{meister}$ was found to be 2806.424854.

A series of empirical kd values is listed for regular polygrams of 5 to 16 points in Table One.

Serial	n	kd	$\log_2(n)$	$\log_2(kd)$
1	5	1.777780000000	2.32192809	0.83007680
2	6	1.818181818	2.58496250	0.86249648
3	7	1.846153847	2.80735492	0.88452278
4	8	1.866666667	3.00000000	0.90046433
5	9	1.882353	3.16992500	0.91253720
6	10	1.894737	3.32192809	0.92199761
7	11	1.90477	3.45943162	0.92961680
8	12	1.913043478	3.58496250	0.93586966
9	13	1.92	3.70043972	0.94110631
10	14	1.926	3.80735492	0.94560770
11	15	1.93104	3.90689060	0.94937805
12	16	1.935484	4.00000000	0.95269438
13	32	1.968254	5.00000000	0.97691641
14	63	1.984	5.97727992	0.98841203
15	129	1.992218	7.01122726	0.99437552
16	256	1.99608625	8.00000000	0.99717406
17	511	1.998041137500	8.99717948	0.99858629
18	1025	1.99902395	10.00140819	0.99929576
19	2048	1.9995116	11.00000000	0.99964765
20	4096	1.99975583	12.00000000	0.99982386
21	8192	1.999877925	13.00000000	0.99991194

Table One
The Subscript Computation Parameter kd
Listed for Selected n Between 5 and 8192

It is clear that for n tending to infinity, kd converges to two and the Px_j subscript j nears *but does not attain* floor(i+1).

I applied an EXCEL® sextic regression to a plot of (n, kd_n). It may be the case, however, that the computation of (Px_j, Py_j) is susceptible of a more analytic treatment.

A similar cubic regression was indicative but inadequate over the stated range.

Table Two presents the sextic Equation Thirty coefficients to twelve-decimal accuracy along with the R^2 Regressive Determination Coefficient of 0.999999359435. The body of the tabulation shows the regenerate value of kd for the cases of n = 5...16.

Figure Three shows the relevant kd data plot along with the fitted sextic regression curve, its equation and its regression coefficient.

Table Two
Sextic Regression Coefficients for (n, kd_n)
With Selected Comparative Regenerations

Regression Coefficient, R^2

	Mantissa	Exponent	Coefficient Value
R^2	9.999993594349	-1	0.9999993594349
0 c_0	9.937445792995	-1	0.9937445792995
1 c_1	3.553563202852	-1	0.3553563202852
2 c_2	-6.518031997460	-2	-0.06518031997460
3 c_3	6.873818630732	-3	0.006873818630732
4 c_4	-4.206004810974	-4	-0.0004206004810974
5 c_5	1.389398430529	-5	0.00001389398430984
6 c_6	-1.917719658628	-7	-0.0000001917719658777
Subscript Parameter	1.290387499972		
Percentage Specific Defect (H_{ped})			0.0000000002392

Regression Terms

Number of Points

	5	6	7	8	9	10	11	12	13	14	15	16
R^2	0.9937445579	0.9937445579	0.9937445579	0.9937445579	0.9937445579	0.9937445579	0.9937445579	0.9937445579	0.9937445579	0.9937445579	0.9937445579	0.9937445579
c_0	1.776781601	2.132137922	2.487494242	2.842850562	3.198206883	3.553563203	3.908919523	4.264275843	4.619632164	4.974988484	5.330344804	5.685701125
c_1	-1.629502999	-2.346491519	-3.193835679	-4.171540478	-5.279605918	-6.518031997	-7.886818717	-9.385966076	-11.01547408	-12.77534272	-14.66557199	-16.68616191
c_2	0.859227329	1.484744824	2.35771979	3.519395139	5.011013782	6.873818631	9.149052598	11.87795859	15.10177953	18.86175832	23.19913788	28.15516111
c_3	-0.262875301	-0.545098224	-1.009861755	-1.722779571	-2.759559756	-4.206004811	-6.158011644	-8.721571576	-12.01277034	-16.15778808	-21.29289936	-27.56447313
c_4	0.043418701	0.108039622	0.233516194	0.455278078	0.820425879	1.389398431	2.237640066	3.452726903	5.158739115	7.472518215	10.55074433	14.56889849
c_5	-0.002996437	-0.008947313	-0.02256178	-0.05027187	-0.101915485	-0.191771966	-0.339735736	-0.577628022	-0.925646651	-1.443953921	-2.184402549	-3.217399694

Number of Points

	5	6	7	8	9	10	11	12	13	14	15	16
Subscript Parameter	1.777792477504	1.818129891780	1.846215591974	1.866676439063	1.882309963180	1.894716069116	1.904790670003	1.913081245188	1.920004322285	1.925924883415	1.931097695631	1.935470565523
Actual kd	1.777780000000	1.818181818200	1.846153847000	1.866666666700	1.882353000000	1.894737000000	1.904770000000	1.913043478265	1.920000000000	1.926000000000	1.931040000000	1.935484000000
Percentage Specific Defect	0.000701633698	-0.002855953121	0.003344519420	0.005235519454	-0.002286230977	-0.003110468546	0.001085170532	0.001974180054	0.000225119002	-0.003900134208	0.002988780914	-0.000694114593

Number of Points

	5	6	7	8	9	10	11	12	13	14	15	16

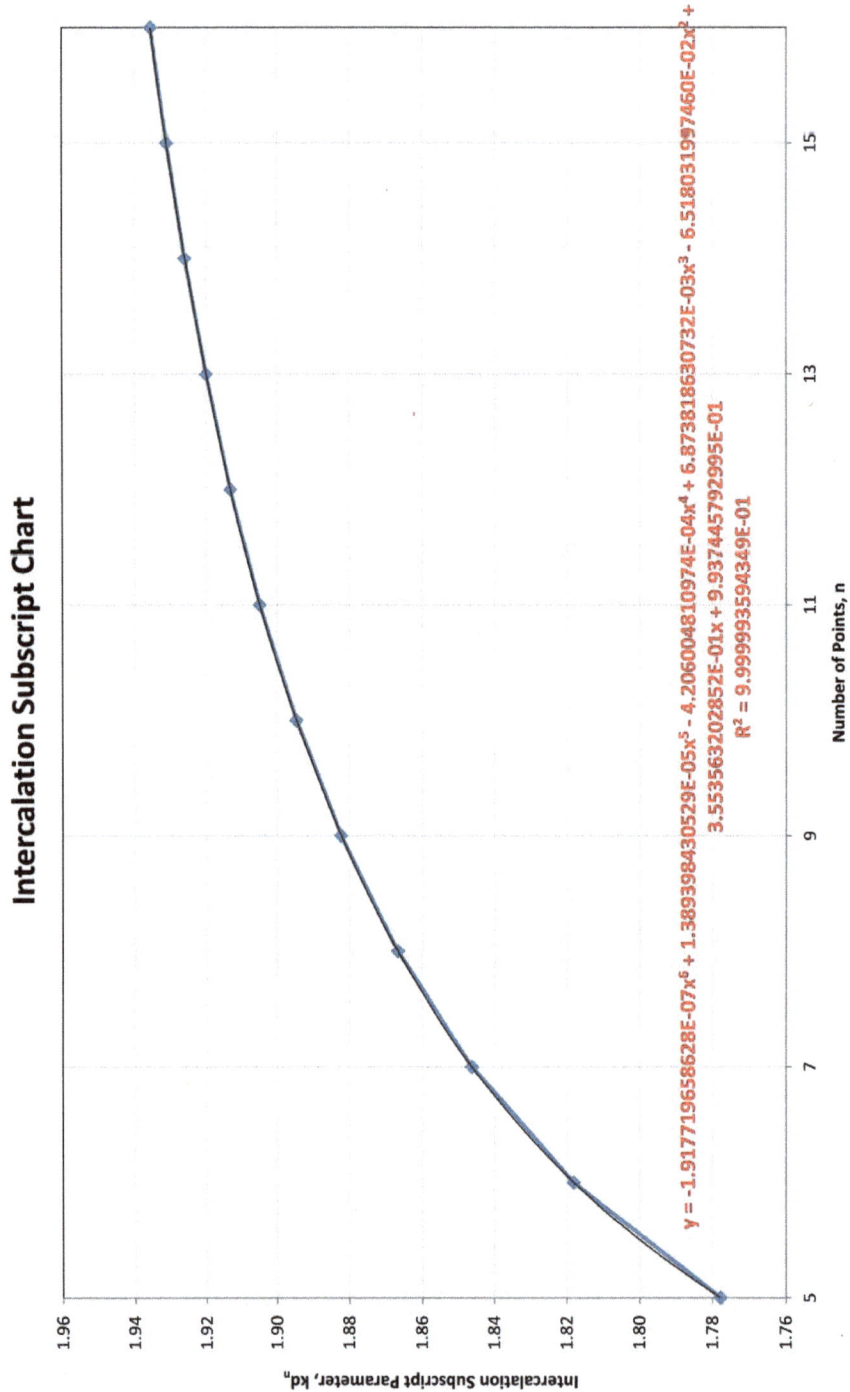

Figure Three
Plot of (n, kdₙ) with a Sextic Regression and its Coefficient of Determination
for Selected n Between 5 and 8192

Appendix One tabulates the calculated properties for polygrams of selected Vertex Counts n, and it is clear that several of the computed elements converge as n tends to infinity.

PART TWO
THE MENSURATION OF IRREGULAR POLYGRAMS
OF THE {n,2} PATTERN

For the sake of introduction we initially shall study the case of an irregular eight-pointed octagram (forgive the pleonasm) whose angular displacements are conventionally reckoned from the easternmost edge of the radial circumference.

The general situation is illustrated by Figure Four.

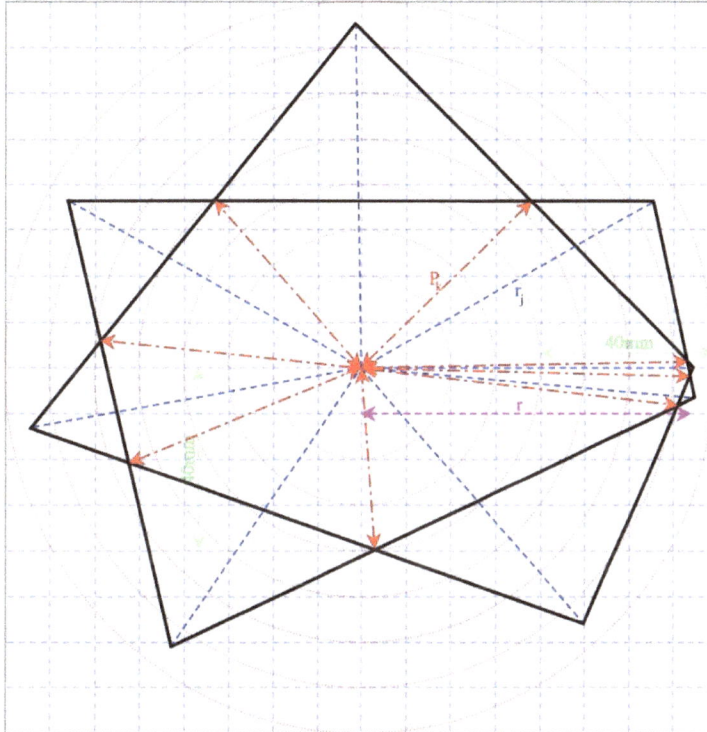

Figure Four
An Irregular Octagram (I1) of
n = 8 and r = 75

Using degrees for convenience, the (eastern) Initial Vertex is by definition αdeg_0 = 0. The eight remaining Sector Angles, αdeg_j, are αdeg_2 = 29, αdeg_4 = 62, αdeg_6 = 60, αdeg_8 = 39, αdeg_{10} = 45, αdeg_{12} = 77, αdeg_{14} = 43, αdeg_{16} = 5.

In terms of Mathcad Express® conventions array subscriptions involve:-

$$jn = 2n \qquad\qquad j = 0,2 \dots jn \qquad\qquad k = 1,3 \dots jn$$

Accordingly j controls the even circumferential radii points at r and k the odd internal P_k positions. In addition, we shall take the opportunity to define $\varepsilon=10^{-10}$ as an arbitrary divisor infinitesimal designed to prevent arithmetic division-by-zero.

Basic angular definitions include:-

$$\alpha lapse_j = \sum_{j=0}^{j} \alpha deg_j$$

Equation 36

$$\alpha_j = \pi \cdot \frac{\alpha lapse_j}{180}$$

Equation 37

$$\beta_j = \pi \cdot \frac{\alpha deg_j}{180}$$

Equation 38

α_j is the Lapsed Angle in radians swept from α_0, and β_j is the jth Sector Angle in radians. Obviously, $\alpha lapse_{jn}$ is 360°; α_{jn} is 2π; and $\Sigma\beta_{jn}$ is 2π.

The Linear Equation Approach to the Intersection Radii P_k

The Apical Co-ordinates of the polygram points are defined by:-

$$x_j = r \cdot \cos \alpha_j \qquad \textbf{Equation 39a}$$
$$y_j = r \cdot \sin \alpha_j \qquad \textbf{Equation 39a}$$

In the case of irregular polygrams specified in our manner the n+1 Circumferential Radii, Q_j, are the constant values r = 75 defined as:-

$$Q_j = \sqrt{x_j{}^2 + y_j{}^2}$$

Equation 40

To establish the co-ordinates (Px_k, Py_k) at the odd positions we will first compute the Grades, b_k, and the Intercepts, a_k, of the first-order algebraic polynomials (i.e. straight lines) of the chords between alternate vertices. In the Express® idiom we may use the following equations:-

$$b_k = \frac{y_{mod(k+3,jn)} - y_{mod(k-1,jn)}}{\left(x_{mod(k+3,jn)} - x_{mod(k-1,jn)}\right) + \varepsilon}$$

Equation 41

$$a_k = y_{mod(k-1,jn)} - b_k \cdot x_{mod(k-1,jn)}$$

Equation 42

The Cartesian co-ordinates of the Intersectoral Radial Lineaments, Px_k and Py_k, are then given by:-

$$Px_{mod(k+2,jn)} = \frac{a_{mod(k+2,jn)} - a_k}{b_k - b_{mod(k+2,jn)}}$$

Equation 43

$$Py_{mod(k+2,jn)} = \frac{b_k \cdot a_{mod(k+2,jn)} - b_{mod(k+2,jn)} \cdot a_k}{b_k - b_{mod(k+2,jn)}}$$

Equation 44

from which it follows that:-

$$P_k = \sqrt{Px_k{}^2 + Py_k{}^2}$$

Equation 45

Tables in Appendix Two present the angular data and computed results respectively for the n = 8 and r = 75 case of Irregular Octagram (I1); for the Irregular Octagram n = 8 r = 80 (I2); and the Irregular Septagram (heptagram) n = 7 r = 80 (I3). Each Appendix Two table is followed by a relevant diagram of the basic figure geometry on a centimeter sketch grid.

PART THREE

MATRIX DEVELOPMENTS FOR AREA COMPUTATION

Some Matrix Approaches to Polygram Mensuration

The intercalation of alternating apical and intersectoral co-ordinates can be epitomised by the rectangular matrix H whose individual elements may be addressed as $H_{iu,iv}$, for example.

Whatever engine of alternation is employed, such a matrix treatment may facilitate later generalisations beyond the scope of our present disquisition.

Figure Seven shows, expressed in the Mathcad Express® idiom, two variants of matrix H as it applies to the irregular octagram I1, together with the equivalent numerical representation:-

NOTE: Mathcad Matrix is rated relative to the EXCEL Meister tabulation

$$
H(ii,jj) =
\begin{bmatrix}
x_0 & y_0 \\
Px_1 & Py_1 \\
x_2 & y_2 \\
Px_3 & Py_3 \\
x_4 & y_4 \\
Px_5 & Py_5 \\
x_6 & y_6 \\
Px_7 & Py_7 \\
x_8 & y_8 \\
Px_9 & Py_9 \\
x_{10} & y_{10} \\
Px_{11} & Py_{11} \\
x_{12} & y_{12} \\
Px_{13} & Py_{13} \\
x_{14} & y_{14} \\
Px_{15} & Py_{15} \\
x_{16} & y_{16}
\end{bmatrix}
\quad
H =
\begin{bmatrix}
x_0 & y_0 \\
Px_1 & Py_1 \\
x_2 & y_2 \\
Px_3 & Py_3 \\
x_4 & y_4 \\
Px_5 & Py_5 \\
x_6 & y_6 \\
Px_7 & Py_7 \\
x_8 & y_8 \\
Px_9 & Py_9 \\
x_{10} & y_{10} \\
Px_{11} & Py_{11} \\
x_{12} & y_{12} \\
Px_{13} & Py_{13} \\
x_{14} & y_{14} \\
Px_{15} & Py_{15} \\
x_{16} & y_{16}
\end{bmatrix}
\quad
H =
\begin{bmatrix}
65.596478 & 36.360722 \\
37.999061 & 36.360722 \\
-1.308930 & 74.988577 \\
-33.151277 & 36.360722 \\
-65.596478 & 36.360722 \\
-58.497793 & 5.612939 \\
-73.860581 & -13.023613 \\
-52.496917 & -20.379713 \\
-43.018233 & -61.436403 \\
3.603126 & -39.696507 \\
50.184795 & -55.735862 \\
71.401926 & -8.081407 \\
74.714602 & -6.536681 \\
73.866386 & -2.546138 \\
75.000000 & 0.000000 \\
72.882989 & 2.080380 \\
65.596478 & 36.360722
\end{bmatrix}
$$

Mathcad Matrix Expression for I1 Alternating Vertices (Format One)	Mathcad Matrix Expression for I1 Alternating Vertices (Format Two)	EXCEL Worksheet Tabulation Showing Values Substitutions for I1 Alternating Vertices

Figure Seven
Variants of the Vertices Matrix H for Irregular Octagram I1

In terms of Equations Thirty-Six through Forty-Five we may now note by way of example that, in the Mathcad® idiom:-

$$G = H_{iu,iv} = 36.360722$$
Equation 46

Note that for the Mathcad matrix iu = 2 and iv = 1, and for the EXCEL version of H iu = 16 and iv = 1 due to rotation.
But to continue with Mathcad syntax:-

$$X = submatrix(H, 0,0,0,1) = [75 \quad 0]$$
Equation 47

In terms of H we may now express several Arial Equations for the Polygram in terms of H, where A2 is the (exact) Estimate of Area in terms of Meister's Rule; A3 is the Area by Sum of Component Triangles; and A4 is the Area by the Sum of Component Triangles assessed using Heron's Formula. All these Areas A2, A3 and A4 should of course be identical, and for the I1 irregular polygram the numerical value should be 12104.365572.

Meister Area Estimate

The Meister Rule Area of I1, A2, is given by:-

$$A2 = \frac{1}{2} \cdot \left| \sum_{i=0}^{2n-1} H_{i,0} \cdot H_{i+1,1} + H_{2n,0} \cdot H_{0,1} - \sum_{i=0}^{2n-1} H_{i+1,0} \cdot H_{i,1} - H_{0,0} \cdot H_{2n,1} \right|$$
Equation 48

Sum of Component Triangles Areal Estimate

For A3 we require the Functional Lapse Angle, γ_i, and its corresponding Functional Interradial Angle, δ_i, according to:-

$$\gamma_i = angle(H_{i,0}, H_{i,1}) \qquad \text{**Equation 49**}$$

$$\gamma deg_i = 360 \cdot \frac{\gamma_i}{2\pi} \qquad \text{**Equation 50**}$$

for i = 0...2n.
and:-

$$\delta_l = \gamma_{l+1} - \gamma_l \qquad \text{**Equation 51**}$$

$$\delta deg_l = 360 \cdot \frac{\delta_l}{2\pi} \qquad \text{**Equation 52**}$$

for l = 0...2n-1.

We may then specify:-

$$R_i = \sqrt{H_{i,0}{}^2 + H_{i,1}{}^2}$$

Equation 53

$$t_l = \frac{1}{2} \cdot R_l \cdot R_{l+1} \cdot \sin \delta_l$$

Equation 54

$$A3 = \sum_{l=0}^{2n-1} t_l$$

Equation 55

Heronian Sum of Component Triangles Areal Estimate

For A4 we need the following equations:-

$$d_l = \sqrt{R_l{}^2 + R_{l+1}{}^2 - 2 \cdot R_l \cdot R_{l+1} \cdot \cos \delta_l}$$

Equation 56

$$s_l = \frac{R_l + R_{l+1} + d_l}{2}$$

Equation 57

$$h_l = \sqrt{s_l \cdot (s_l - R_l) \cdot (s_l - R_{l+1}) \cdot (s_l - d_l)}$$

Equation 58

$$A4 = \sum_{l=0}^{2n-1} h_l$$

Equation 59

In regard to I1, the irregular octagram with the 75-unit apical radius, comparisons of Percentage Specific Defects involving A2, A3 and A4 demonstrate that they all lie within 10^{-13} of each other, which of course implies a common value of 12104.365572 to six places of decimals.

Subscriptive Predicates

Further to discuss the use of simple matrices in area computation we need to (re)specify the following subscript definitions:-

$$i = 0 \ldots 2n$$
Equation 60

$$l = 0 \ldots 2n - 1$$
Equation 61

$$K(l) = ent\left(\frac{l+1}{2} + 3\right)$$
Equation 62

$$L(l) = ent\left(\frac{l}{2}\right) + ent\left(\frac{l}{2} + 1\right)$$
Equation 63

$ent()$ is the truncate-to-integer entier function, rendered in MathCad Express® syntax as *floor*(), and in EXCEL® syntax as INT().

Component Triangle Area Studies

Appendix Three presents simple matrices of several key polygram series, where necessary defined below.

The first two tables show the Formula and Number matrices of i, l, γdeg_l, γ_l, δdeg_l, and δ_l: All of which are simple $1 \times (2n-1)$ columnar matrices, i.e. series.

The only point of difference between i and l is that l is one element shorter.

The third and fourth tabulations show the Formula and Number matrices of R and ta_l. The series variable ta_l is the same thing as t_l and τ_l: The Area of a Component Triangle defined by:

$$\tau_l = \frac{1}{2} \cdot R_{l+1} R_l \sin \delta_l = \frac{1}{2} \cdot R_{2.K(l)-6} \cdot R_{L(l).} \cdot \sin \delta_l$$
Equation 64

Note that K(l) and L(l) are not redundant: They have to operate the repetition of l values necessary to calculation of τ_l.

To clarify the meaning of this we may observe that:-

$$l = \begin{bmatrix} 0 \\ 1 \\ 2 \\ 3 \\ 4 \\ 5 \\ 6 \\ 7 \\ 8 \\ 9 \\ 10 \\ 11 \\ 12 \\ 13 \\ 14 \\ 15 \end{bmatrix} \qquad Q_{sub} = K(l) = 2.(\text{floor}((l+1)/2+3)) - 6 = \begin{bmatrix} 0 \\ 2 \\ 2 \\ 4 \\ 4 \\ 6 \\ 6 \\ 8 \\ 8 \\ 10 \\ 10 \\ 12 \\ 12 \\ 14 \\ 14 \\ 16 \end{bmatrix} \qquad P_{sub} = L(l) = \text{floor}(l/2)+\text{floor}(l/2+1) = \begin{bmatrix} 1 \\ 1 \\ 3 \\ 3 \\ 5 \\ 5 \\ 7 \\ 7 \\ 9 \\ 9 \\ 11 \\ 11 \\ 13 \\ 13 \\ 15 \\ 15 \end{bmatrix}$$

Figure Eight
Illustrative Compound Subscripts Tabulation

From Equation Sixty-Four it follows that the total Polygram Area, A_τ, is yielded by:-

$$A_\tau = \sum_{l=0}^{2n-1} \tau_l$$
$$= \sum_{l=0}^{2n-1} \frac{1}{2} . R_{2.K(l)-6} . R_{L(l).} . \sin \delta_l$$
$$= \sum_{l=0}^{2n-1} \frac{1}{2} . R_{2.ent\left(\frac{l+1}{2}+3\right)-6} . R_{ent\left(\frac{l}{2}\right)+ent\left(\frac{l}{2}+1\right)} . \sin \delta_l$$

Equation 65

For the sake of efficiency one could of course cast Equation Sixty-Five as:-

$$A_\tau = \frac{1}{2} \sum_{l=0}^{2n-1} R_{2.ent\left(\frac{l+1}{2}+3\right)-6} . R_{ent\left(\frac{l}{2}\right)+ent\left(\frac{l}{2}+1\right)} . \sin \delta_l$$

Equation 66

This latter expression may not itself be the slickest rendition. At all times, A_τ should be the same as the Meister and Heronian areal estimates and in the trialled cases (I1, I2, I3, I4, I5, R1, R2, R3, R4 and R5) this obtained.

Because the special form of our {n,2} irregular polygons assumes that the apical radii $Q_{K(l)}$ are constant, i.e. they are r, we are able to simplify τ_l as:-

$$\tau_l = \frac{1}{2} \cdot r \cdot P_{floor\left(\frac{l}{2}\right)+floor\left(\frac{l}{2}+1\right)} \cdot \sin \delta_l$$
Equation 67

so that:-

$$A_\tau = \frac{1}{2} \cdot r \cdot \sum_{l=0}^{2n-1} P_{L(l)} \cdot \sin \delta_l$$
$$= \frac{1}{2} \cdot r \cdot \sum_{l=0}^{2n-1} P_{floor\left(\frac{l}{2}\right)+floor\left(\frac{l}{2}+1\right)} \cdot \sin \delta_l$$
Equation 68

The Matrices T, U, V and W and their relations to Polygram Area

The fifth and sixth tabulations of Appendix Three respectively display the series matrices T, U, V and W for the case of Polygram I1.

T contains the series of Component Triangle Areas τ_l such that:-

$$T = [\tau_l]_0^{2n-1}$$
Equation 69

where the RHS expression is understood to indicate the series $\tau_0 \ldots \tau_{2n-1}$, the Family of Polygram Component Triangles.

Similarly:-

$$U = \left[P_{L(l)} \cdot \sin \delta_l \right]_0^{2n-1} \qquad \textbf{Equation 70}$$
The Family of Sector Transforms

$$V = \left[P_{L(l)} \right]_0^{2n-1} \qquad \textbf{Equation 71}$$
The Family of Chord Intersection Radii

$$W = [\sin \delta_l]_0^{2n-1} \qquad \textbf{Equation 72}$$
The Family of Inter-P Angle Sines

From the above, the following facts follow:-

$$\sum V = 2 \cdot \sum_k P_k$$
Equation 73

$$A_\tau = \sum T = \frac{1}{2}.r.\sum U = \frac{r}{2}.V.W$$
Equation 74

The Expectation of Area

For the series represented by the uni-columnar matrices U, V and W it is possible to define the following arithmetic means:-

$$U_\mu = \frac{\sum U}{2n}$$
Equation 75

$$V_\mu = \frac{\sum V}{2n}$$
Equation 76

$$W_\mu = \frac{\sum W}{2n}$$
Equation 77

as well as the more primitive mean of the Chord Intersectional Radii, P_k, as:-

$$P_\mu = \frac{\sum_k P_k}{n}$$
Equation 78

In terms of these means the Polygram Area, A_{amean}, may be approximated as:-

$$A_{amean} = rnP_\mu W_\mu$$
Equation 79

In the case of Polygram I1, A_{amean} is 13135.716197 whereas the exact Area A_τ is 12104.365572. To assess the Percentage Specific Defect, $PSD(A_\tau, A_{amean})$, a measure of computational accuracy, we may define:-

$$PSD(A_\tau, A_{amean}) = 100.\left(\frac{A_\tau - A_{amean}}{A_\tau}\right)$$
Equation 80

For Polygram I1, $PSD(A_\tau, A_{amean})$ is -8.520485, indicating an 8½% overestimate of the area by Equation Seventy-Nine.
The interest of such procedures of approximation rests on their efficacy at predicting area from limited numbers of scans or transects; or for the vicinitation of starting values in refining computation.

Table Three presents averages calculated for the five regular and five irregular polygrams investigated in detail.

Regular Polygrams

Folder Code	Diagram Code	Number of points, n	Apical Radius, r	%SpDef (P_μ, W_μ)
RA	R3	7	1	0
RB	R2	8	1	0
RC	(none)			
RD	R4	5	1	0
RE	R5	16	1	0

Irregular Polygrams

Folder Code	Diagram Code	Number of points, n	Apical Radius, r	%SpDef (P_μ, W_μ)
IA	I3	7	80	-8.130164
IB	I2	8	80	-5.803603
IC	I1	8	75	-8.520485
ID	I4	5	75	-4.342749
IE	I5	16	75	-1.52074

Table Three
Polygram Area Averages

For regular polygram U-matrix arithmetic means it can be seen that Equation Seventy-Nine outcomes are exact, but for irregular polygrams the estimates of area from arithmetic means very from eight and a bit percent overestimates to 1½ percent at sixteen points (an irregular hexadecagram). There is a suggestion that the deviation from truth becomes progressively negligible with increasing vertex-count n.

The AGM:- A Failed Idea

I observed that the Arithmetic Mean of the Chord intersection Radii was:-

$$P_\mu = \frac{\sum P}{n}$$

Equation 81

and that the Arithmetic Mean of the Central Angles was:-

$$W_\mu = \frac{\sum \sin \delta_l}{2n}$$

Equation 82

Also, I considered that a plausible approximation of the Polygram Area might be yielded by the relevant circle area:-

$$B_{av} = \pi . P_\mu{}^2$$

Equation 83

Whilst:-

$$A_{av} = 2n.\frac{1}{2}r.P_\mu.W_\mu = 2n.\frac{1}{2}r.\frac{\sum P}{n}.\frac{\sum \sin \delta_l}{2n} = r.\sum P.\frac{\sum \sin \delta_l}{2n} = r.W_\mu.\sum P$$

Equation 84

For the purposes of computing the Arithmetic-Geometric Mean (AGM) the Initial Upper Bound Arithmetic Mean, AM_0, is formulable as:-

$$AM_0 = \frac{A_{av} + B_{av}}{2}$$

Equation 85

Whilst the Initial Lower Bound Geometric Mean, GM_0, is given by:-

$$GM_0 = \sqrt{A_{av}. B_{av}}$$

Equation 86

There is little point in boring you with further tedious details because I was guilty of a mis-conception.

References

1   "Generalized Star Polygons and Star Polygrams"

     Eidur Sveinn Gunnarsson and Karl Thorlaksson
     School of Computer Science, Reykjavik University, Reykjavik, Iceland
     https://skemman.is/bitstream/1946/26421/1/
          generalized_star_polygons_and_star_polygrams.pdf

2   Wikipedia contributors. (2018, September 19). Regular polygon.
     In *Wikipedia, The Free Encyclopedia*. Retrieved 13:25, October 7, 2018,
     from https://en.wikipedia.org/w/index.php?title=
     Regular_polygon&oldid=860286235

3   Wikipedia contributors. (2018, October 7). Linear equation.
     In *Wikipedia, The Free Encyclopedia*. Retrieved 13:24, October 7, 2018,
     from https://en.wikipedia.org/w/index.php?title=
     Linear_equation&oldid=862870878

4   Wikipedia contributors. (2018, August 30). Line–line intersection.
     In *Wikipedia, The Free Encyclopedia*. Retrieved 13:23, October 7, 2018,
     from https://en.wikipedia.org/w/index.php?title=
     Line%E2%80%93line_intersection&oldid=857265238

5   Wikipedia contributors. (2018, March 29). Shoelace formula.
     In *Wikipedia, The Free Encyclopedia*. Retrieved 13:22, October 7, 2018,
     from https://en.wikipedia.org/w/index.php?title=
     Shoelace_formula&oldid=833149289

6   Wikipedia contributors. (2018, July 25). Heron's formula.
     n *Wikipedia, The Free Encyclopedia*. Retrieved 13:25, October 7, 2018,
     from https://en.wikipedia.org/w/index.php?title=
     Heron%27s_formula&oldid=851939741

APPENDIX ONE

LIST OF {n,2} POLYGRAM PROPERTIES
FOR SELECTED VERTEX COUNTS n

FIGURE SIDE RELATIONS

Serial	Number of Sides n	Apical Radius r	Sector Angle α	Half Sector Angle β	Peripheral Angle δ	Right Tending Angle η	Inter Sectoral Radius P	Analytic Area A	Area Parameter A/P	s =cos(2π/n)	t =cos(π/n)	u =s/t	Figure Side Ratio Parameter φ$_n$
1	0	1	#DIV/0!	#DIV/0!	#DIV/0!	#DIV/0!	#DIV/0!	#DIV/0!	#DIV/0!	#DIV/0!	#DIV/0!	#DIV/0!	#DIV/0!
2	1	1	6.283185	3.141593	-4.712389	4.712389	-1	-1.2E-16	1.2251E-16	1	-1	-1	#DIV/0!
3	2	1	3.141593	1.570796	-1.570796	3.141593	-8.2E+15	-3.3E+16	4	-1	6.13E-17	-1.6E+16	-2
4	3	1	2.094395	1.047198	-0.523599	2.617994	-1	-2.59808	2.59807621	-0.5	0.5	-1	-1
5	4	1	1.570796	0.785398	0	2.356194	0	2.45E-16	#DIV/0!	6.12574E-17	0.707107	8.66E-17	1.2251E-16
6	5	1	1.256637	0.628319	0.3141593	2.199115	0.381966	1.12257	2.93892626	0.309016994	0.809017	0.381966	0.61803399
7	6	1	1.047198	0.523599	0.5235988	2.094395	0.57735	1.732051	3	0.5	0.866025	0.57735	1
8	7	1	0.897598	0.448799	0.6731984	2.019595	0.692021	2.101798	3.03718617	0.623489802	0.900969	0.692021	1.2469796
9	8	1	0.785398	0.392699	0.7853982	1.963495	0.765367	2.343146	3.06146746	0.707106781	0.92388	0.765367	1.41421356
10	9	1	0.698132	0.349066	0.8726646	1.919862	0.815207	2.509356	3.07818129	0.766044443	0.939693	0.815207	1.53208889
11	10	1	0.628319	0.314159	0.9424778	1.884956	0.850651	2.628656	3.09016994	0.809016994	0.951057	0.850651	1.61803399
12	11	1	0.571199	0.285599	0.9995977	1.856396	0.876769	2.717158	3.09905813	0.841253533	0.959493	0.876769	1.68250707
13	12	1	0.523599	0.261799	1.0471976	1.832596	0.896575	2.78461	3.10582854	0.866025404	0.965926	0.896575	1.73205081
14	13	1	0.483322	0.241661	1.0874744	1.812457	0.911956	2.837189	3.11110364	0.885456026	0.970942	0.911956	1.77091205
15	14	1	0.448799	0.224399	1.1219974	1.795196	0.924139	2.878964	3.11529308	0.900968868	0.974928	0.924139	1.80193774
16	15	1	0.418879	0.20944	1.1519173	1.780236	0.933955	2.912701	3.11867536	0.913545458	0.978148	0.933955	1.82709092
17	16	1	0.392699	0.19635	1.1780972	1.767146	0.941979	2.940337	3.12144515	0.923879533	0.980785	0.941979	1.84775907
18	17	1	0.369599	0.1848	1.2011972	1.755596	0.948624	2.963258	3.1237418	0.932472229	0.982973	0.948624	1.86494446
19	18	1	0.349066	0.174533	1.2217305	1.745329	0.954189	2.982477	3.1256672	0.939692621	0.984808	0.954189	1.87938524
20	19	1	0.330694	0.165347	1.2401024	1.736143	0.958895	2.998751	3.12729722	0.945817242	0.986361	0.958895	1.89163448
21	20	1	0.314159	0.15708	1.2566371	1.727876	0.962912	3.012651	3.1286893	0.951056516	0.987688	0.962912	1.90211303
22	32	1	0.19635	0.098175	1.3744468	1.668971	0.985531	3.091165	3.13654849	0.98078528	0.995185	0.985531	1.96157056
23	64	1	0.098175	0.049087	1.4726216	1.619884	0.996385	3.128979	3.14033116	0.995184727	0.998795	0.996385	1.99036945
24	128	1	0.049087	0.024544	1.5217089	1.59534	0.999096	3.138439	3.14127725	0.998795456	0.999699	0.999096	1.99759091
25	256	1	0.024544	0.012272	1.5462526	1.583068	0.999774	3.140804	3.1415138	0.999698819	0.999925	0.999774	1.99939764
26	8192	1	0.000767	0.000383	1.5700293	1.57118	1	3.141592	3.14159258	0.999999706	1	1	1.99999941

APPENDIX TWO

**TABULATIONS OF DATA AND RESULTS FOR
IRREGULAR POLYGRAMS
TOGETHER WITH
THEIR GEOMETRIC DIAGRAMS**

PROJECT TITLE: On the Quadrature of Polygrams
FIGURE: I1
FIGURE TITLE: Irregular Eight-Pointed Sideroid (Octagram)

NUMBER OF VERTICES: 8
NORMATIVE RADIUS: 75

Chord Angle Adjustor: 0

POLYGRAM AREA ESTIMATORS:

Heron's Formula	12104.36557
Formula ½A=d₁SabSin(θ)	12104.36557
Dart Formula	12104.36557
Meister's Rule	12104.36557
Counted Squares	Not Assessed

Fundamental Data and Geometrical Derived Variables For the Irregular Polygram I1

Point Serial i	Point Radius r_i	Inter Vertex Angle β^o (Degrees)	Lapse Angle α^o (Degrees)	Lapse Angle α (Radians)	Lapse Angle Cosine	Lapse Angle Sine	Vertex Terminus x_i	Vertex Terminus y_i	Inter Cept c_i	Grade a_i	Inter Cept d_i	Grade b_i	Chord Inter Section u_i	Chord Inter Section v_i	Calc Inter Section Radius P_i	Meas Inter Section Radius Pm_i	Percentage Specific Defect $(P_i - Pm_i)$	Sector Constant (1 # TC+ve)	Chord Intersect Lapse Angle γ	Chord Intersect Lapse Angle γ	Adjusted Chord Inter Angle δ^o	Meas Chord Intersect Inter Angle δ^o_m	Percentage Specific Defec $(\delta^o - \delta^o m_i)$
1	75	29	29	0.50614548	0.87462	0.48481	65.596648	36.36072	73.702	-0.983	36.361	-5E-17	37.99906	36.36072	52.5930671	52.61586	-0.0433373	0	0.763369	43.73783	43.73783	41.5	5.11646.375
2	75	62	91	1.58824962	-0.01745	0.9999848	-1.30893	74.98858	76.576	1.2131	36.361	-5E-17	-33.1513	36.36072	49.2047687	49.42702	-0.45168197	0	2.310056	132.3565	88.61864	88.6	0.021034151
3	75	60	151	2.63544717	-0.87462	0.48481	-65.5965	36.36072	-247.8	-4.331	76.576	1.2131	-58.4978	5.612939	58.7664606	59.39214	-1.06469186	0	3.045934	174.5192	42.16271	42.2	-0.08845447
4	75	39	190	3.31612558	-0.98481	-0.17365	-73.8606	-13.0236	-38.46	-0.344	-247.8	-4.331	-52.4969	-20.3797	39.8596935	40.2591	-1.00203896	1	3.511892	201.2166	26.69743	27	-1.13334267
5	75	45	235	4.10152374	-0.57358	-0.81915	-43.0182	-61.4364	-41.38	0.4663	-38.46	-0.344	3.603126	-39.6965	71.857805	71.7489	0.151560761	1	4.802908	275.1863	73.96974	73	1.310992747
6	75	77	312	5.44542727	0.669131	-0.74314	50.1848	-55.7359	-168.5	2.246	-41.38	0.4663	71.40193	-8.08141	73.9102554	74.24018	-0.44638259	1	6.170483	353.5426	78.35629	79	-0.82153828
7	75	43	355	6.19591884	0.996195	-0.08716	74.7146	-6.53668	344.97	-4.705	-168.5	2.246	73.86639	-2.54614	72.9126747	73.44297	-0.72729904	1	6.248729	358.0258	4.483191	5	-11.5277034
8	75	5	360	6.28318531	1	-2.5E-16	75	-1.8E-14	73.702	-0.983	344.97	-4.705	72.88299	2.08038					6.311722	361.635	3.609189	3	16.87883585
Total		360	1723	30.07023	1.089489	0.146366	81.71165	10.97746							475.418657	477.7281	-4.09524226		33.16509	1900.22	361.635	359.3	9.756312479
Mean		45	215.375	3.75900288	0.136186	0.018296	10.21396	1.372183							59.4273321	59.71601	-0.51190528		4.145637	237.5275	45.20438	44.9125	1.219539306
Norm		45	360	3.14159265													0			180	22.5	45	0

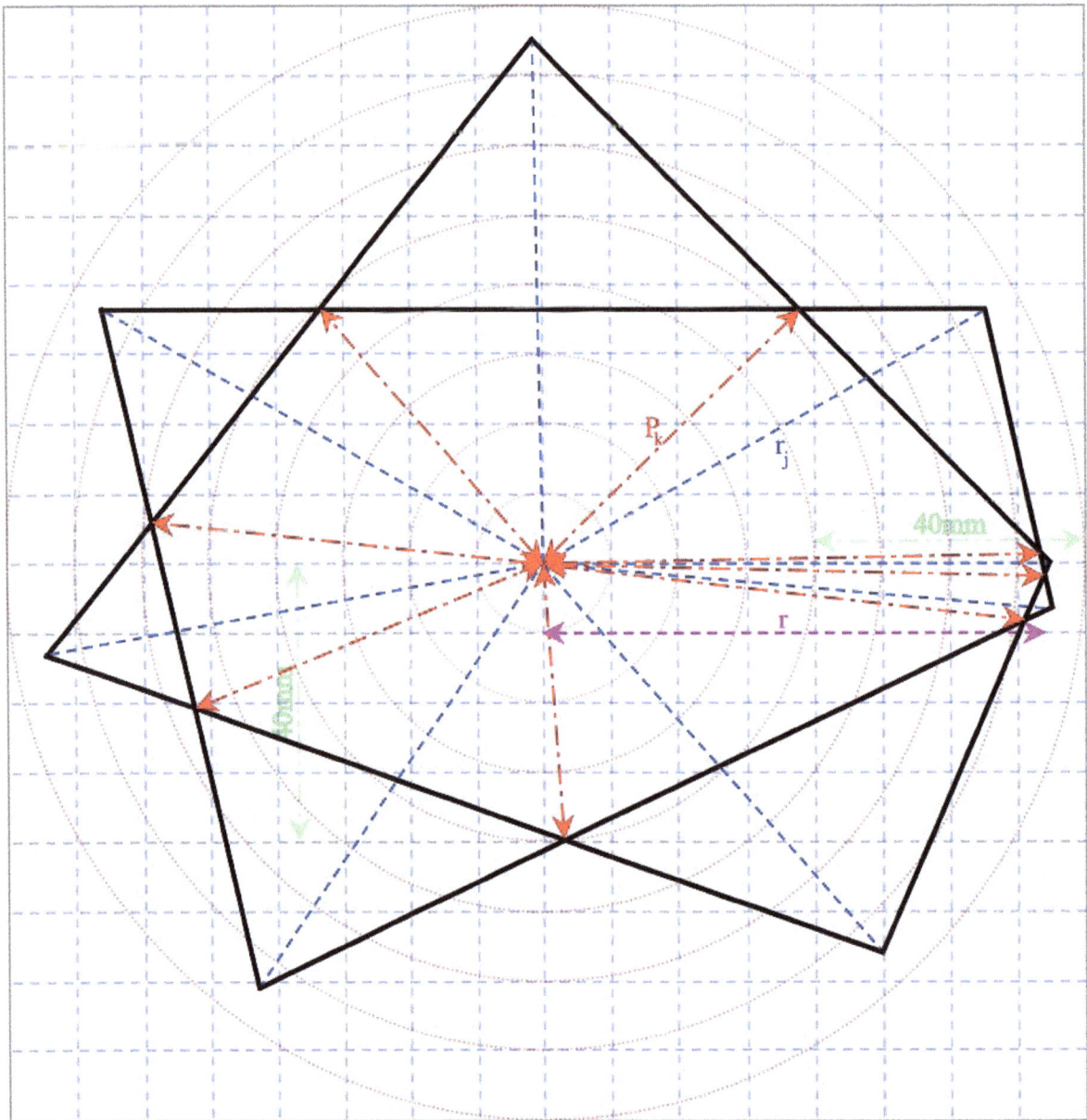

The Main Geometric Features
of the
Irregular {n,2} Polygram
I1

PROJECT TITLE: On the Quadrature of Polygrams
FIGURE: I2
FIGURE TITLE: Irregular Eight-Pointed Sideroid (Octagram)

NUMBER OF VERTICES: 8
NORMATIVE RADIUS: 80

Chord Angle Adjustor: 0

POLYGRAM AREA ESTIMATORS:

Heron's Formula	14231.33716
Formula ΣA=0.5abSin(θ)	14231.33716
Dart Formula	14231.33716
Meister's Rule	14231.33716
Counted Squares	Not Assessed

Fundamental Data and Geometrical Derived Variables For the Irregular Polygram I2

Point Serial i	Point Radius r_i	Inter Vertex Angle $\beta°$ (Degrees)	Lapse Angle $\alpha°$ (Degrees)	Lapse Angle α (Radians)	Lapse Angle Cosine	Lapse Angle Sine	Vertex Terminus x_i	Vertex Terminus y_i	Inter Cept c_i	Grade a_i	Inter Cept d_i	Grade b_i	Chord Inter Section u_i	Chord Inter Section v_i	Calc Inter Section Radius P_i	Meas Inter Section Radius Pm_i	Percentage Specific Defect $(P_i\text{-}Pm_i)$	Sector Constant (1 if TC+ve)	Chord Intersect Lapse Angle γ	Chord Intersect Lapse Angle $\gamma°$	Adjusted Chord Intersect Inter Angle $\delta°$	Meas Chord Intersect Inter Angle $\delta°_m$	Percentage Specific Defect $(\delta°_i\text{-}\delta°_m)$
1	80	39	39	0.680674941	0.77146	0.62932	62.17168	50.34563	81.409	-1.018	44.906	0.0875	33.03086	47.79614	58.0991307	58.23806	-0.23912311	0	0.966083	55.3525	55.3525	48.5	12.37974946
2	80	50	89	1.55334303	0.017452	0.999848	1.396193	79.98782	78.616	0.9827	44.906	0.0875	-37.6554	41.61189	56.1202373	56.14746	-0.04851166	0	2.306323	132.1426	76.7900	78	-1.57562305
3	80	62	151	2.63544717	-0.87462	0.48481	-69.9696	38.78477	-53.1	-8.144	-531.1	-8.144	-66.8001	12.9715	68.0478796	66.99866	1.541890105	0	2.949796	169.0108	36.86827	37	-0.35729679
4	80	29	180	3.14159265	-1	1.23E-16	-80	9.8E-15	-46.19	-0.577	-531.1	-0.577	-64.0788	-9.19214	64.7347047	64.4103	0.501132618	1	3.284071	188.1634	19.15257	18.8	1.840871929
5	80	43	223	3.89208423	-0.73135	-0.682	-58.5083	-54.5599	-39.97	0.2493	-46.19	0.2493	-7.51914	-41.8468	42.5170064	42.60836	-0.21485854	1	4.534604	259.8137	71.65023	72.6	-1.32556306
6	80	77	300	5.23598776	0.5	-0.86603	40	-69.282	-138.6	1.7321	-39.97	1.7321	66.49385	-23.3933	70.488868	70.68209	-0.27411549	1	5.944897	340.6175	80.80386	81	-0.24773959
7	80	45	345	6.02138592	0.965926	-0.25882	77.27407	-20.7055	342.84	-4.705	-138.6	-4.705	74.79079	-9.02262	75.3330582	75.36104	-0.03714954	1	6.163127	353.1212	12.50367	12	4.028168048
8	80	15	360	6.28318531	1	-2.5E-16	80	-2E-14	81.409	-1.018	342.84	-1.018	70.90593	9.254191	71.5072813	71.2794	0.31867932	1	6.412965	367.4358	14.31466	15	-4.78765889
Total		360	1687	29.4437045	0.654551	0.307135	52.36406	24.57079							506.848166	505.7254	1.547943704	0	32.56187	1865.658	367.4358	362.9	9.959919071
Mean		45	210.875	3.68046306	0.081819	0.038392	6.545508	3.071349							63.3560208	63.21567	0.193492963		4.070233	233.2072	45.92948	45.3625	1.244988884
Norm		45	360	3.14159265													0			180	22.5	360	0

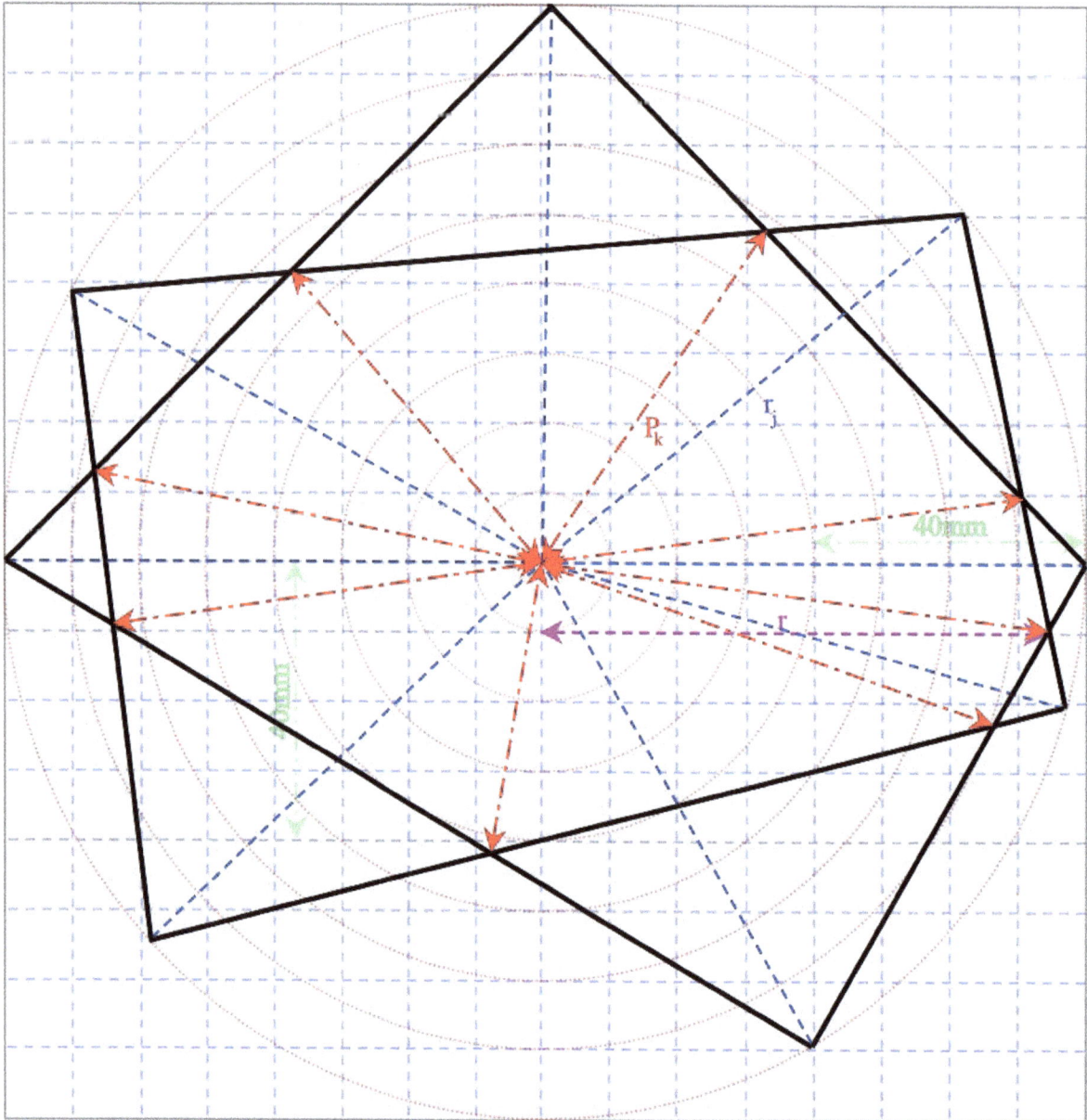

The Main Geometric Features
of the
Irregular {n,2} Polygram
I2

PROJECT TITLE: On the Quadrature of Polygrams
FIGURE: I3
FIGURE TITLE: Irregular Seven-Pointed Sideroid (Septagram)

NUMBER OF VERTICES: 7
NORMATIVE RADIUS: 80

Chord Angle Adjustor: 46.7440136

POLYGRAM AREA ESTIMATORS:
Heron's Formula	12286.62351
Formula ΣA=0.5absSin(δ)	12286.62351
Dart Formula	12286.62351
Meister's Rule	12286.62351
Counted Squares	Not Assessed

Point Serial i	Point Radius r_i	Inter Vertex Angle $\beta°$ (Degrees)	Lapse Angle $\alpha°$ (Degrees)	Lapse Angle α (Radians)	Lapse Angle Cosine	Lapse Angle Sine	Vertex Terminus x_i	Vertex Terminus y_i	Inter Cept c_i	Grade a_i	Inter Cept d_i	Grade b_i	Chord Inter Section u_i	Chord Inter Section v_i	Calc Inter Section Radius P_i	Meas Inter Section Radius Pm_i	Percentage Specific Defect $(P_i\text{-}Pm)$	Sector Constant (1 if TC+ve)	Chord Intersect Lapse Angle γ	Chord Intersect Lapse Angle $\gamma°$	Adjusted Chord Intersect Inter Angle $\delta°$	Meas Chord Intersect Inter Angle $\delta°_m$	Percentage Specific Defect $(\delta°\text{-}\delta°m)$
1	80	91	91	1.58824962	-0.01745	0.9998848	-1.39619	79.98782	33.958	-0.424	82.266	1.6319	-23.4925	43.92995	49.8170378	49.76784	0.0987607	0	2.061873	118.1366	71.39258	71	0.549851748
2	80	43	134	2.3387412	-0.69466	0.71934	-55.5727	57.54718	-193.1	-4.511	82.266	1.6319	-44.8333	9.10493	45.7484732	45.38827	0.787359127	0	2.941233	168.5203	50.38366	52	-3.20806596
3	80	72	206	3.59537826	-0.89879	-0.43837	-71.9035	-35.0697	-193.1	-0.499	-193.1	-4.511	-30.459	-55.7332	63.5132815	63.20515	0.485138004	1	4.212225	241.3427	72.82247	72.6	0.305492009
4	80	45	251	4.38077642	-0.32557	-0.94552	-26.0455	-75.6415	-69.39	0.2401	-193.1	-0.499	-2.07259	-69.8861	69.9168373	69.4759	0.630656728	1	4.682741	268.3013	26.95857	27	-0.15346523
5	80	30	281	4.9043752	0.190809	-0.98163	15.26472	-78.5302	-97.05	1.2131	-69.39	0.2401	28.42622	-62.564	68.7190184	68.38101	0.491870812	1	5.138858	294.4349	26.13358	27	-3.31533219
6	80	35	316	5.51524044	0.71934	0.69466	57.54718	-55.5727	76.777	-2.3	-97.05	1.2131	49.48123	-37.0222	61.7983615	62.11026	-0.50470685	1	5.640832	323.1959	28.761	28.5	0.907431032
7	80	44	360	6.28318531	1	-2.5E-16	80	-2E-14	33.958	-0.424	76.777	-2.3	22.83222	24.26628	33.3191047	33.04584	0.82013065	1	7.099022	406.744	83.54813	84	-0.54084661
Total		360	1639	28.6059464	-1.34099	-0.02632	-2.10593	-107.279							392.832114	391.3743	2.809209169	0	31.77678	1820.676	360	362.1	-5.45524519
Mean		51.4285714	234.142857	4.08656378	-0.19157	-0.00376	-0.30085	-15.3256							56.1188735	55.91061	0.401315596		4.539541	260.0965	51.4285714	51.7285714	-0.75929217
Norm		51.4285714	360	3.14159265																	180	25.71429	360	0

Fundamental Data and
Geometrical Derived Variables
For the Irregular Polygram I3

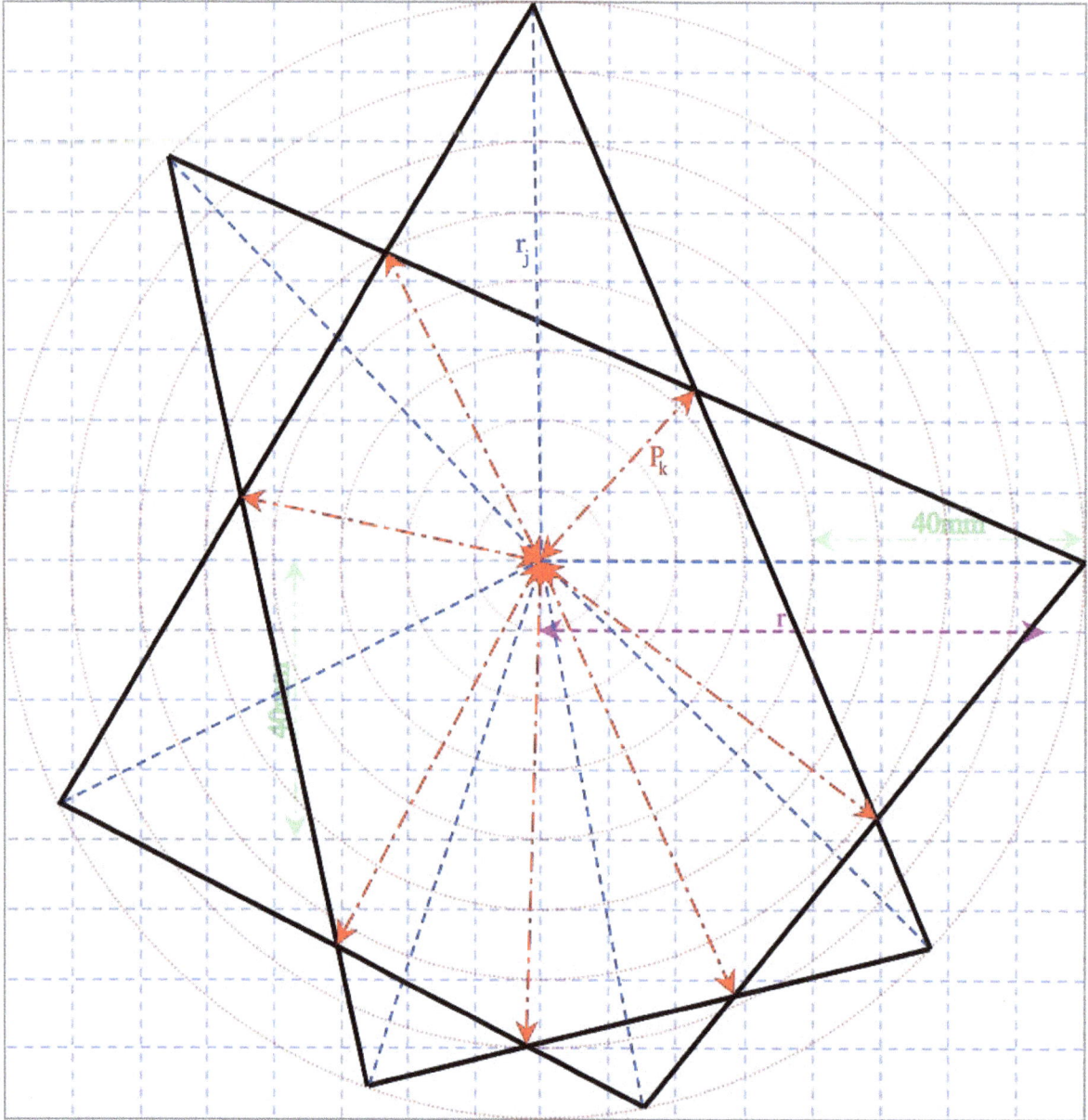

The Main Geometric Features
of the
Irregular {n,2} Polygram
I3

APPENDIX THREE

TABULATIONS OF FORMULAE AND VALUES
FOR SELECTED DERIVED
SERIES AND MATRICES

Internal Formulae for the Angular Series of the Irregular Polygram I1

$$
i = \begin{bmatrix} 0 \\ 1 \\ 2 \\ 3 \\ 4 \\ 5 \\ 6 \\ 7 \\ 8 \\ 9 \\ 10 \\ 11 \\ 12 \\ 13 \\ 14 \\ 15 \\ 16 \end{bmatrix}
\qquad
I = \begin{bmatrix} 0 \\ 1 \\ 2 \\ 3 \\ 4 \\ 5 \\ 6 \\ 7 \\ 8 \\ 9 \\ 10 \\ 11 \\ 12 \\ 13 \\ 14 \\ 15 \end{bmatrix}
$$

$$
\gamma deg_I = \begin{bmatrix}
360 \cdot \gamma_0 / 2\pi \\
360 \cdot \gamma_1 / 2\pi \\
360 \cdot \gamma_2 / 2\pi \\
360 \cdot \gamma_3 / 2\pi \\
360 \cdot \gamma_4 / 2\pi \\
360 \cdot \gamma_5 / 2\pi \\
360 \cdot \gamma_6 / 2\pi \\
360 \cdot \gamma_7 / 2\pi \\
360 \cdot \gamma_8 / 2\pi \\
360 \cdot \gamma_9 / 2\pi \\
360 \cdot \gamma_{10} / 2\pi \\
360 \cdot \gamma_{11} / 2\pi \\
360 \cdot \gamma_{12} / 2\pi \\
360 \cdot \gamma_{13} / 2\pi \\
360 \cdot \gamma_{14} / 2\pi \\
360 \cdot \gamma_{15} / 2\pi \\
360 \cdot \gamma_0 / 2\pi
\end{bmatrix}
\qquad
\gamma_I = \begin{bmatrix}
\text{angle}(H_{0,0}, H_{0,1}) \\
\text{angle}(H_{1,0}, H_{1,1}) \\
\text{angle}(H_{2,0}, H_{2,1}) \\
\text{angle}(H_{3,0}, H_{3,1}) \\
\text{angle}(H_{4,0}, H_{4,1}) \\
\text{angle}(H_{5,0}, H_{5,1}) \\
\text{angle}(H_{6,0}, H_{6,1}) \\
\text{angle}(H_{7,0}, H_{7,1}) \\
\text{angle}(H_{8,0}, H_{8,1}) \\
\text{angle}(H_{9,0}, H_{9,1}) \\
\text{angle}(H_{10,0}, H_{10,1}) \\
\text{angle}(H_{11,0}, H_{11,1}) \\
\text{angle}(H_{12,0}, H_{12,1}) \\
\text{angle}(H_{13,0}, H_{13,1}) \\
\text{angle}(H_{14,0}, H_{1,14}) \\
\text{angle}(H_{15,0}, H_{15,1}) \\
\text{angle}(H_{0,0}, H_{0,1})
\end{bmatrix}
$$

$$
\delta deg_I = \begin{bmatrix}
360 \cdot \delta_0 / 2\pi \\
360 \cdot \delta_1 / 2\pi \\
360 \cdot \delta_2 / 2\pi \\
360 \cdot \delta_3 / 2\pi \\
360 \cdot \delta_4 / 2\pi \\
360 \cdot \delta_5 / 2\pi \\
360 \cdot \delta_6 / 2\pi \\
360 \cdot \delta_7 / 2\pi \\
360 \cdot \delta_8 / 2\pi \\
360 \cdot \delta_9 / 2\pi \\
360 \cdot \delta_{10} / 2\pi \\
360 \cdot \delta_{11} / 2\pi \\
360 \cdot \delta_{12} / 2\pi \\
360 \cdot \delta_{13} / 2\pi \\
360 \cdot \delta_{14} / 2\pi \\
360 \cdot \delta_{15} / 2\pi \\
360 \cdot \delta_0 / 2\pi
\end{bmatrix}
\qquad
\delta_I = \begin{bmatrix}
\gamma_1 - \gamma_0 \\
\gamma_2 - \gamma_1 \\
\gamma_3 - \gamma_2 \\
\gamma_4 - \gamma_3 \\
\gamma_5 - \gamma_4 \\
\gamma_6 - \gamma_5 \\
\gamma_7 - \gamma_6 \\
\gamma_8 - \gamma_7 \\
\gamma_9 - \gamma_8 \\
\gamma_{10} - \gamma_9 \\
\gamma_{11} - \gamma_{10} \\
\gamma_{12} - \gamma_{11} \\
\gamma_{13} - \gamma_{12} \\
\gamma_{14} - \gamma_{13} \\
\gamma_{15} - \gamma_{14} \\
\gamma_{16} - \gamma_{15} \\
\gamma_1 - \gamma_0
\end{bmatrix}
$$

Internal Values for the Angular Series of the Irregular {n,2} Polygram II

INTERNAL VALUES FOR THE ANGULAR SERIES

i	l	γdeg_l	γ_l	δdeg_l	δ_l
0	0	0.	0	1.635013	0.02853636
1	1	1.635013	0.02853636	27.364987	0.47760912
2	2	29.	0.50614548	14.737829	0.25722364
3	3	43.737829	0.76336912	47.262171	0.8248805
4	4	91.	1.58824962	41.356469	0.72180655
5	5	132.356469	2.31005617	18.643531	0.325391
6	6	151.	2.63544717	23.519175	0.41048704
7	7	174.519175	3.04593421	15.480825	0.27019137
8	8	190.	3.31612558	11.216601	0.19576662
9	9	201.216601	3.518922	33.783399	0.58963155
10	10	235.	4.10152374	40.186344	0.70138402
11	11	275.186344	4.80290776	36.813656	0.64251951
12	12	312.	5.44542727	41.542633	0.72505573
13	13	353.542633	6.17048299	1.457367	0.02543585
14	14	355.	6.19591884	3.025824	0.05281059
15	15	358.025824	6.24872944	1.974176	0.03445587
16		360.	0	1.635013	0.02853636

$$
i = \begin{bmatrix} 0 \\ 1 \\ 2 \\ 3 \\ 4 \\ 5 \\ 6 \\ 7 \\ 8 \\ 9 \\ 10 \\ 11 \\ 12 \\ 13 \\ 14 \\ 15 \\ 16 \end{bmatrix}
\quad
l = \begin{bmatrix} 0 \\ 1 \\ 2 \\ 3 \\ 4 \\ 5 \\ 6 \\ 7 \\ 8 \\ 9 \\ 10 \\ 11 \\ 12 \\ 13 \\ 14 \\ 15 \end{bmatrix}
\quad
R_i = \begin{bmatrix}
(H_{14,0}{}^2 + H_{14,1}{}^2)^{0.5} \\
(H_{15,0}{}^2 + H_{15,1}{}^2)^{0.5} \\
(H_{16,0}{}^2 + H_{16,1}{}^2)^{0.5} \\
(H_{1,0}{}^2 + H_{1,1}{}^2)^{0.5} \\
(H_{2,0}{}^2 + H_{2,1}{}^2)^{0.5} \\
(H_{3,0}{}^2 + H_{3,1}{}^2)^{0.5} \\
(H_{4,0}{}^2 + H_{4,1}{}^2)^{0.5} \\
(H_{5,0}{}^2 + H_{5,1}{}^2)^{0.5} \\
(H_{6,0}{}^2 + H_{6,1}{}^2)^{0.5} \\
(H_{7,0}{}^2 + H_{7,1}{}^2)^{0.5} \\
(H_{8,0}{}^2 + H_{8,1}{}^2)^{0.5} \\
(H_{9,0}{}^2 + H_{9,1}{}^2)^{0.5} \\
(H_{10,0}{}^2 + H_{10,1}{}^2)^{0.5} \\
(H_{11,0}{}^2 + H_{11,1}{}^2)^{0.5} \\
(H_{12,0}{}^2 + H_{12,1}{}^2)^{0.5} \\
(H_{13,0}{}^2 + H_{13,1}{}^2)^{0.5} \\
(H_{14,0}{}^2 + H_{14,1}{}^2)^{0.5}
\end{bmatrix}
\quad
ta_l = \begin{bmatrix}
0.5 * R_0 R_1 . \mathrm{Sin}(\delta_0) \\
0.5 * R_1 R_2 . \mathrm{Sin}(\delta_1) \\
0.5 * R_2 R_3 . \mathrm{Sin}(\delta_2) \\
0.5 * R_3 R_4 . \mathrm{Sin}(\delta_3) \\
0.5 * R_4 R_5 . \mathrm{Sin}(\delta_4) \\
0.5 * R_5 R_6 . \mathrm{Sin}(\delta_5) \\
0.5 * R_6 R_7 . \mathrm{Sin}(\delta_6) \\
0.5 * R_7 R_8 . \mathrm{Sin}(\delta_7) \\
0.5 * R_8 R_9 . \mathrm{Sin}(\delta_8) \\
0.5 * R_9 R_{10} . \mathrm{Sin}(\delta_9) \\
0.5 * R_{10} R_{11} . \mathrm{Sin}(\delta_{10}) \\
0.5 * R_{11} R_{12} . \mathrm{Sin}(\delta_{11}) \\
0.5 * R_{12} R_{13} . \mathrm{Sin}(\delta_{12}) \\
0.5 * R_{13} R_{14} . \mathrm{Sin}(\delta_{13}) \\
0.5 * R_{14} R_{15} . \mathrm{Sin}(\delta_{14}) \\
0.5 * R_{15} R_{16} . \mathrm{Sin}(\delta_{15}) \\
0.5 * R_0 R_1 . \mathrm{Sin}(\delta_{16})
\end{bmatrix}
$$

rotated
to conform
with
MATHCAD®
output

The Matrix of
Heronian
Subtriangles ha_l
identifies with
ta_l

INTERNAL FORMULAE FOR THE SERIES R_i AND ta_i

**Internal Formulae for the
Component Triangle Series of the
Irregular {n,2} Polygram
I1**

$$
i = \begin{bmatrix} 0 \\ 1 \\ 2 \\ 3 \\ 4 \\ 5 \\ 6 \\ 7 \\ 8 \\ 9 \\ 10 \\ 11 \\ 12 \\ 13 \\ 14 \\ 15 \\ 16 \end{bmatrix}
\qquad
l = \begin{bmatrix} 0 \\ 1 \\ 2 \\ 3 \\ 4 \\ 5 \\ 6 \\ 7 \\ 8 \\ 9 \\ 10 \\ 11 \\ 12 \\ 13 \\ 14 \\ 15 \end{bmatrix}
\qquad
R_i = \begin{bmatrix} 75. \\ 72.912675 \\ 75. \\ 52.593067 \\ 75. \\ 49.204769 \\ 75. \\ 58.766461 \\ 75. \\ 56.313932 \\ 75. \\ 39.859694 \\ 75. \\ 71.857805 \\ 75. \\ 73.910255 \\ 75. \end{bmatrix}
\qquad
ta_l = \begin{bmatrix} 78.014249 \\ 1256.806242 \\ 501.730994 \\ 1448.544589 \\ 1219.186724 \\ 589.865462 \\ 879.416480 \\ 588.213777 \\ 410.778952 \\ 1174.261254 \\ 964.518323 \\ 895.668882 \\ 1787.042043 \\ 68.533779 \\ 146.303633 \\ 95.480189 \\ 78.014249 \end{bmatrix}
$$

rotated
to conform
with
MATHCAD®
output

The Matrix
of
Heronian
Subtriangles
ha_l
identifies
with
ta_l

INTERNAL VALUES FOR THE SERIES R_i AND ta_i

**Internal Values for the
Component Triangle Series of the
Irregular {n,2} Polygram
I1**

$$i = \begin{bmatrix} 0 \\ 1 \\ 2 \\ 3 \\ 4 \\ 5 \\ 6 \\ 7 \\ 8 \\ 9 \\ 10 \\ 11 \\ 12 \\ 13 \\ 14 \\ 15 \\ 16 \end{bmatrix} \quad l = \begin{bmatrix} 0 \\ 1 \\ 2 \\ 3 \\ 4 \\ 5 \\ 6 \\ 7 \\ 8 \\ 9 \\ 10 \\ 11 \\ 12 \\ 13 \\ 14 \\ 15 \end{bmatrix}$$

$$L(l) = \begin{bmatrix} \mathrm{ent}(l_0/2)+\mathrm{ent}(l_0/2+1) \\ \mathrm{ent}(l_1/2)+\mathrm{ent}(l_1/2+1) \\ \mathrm{ent}(l_2/2)+\mathrm{ent}(l_2/2+1) \\ \mathrm{ent}(l_3/2)+\mathrm{ent}(l_3/2+1) \\ \mathrm{ent}(l_4/2)+\mathrm{ent}(l_4/2+1) \\ \mathrm{ent}(l_5/2)+\mathrm{ent}(l_5/2+1) \\ \mathrm{ent}(l_6/2)+\mathrm{ent}(l_6/2+1) \\ \mathrm{ent}(l_7/2)+\mathrm{ent}(l_7/2+1) \\ \mathrm{ent}(l_8/2)+\mathrm{ent}(l_8/2+1) \\ \mathrm{ent}(l_9/2)+\mathrm{ent}(l_9/2+1) \\ \mathrm{ent}(l_{10}/2)+\mathrm{ent}(l_{10}/2+1) \\ \mathrm{ent}(l_{11}/2)+\mathrm{ent}(l_{11}/2+1) \\ \mathrm{ent}(l_{12}/2)+\mathrm{ent}(l_{12}/2+1) \\ \mathrm{ent}(l_{13}/2)+\mathrm{ent}(l_{13}/2+1) \\ \mathrm{ent}(l_{14}/2)+\mathrm{ent}(l_{14}/2+1) \\ \mathrm{ent}(l_{15}/2)+\mathrm{ent}(l_{15}/2+1) \end{bmatrix}$$

$$\sin(\delta_l) = \begin{bmatrix} \sin(\delta_0) \\ \sin(\delta_1) \\ \sin(\delta_2) \\ \sin(\delta_3) \\ \sin(\delta_4) \\ \sin(\delta_5) \\ \sin(\delta_6) \\ \sin(\delta_7) \\ \sin(\delta_8) \\ \sin(\delta_9) \\ \sin(\delta_{10}) \\ \sin(\delta_{11}) \\ \sin(\delta_{12}) \\ \sin(\delta_{13}) \\ \sin(\delta_{14}) \\ \sin(\delta_{15}) \end{bmatrix}$$

$$T = \begin{bmatrix} \tau_0 \\ \tau_1 \\ \tau_2 \\ \tau_3 \\ \tau_4 \\ \tau_5 \\ \tau_6 \\ \tau_7 \\ \tau_8 \\ \tau_9 \\ \tau_{10} \\ \tau_{11} \\ \tau_{12} \\ \tau_{13} \\ \tau_{14} \\ \tau_{15} \end{bmatrix}$$

τ_l is the same as ta_i, the series of Heronian SubTriangles

$$U = \begin{bmatrix} P_{L(0)} \cdot \sin(\delta_0) \\ P_{L(1)} \cdot \sin(\delta_1) \\ P_{L(2)} \cdot \sin(\delta_2) \\ P_{L(3)} \cdot \sin(\delta_3) \\ P_{L(4)} \cdot \sin(\delta_4) \\ P_{L(5)} \cdot \sin(\delta_5) \\ P_{L(6)} \cdot \sin(\delta_6) \\ P_{L(7)} \cdot \sin(\delta_7) \\ P_{L(8)} \cdot \sin(\delta_8) \\ P_{L(9)} \cdot \sin(\delta_9) \\ P_{L(10)} \cdot \sin(\delta_{10}) \\ P_{L(11)} \cdot \sin(\delta_{11}) \\ P_{L(12)} \cdot \sin(\delta_{12}) \\ P_{L(13)} \cdot \sin(\delta_{13}) \\ P_{L(14)} \cdot \sin(\delta_{14}) \\ P_{L(15)} \cdot \sin(\delta_{15}) \end{bmatrix}$$

$$V = \begin{bmatrix} P_{L(0)} \\ P_{L(1)} \\ P_{L(2)} \\ P_{L(3)} \\ P_{L(4)} \\ P_{L(5)} \\ P_{L(6)} \\ P_{L(7)} \\ P_{L(8)} \\ P_{L(9)} \\ P_{L(10)} \\ P_{L(11)} \\ P_{L(12)} \\ P_{L(13)} \\ P_{L(14)} \\ P_{L(15)} \end{bmatrix}$$

$$W = \begin{bmatrix} \sin(\delta_0) \\ \sin(\delta_1) \\ \sin(\delta_2) \\ \sin(\delta_3) \\ \sin(\delta_4) \\ \sin(\delta_5) \\ \sin(\delta_6) \\ \sin(\delta_7) \\ \sin(\delta_8) \\ \sin(\delta_9) \\ \sin(\delta_{10}) \\ \sin(\delta_{11}) \\ \sin(\delta_{12}) \\ \sin(\delta_{13}) \\ \sin(\delta_{14}) \\ \sin(\delta_{15}) \end{bmatrix}$$

W is the Matrix of Sines of δ_l

INTERNAL FORMULAE FOR THE MATRICES L(l), T, U, V, AND W = array of sines

Internal Formulae for the L, T, U, V and W Matrices of the Irregular Polygram I1

Internal Values for the L, T, U, V and W Matrices of the Irregular {n,2} Polygram I1

INTERNAL VALUES FOR THE MATRICES L(l), T, U, V, AND W = array of sines

$i =$	$l =$	$L(l) =$	$\sin(\delta_l) =$	$T =$	$U =$	$V =$	$W =$
0	0	1	0.02853249	78.014249	2.080380	72.912675	0.02853249
1	1	1	0.45965716	1256.806242	33.514833	72.912675	0.45965716
2	2	3	0.25439652	501.730994	13.379493	52.593067	0.25439652
3	3	3	0.73446669	1448.544589	38.627856	52.593067	0.73446669
4	4	5	0.66074177	1219.186724	32.511646	49.204769	0.66074177
5	5	5	0.31967929	589.865462	15.729746	49.204769	0.31967929
6	6	7	0.39905596	879.416480	23.451106	58.766461	0.39905596
7	7	7	0.26691587	588.213777	15.685701	58.766461	0.26691587
8	8	9	0.19451857	410.778952	10.954105	56.313932	0.19451857
9	9	9	0.55605482	1174.261254	31.313633	56.313932	0.55605482
10	10	11	0.64527562	964.518323	25.720489	39.859694	0.64527562
11	11	11	0.59921443	895.668882	23.884504	39.859694	0.59921443
12	12	13	0.66317715	1787.042043	47.654454	71.857805	0.66317715
13	13	13	0.02543311	68.533779	1.827567	71.857805	0.02543311
14	14	15	0.05278605	146.303633	3.901430	73.910255	0.05278605
15	15	15	0.03444905	95.480189	2.546138	73.910255	0.03444905
16							

Σ			5.894355	12104.365572	322.783082	$2*\Sigma P$ 950.837314	5.894355
						ΣP 475.418657	

W is the Matrix of Sines of δ_l